Numerical Simulation - Advanced Techniques for Science and Engineering

Edited by Ali Soofastaei

Published in London, United Kingdom

Numerical Simulation - Advanced Techniques for Science and Engineering
http://dx.doi.org/10.5772/intechopen.100786
Edited by Ali Soofastaei

Contributors
Prasenjit Singha, Sunil Yadav, Soumya Ranjan Mohanty, Abhishek Tiwari, Ajay Kumar Shukla, Muzammil Arshad, Umurdin Dalabaev, Malika Raximberdiyevna Ikramova, Takaaki Uda, Takuya Yokota, Yasuhito Noshi, Arash Mohammadi, Amadou Coulibaly, Bayo J. Omotosho, Amoro Coulibaly, Mouhamadou B. Sylla, Abdoulaye Ballo, Inci Zaim Gokbay, Sacide Pehlivan, Yasemin Oyacı, Chiş Timur, Jugastreaanu Cristina, Tabatabai Seyed Mehdi, Renata Rădulescu, Tao Wang, Bharat Ramanathan, Mayane Batista Lima, KP Mredula Pyarelal, Pavel Loskot, Mykola Ivanovich Yaremenko, Shazali Abdalla Fadul, Ali Soofastaei

First published in London, United Kingdom, 2023 by IntechOpen
IntechOpen is the global imprint of INTECHOPEN LIMITED, registered in England and Wales, registration number: 11086078, 5 Princes Gate Court, London, SW7 2QJ, United Kingdom

British Library Cataloguing-in-Publication Data
A catalogue record for this book is available from the British Library

Additional hard and PDF copies can be obtained from orders@intechopen.com

Numerical Simulation - Advanced Techniques for Science and Engineering
Edited by Ali Soofastaei
p. cm.
Print ISBN 978-1-80356-953-6
Online ISBN 978-1-80356-954-3
eBook (PDF) ISBN 978-1-80356-955-0

For EU product safety concerns:
IN TECH d.o.o., Prolaz Marije Krucifikse Kozulić 3, 51000 Rijeka, Croatia,
info@intechopen.com or visit our website at intechopen.com.

Meet the editor

Dr. Ali Soofastaei is a global artificial intelligence (AI) projects leader, an international keynote speaker, and a professional author. He completed his Ph.D. and was a postdoctoral research fellow at the University of Queensland, Australia, in the field of AI applications in mining engineering, where he led a revolution in the use of deep learning and AI methods to increase energy efficiency, reduce operation and maintenance costs, and reduce greenhouse gas emissions in surface mines. As a scientific supervisor, he has provided practical guidance to undergraduate and postgraduate students in mechanical and mining engineering and information technology for many years. Dr. Soofastaei has more than 15 years of academic experience as an assistant professor and leader of global research activities. Results from his research and development projects have been published in international journals and keynote presentations. He has presented his practical achievements at conferences in the United States, Europe, Asia, and Australia. He has been involved in industrial research and development projects in several industries, including oil and gas (Royal Dutch Shell), steel (Danieli), and mining (BHP, Rio Tinto, Anglo American, and Vale). His extensive practical experience in the industry has equipped him to work with complex industrial problems in highly technical and multi-disciplinary teams.

Contents

Preface

The relentless pursuit of understanding the complexities of our world has always been a driving force in the evolution of human knowledge. From the early days of mathematical breakthroughs to the current age of advanced computing, our quest to analyze, predict, and control natural and humanmade phenomena has led us to develop increasingly sophisticated tools and techniques. The advent of numerical simulation has been a key milestone in this journey, empowering scientists and engineers to tackle problems once thought out of reach.

Numerical Simulation – Advanced Techniques for Science and Engineering provides a comprehensive and accessible introduction to this ever-expanding field. It equips readers with the knowledge and skills to understand, design, and implement numerical simulations in various disciplines, such as engineering, physics, chemistry, and biology.

This book caters to a wide readership, including students, researchers, and professionals in science and engineering. It begins with the history and principles of numerical simulation, then explores mathematical modeling, numerical methods, and computational algorithms. Further chapters delve into various applications, illustrating the potential of numerical simulations. The book emphasizes the significance of accuracy, stability, and efficiency in simulations, striving to blend theory with practice. It includes examples, exercises, software, and code snippets, encouraging readers to apply what they learn to practical scenarios.

This book contains different use cases for numerical simulation in different industries. Practical examples and scientific details support all the information. The chapters contain enough information for beginners to get familiar with the high technology and science applications to solve business problems and more detailed technical information for advanced readers.

Chapter 1: “Introductory Chapter: Numerical Simulation”

Chapter 1 is about numerical simulation, a computational technique used across numerous scientific and engineering disciplines to study complex systems that are otherwise difficult to explore physically. Numerical simulation creates approximate models of real-world behavior, helping to predict outcomes under various scenarios. This method involves discretizing continuous variables into finite points and converting differential equations into algebraic ones that computers can solve. Its diverse applications include weather prediction, fluid flow simulation in aerodynamics, stress analysis in mechanical structures, and biological system simulation in medicine. While computationally intensive, the development of high-performance computing has significantly broadened the use of numerical simulation, establishing it as a crucial tool in contemporary research and development.

Chapter 2: "Mathematical Basics as a Prerequisite to Artificial Intelligence in Forensic Analysis"

This chapter delves into the confluence of mathematics, statistics, and AI, emphasizing their application in image processing and forensics. It underscores the importance of math in these areas and the progression of image processing techniques. Key notions like neural networks are examined, setting the groundwork for understanding artificial neural networks. The chapter tackles hurdles in understanding prerequisites and explores niche areas like steganographic security and image forensic detection. It proposes a score-based likelihood ratio over traditional statistical methods. The chapter is divided into two sections, tackling math prerequisites for image processing and connecting these to forensic sciences, facilitating an efficient overview of related concepts across multiple specializations.

Chapter 3: "Computer Vision: Anthropology of Algorithmic Bias in Facial Analysis Tool"

This chapter examines bias in Computer Vision (CV), notably in autonomous machines, due to human influence during data labeling for machine learning, leading to an unequal representation of social groups. Referencing Russell and Lee, it emphasizes the need for broad and relevant datasets for effective recognition, highlighting those biases from disproportionate representation can result in algorithmic decisions that marginalize specific social groups. The authors scrutinize Amazon's facial recognition tool's identification and categorization of non-conventional or dissenting genders, questioning these machines' capacity to recognize beyond binary gender classifications.

Chapter 4: "Numerical Simulations and Validation of Engine Performance Parameters Using Chemical Kinetics"

This chapter promotes computer modeling and simulations for enhancing fuels and engines, critiquing the conventional dependence on global reactions in combustion simulations due to its impact on engine performance predictions. The authors suggest a refined combustion model combining a 3D turbulent Navier–Stokes solver with detailed kinetic reactions and fluid dynamics for improved accuracy. They advocate for reduced chemical reaction mechanisms to expedite simulations, aiding efficient engine performance analysis. The chapter underscores sensitivity analysis and the computational singular perturbation method to hone the reaction mechanism. An interface for surrogate fuels study is proposed, emphasizing the need for simulation validation through experimental data. Comprehensive studies are urged for validating performance parameters across all mixtures. The necessity of a standard reduced mechanisms library for varied engines and combustion systems is highlighted, ending with a validated reduced reaction mechanism for premixed and direct injection spark ignition engines.

Chapter 5: "Bayesian Methods and Monte Carlo Simulations"

This chapter discusses Bayesian methods and tools for studying probabilistic models of linear and non-linear stochastic systems. They allow the tracking of probability distribution changes using Bayes' theorem and the chain rule, but their complexity often requires numerical statistical and causal inference methods. The chapter introduces various Bayesian techniques for managing intractable distributions, including sampling, filtering, approximation, and likelihood-free methods, explaining their principles and key challenges. These methods find applications in various areas: Bayesian experiment design maximizes information gain and is usually combined with optimal model selection; Bayesian hypothesis testing improves data-driven decision-making; Bayesian machine learning treats data labels as random variables; and a Bayesian optimization is a powerful tool for configuring and optimizing large-scale complex systems. The chapter discusses Bayesian Monte-Carlo simulations, proposing that augmented Monte-Carlo simulations can better explain capability and information efficiency.

Chapter 6: "Numerical Simulation on Sand Accumulation behind Artificial Reefs and Enhancement of Windblown Sand to Hinterland"

This study investigates the impact of artificial reefs on Kimigahama Beach in Chiba Prefecture, Japan. Due to their wave-sheltering effect, the reefs formed Salients, or protrusions, in the shoreline, leading to a significant amount of fine sand being transported inland by the wind. To analyze these effects, the study employed a model that combines the Beta-Geometric (BG) model (for predicting three-dimensional beach changes due to waves) with a cellular automaton method. This model was used to forecast shoreline changes after the installation of reefs, beach changes induced by windblown sand, beach changes after reef removal, and the impact of beach nourishment. The findings indicate that constructing wave-sheltering structures like artificial reefs on fine sand coasts speeds up the wind-driven landward sand transport. The model used was successful in predicting such effects.

Chapter 7: "Numerical Simulation of Land and Sea Breeze (LSB) Circulation along the Guinean Coast of West Africa"

This chapter examines the dynamics of land and sea-breeze rotation along the West African Guinean Coast using observed and simulated data. The Weather Research and Forecasting (WRF) model, modified with ERA-Interim (a global atmospheric reanalysis) and Climate Forecast System (CFS) forcing data, was used to simulate the local circulation, displaying accurate results aligned with observed data. The research reveals that pressure gradients, advection, and diffusion forces shape wind rotation direction. An hourly breakdown indicated surface gradient forces dominate the ocean, while diffusion terms impact more on land due to variations in surface roughness caused by landscape and urbanization. The study highlights a connection between urbanization and local circulation in major cities along the Guinean Coast.

Chapter 8: "Fluid Dynamics Simulation of an NREL-S Series Wind Turbine Blade"

This chapter focuses on the detailed study of the theory, design, modeling, and simulation of a 1.2-MW wind turbine blade that measures 35 meters. Given the wind turbine blade geometry's complexity and unpredictable characteristics, the chapter employs Computational Fluid Dynamics (CFD) to simulate the blade. The design's central focus is the Tip Speed Ratio (TSR), optimally set at 7 for this study. The chapter then juxtaposes the simulation results with those from the Blade Element & Momentum (BEM) theory. Finally, the results from Q Blade and X Foils are compared with a more precise CFD Simulation. The chapter concludes by comparing and evaluating the accuracy of the various methods used in the study.

Chapter 9: "Methods of the Perturbation Theory for Fundamental Solutions to the Generalization of the Fractional Laplaciane"

This chapter delves into the regularity properties of solutions to the fractional Laplacian equation with perturbations, a model significant in various fields of mathematics and physics. The authors utilize semigroup theory to illuminate the dynamics of solutions, highlighting the effects of perturbations. They establish the Harnack inequality for a weak solution to the fractional Laplacian problem, which is a crucial tool for analyzing elliptic and parabolic partial differential equations and provides vital information about the interior regularity of solutions. Furthermore, the authors estimate the oscillation of the solution to the fractional Laplacian, a necessary step for understanding solutions' qualitative behavior and developing numerical methods for the equation. Overall, this chapter provides a comprehensive exploration of the regularity properties of the fractional Laplacian equation with perturbations, aiming to spur further research and development in this crucial field.

Chapter 10: "The Analysis on the Effects of COMT, DRD2, PER3, eNOS, NR3C1 Functional Gene Variants and Methylation Differences on Behavoiral Inclinations in Addicts through the Decision Tree Algorythm"

This chapter presents a study exploring the influence of functional gene variants (COMT, DRD2, PER3, eNOS, and NR3C1) on individuals with substance use disorder (SUD). A decision tree algorithm is used to analyze and compare the impacts of these gene variants, guided by the influences of genetic and epigenetic sequences. This classification system is evaluated through a 10-fold cross-validation considering various factors, such as criminal history, continuity of substance use, previous polysubstance abuse, suicide attempts, and inpatient treatment. Performance criteria are gauged based on accuracy, sensitivity, and precision values, consistent with earlier research. The gene variants branching structure resulting from the tree classification aligns with existing literature. This research highlights the potential of machine learning in predicting the effect of gene variants on behavior, emphasizing the need for more extensive studies that include data from diverse ethnic groups to improve predictive accuracy rates.

Chapter 11: "Mathematical Modeling of a Porous Medium in Diesel Engines"

This chapter discusses the issue of particulate matter (PM) emissions in direct-injection diesel engines. Despite their high-power density and low exhaust emissions, these engines face challenges with PM emissions due to the simultaneous fuel injection and combustion process. This process results in a non-homogeneous mixture in the cylinder, contributing to emissions. The chapter proposes separating fuel injection and combustion processes to create a homogeneous mixture, using porous media in diesel engine combustion chambers as a practical approach to enable stable ultra-lean combustion and reduce emissions. The chapter presents a thorough overview of the mathematical modeling of PM diesel engines, divided into three parts: thermodynamic modeling, zero-dimensional modeling with chemical kinetics, and three-dimensional computational fluid dynamics (CFD) modeling with chemical kinetics.

Chapter 12: "Modeling of Thermal Conductivity in Gas Field Rocks"

This chapter provides a comprehensive examination of the significance of understanding the thermal conductivity of rocks in petroleum engineering, specifically during the initial stages of oil and gas deposit exploitation. The information is essential for devising secondary and tertiary extraction methods, including hot water and steam injection, CO_2 injection, flue gas injection, and initiation of underground combustion. The authors introduce an innovative method to measure the thermal conductivity of rocks, improving understanding of heat transfer processes in subsurface reservoirs and aiding the development of efficient extraction strategies. They also analyze the relationships between thermal conductivity and properties of oil and gas collector rocks, particularly density and porosity, offering insights that could optimize extraction processes. The chapter is a valuable resource blending theory and application, beneficial for researchers, professionals, and students in petroleum engineering and related fields.

Chapter 13: "Simulation Study of Microwave Heating of Hematite and Coal Mixture"

This chapter presents a computational approach to predict the temperature distribution in a hematite ore mixed with 7.5% coal. Using MATLAB 2018a software, a 1D heat conduction equation was solved via an implicit finite difference method. The study focused on a 20 cm x 20 cm square slab, where coal was assumed to be uniformly mixed with the ore. The model considered convective and radiative boundary conditions, microwave heating time, thermal conductivity, heat capacity, carbon percentage, sample dimensions, and other factors like penetration depth, permittivity, and permeability of the ore-coal mixture. The temperature profile derived from this model could optimize the microwave-assisted carbothermal reduction process for hematite. The model was also extended to slabs of varying sizes, and the predictions aligned well with experimental results.

Chapter 14: "Perspective Chapter: Computational Modeling for Predicting the Optical Distortions through the Hypersonic Flow Fields"

This chapter investigates the impact of aero-optical effects (AOE) on interceptor systems equipped with infrared detectors. As these interceptors move at supersonic speeds, they create a variable density field, altering the optical properties, particularly the index of refraction. This alteration can distort the incoming light, causing blur, shift, jitter, intensity loss, and resolution loss, collectively known as AOE. These aberrations can severely degrade the imaging quality of onboard optical sensors, compromising guidance accuracy and potentially leading to mission failure. Given the importance of achieving high guidance accuracy for endo-atmospheric flight vehicles, it is essential to understand the principles of AOE and evaluate these aberrations. Therefore, this chapter studies the influence of supersonic flow fields on optical propagation and imaging, which holds both theoretical value and practical implications in the design of optical systems and restoration of turbulence-degraded images.

Chapter 15: "Moving Node Method for Differential Equations"

This chapter presents a novel method using moving nodes in computing, promising improved accuracy and efficiency in complex numerical calculations. It uses a common fluid mechanics and heat transfer problem to demonstrate the method's proficiency in solving convective-diffusion issues. The goal is to highlight how this innovative method can tackle such problems more effectively than existing approaches. Validation through test examples illustrates the method's benefits, encouraging its broader use in computing technology. This work aims to stimulate fresh insights and further exploration in this crucial field.

Chapter 16: "On the Analytical Properties of Prime Numbers"

Prime numbers, unique in their properties and fundamental to number theory, intrigue mathematicians due to their unpredictability and fundamental role in mathematics. Despite exhaustive research, their seemingly random distribution remains an enigma encapsulated in the Prime Number Theorem. As the Fundamental Theorem of Arithmetic highlights, their multiplicative properties emphasize their role as the "building blocks" of the number system. Furthermore, unresolved mysteries like the Twin Prime Conjecture and the Riemann Hypothesis, which are deeply connected to the distribution of primes, add to the intrigue. Beyond their theoretical interest, primes have significant practical applications, particularly in cryptography, where their complex factorization properties underpin secure data transmission systems like Rivest–Shamir–Adleman (RSA). In this chapter, the authors focus on the prime numbers and their analytical properties.

This book helps readers understand numerical simulation applications in different areas, and we hope it will be a valuable resource for industry professionals and researchers. The chapters discuss the state of the art of critical topics in numerical simulation. Furthermore, their coverage and depth make this book a helpful tool for all managers and engineers interested in the new generation of data analytics applications. Above all, the editor hopes this volume will spur further discussions on all aspects of numerical simulation applications in different industries.

As you embark on this journey, we encourage you to embrace the spirit of curiosity, perseverance, and innovation that has fueled the development of numerical simulation throughout history. We hope this book will be invaluable and ignite a passion for lifelong learning and discovery in numerical simulations.

Happy reading, and best of luck on your journey into the world of numerical simulation!

Dr. Ali Soofastaei
AI Program Leader,
Artificial Intelligence Center,
Vale, Brisbane, Australia

Section 1

Intersecting Mathematics, Statistics, and AI for Improving the Numerical Simulation

Chapter 1

Introductory Chapter: Numerical Simulation

Ali Soofastaei

1. Introduction

A numerical simulation is an influential tool scientists and engineers use to model and analyze complex systems. From the behavior of subatomic particles to the dynamics of the universe, numerical simulation has become an indispensable tool for understanding the world around us.

At its core, numerical simulation involves using mathematical models to describe a system's behavior. These models can predict the system's behavior under different conditions, allowing researchers to test theories and explore new ideas without expensive and time-consuming experimentation.

The process of numerical simulation involves a series of steps. First, a mathematical model is developed that accurately represents the system being studied. This model can take many forms, depending on the nature of the system and the questions being asked. For example, a water flow model through a pipe might be based on fluid mechanics principles. In contrast, a chemical reaction model might be found on principles of thermodynamics and kinetics.

Once the model has been developed, it must be translated into a form that a computer can understand. This often involves writing code in a programming language such as Python or C++, which can be executed on a computer. The code typically includes algorithms that simulate the system's behavior over time based on the mathematical model.

Simulations can be run on various computer systems, from desktop computers to supercomputers. The amount of computational power required depends on the complexity of the model and the size of the system being studied. For example, simulating a single molecule's behavior might only need a few minutes on a desktop computer, while simulating the behavior of an entire ecosystem could require days or even weeks on a supercomputer.

As numerical simulation continues to evolve, it is also essential for researchers to explore new frontiers and develop new approaches and techniques. For example, machine learning and artificial intelligence are opening up new opportunities in numerical simulation by enabling simulations to be run more efficiently and accurately.

The future of numerical simulation is exciting and filled with possibilities. One promising area of research is the development of multiscale simulations, which can model systems at multiple levels of complexity, from the atomic to the macroscopic. As a result, multiscale simulations can provide a complete understanding of complex systems and can be used to design new materials and devices with novel properties.

Over the past few decades, numerical simulation has grown in popularity and sophistication, driven partly by computer technology advances and new simulation methods and algorithms.

Furthermore, integrating numerical simulation with other emerging technologies, such as artificial intelligence, robotics, and quantum computing, is expected to open up new opportunities for research and development. For example, using artificial intelligence in numerical simulation can help identify patterns and relationships in complex datasets, leading to more accurate predictions and better insights.

2. Advantages/Use cases

One of the critical benefits of numerical simulation is that it allows researchers to explore the behavior of a system under a wide range of conditions. For example, a model of climate change might be used to examine the impact of different levels of greenhouse gas emissions. In climate science, numerical simulation has been used to model the Earth's climate system and predict future changes in the climate. These models have helped scientists understand human activities impact on the environment and develop strategies for mitigating the effects of climate change [1].

The numerical simulation also allows researchers to investigate systems that are difficult or impossible to study experimentally. For example, simulating the behavior of subatomic particles in a particle accelerator might be the only way to explore specific aspects of their behavior.

One example of the impact of numerical simulation can be seen in aerospace engineering. Simulations have been used to design and optimize aircraft, spacecraft, and rockets, reducing the need for expensive and time-consuming physical testing. Numerical simulation has also been used to study air flow around wings and other aerodynamic surfaces, leading to improved designs that are more efficient and produce less noise [2].

Another example can be seen in biotechnology, where simulations have been used to study the behavior of proteins and other biological molecules. For example, simulations have been used to predict the behavior of drug molecules in the human body, helping researchers to develop new treatments for diseases such as cancer and Alzheimer's.

Moreover, numerical simulation has also found application in finance, where it is used to model the behavior of financial markets and evaluate risk. For example, Monte Carlo simulation is a popular technique for estimating financial instruments' value and assessing the associated risk. In addition, simulations can test the impact of different market conditions on a portfolio, helping investors make informed decisions about their investments [3].

In materials science, numerical simulation has been used to study the properties and behavior of materials at the atomic and molecular levels. These simulations can provide insight into the fundamental mechanisms that govern the conduct of materi als and can be used to design new materials with specific properties.

Another area where numerical simulation is increasingly being used is in the development of autonomous vehicles. Simulations can be used to test and refine the behavior of self-driving cars, trucks, and drones, allowing developers to ensure that they are safe and reliable before they are deployed on the roads or in the air [4].

In recent years, there has been growing interest in using machine learning and artificial intelligence to enhance numerical simulation capabilities. Machine learning algorithms can improve the accuracy and efficiency of simulations by learning from data and identifying patterns in the behavior of the system being studied.

Researchers must collaborate across disciplines and share their knowledge and expertise to harness numerical simulations powerfully. Many of the most complex and challenging problems facing science and engineering today require input from experts in multiple fields, and collaboration and communication are essential for making progress.

Furthermore, developing numerical simulation tools and algorithms requires significant investment in research and development. Therefore, governments and private industry must continue to invest in developing new simulation methods and algorithms and the hardware and software infrastructure necessary to run simulations efficiently [5].

In addition, efforts must be made to ensure that the benefits of numerical simulation are accessible to all. Access to computational resources can be expensive, and researchers in developing countries and underfunded institutions may need access to the resources they need to conduct simulations. As such, efforts must be made to promote equity and access to resources, to ensure that all researchers have the tools they need to progress in their fields.

Researchers must communicate their findings to the public in an understandable and accessible way. The results of numerical simulations can have far-reaching implications for society, and the public needs to understand the impact of this research. Therefore, researchers must work to communicate their findings clearly and transparently and engage in dialog with the public about the ethical and societal implications of their work [6].

In addition, numerical simulation can address some of the grand challenges facing humanity, such as exploring the mysteries of the universe and discovering new forms of energy. Numerical simulation in astrophysics, for example, has led to groundbreaking discoveries about the nature of black holes and the universe's structure.

Numerical simulation is not limited to academic research and can play a critical role in industrial applications. For example, numerical simulation can be used to design manufacturing processes and optimize production systems. Manufacturers can identify the most efficient and cost-effective strategies by simulating different manufacturing scenarios and evaluating their impact on production output and quality.

Numerical simulation can also be used to design and optimize buildings and infrastructure. For example, simulations can be used to test the behavior of structures under different loading conditions and to evaluate the effectiveness of different design strategies. This can lead to the development of more resilient and sustainable buildings and infrastructure, better able to withstand natural disasters and the effects of climate change [7].

One area of research that is currently gaining momentum is the use of quantum computing in numerical simulation. Quantum computers operate on the principles of quantum mechanics, which enable them to perform specific calculations much faster than classical computers. This makes quantum computers particularly well-suited for simulating the behavior of quantum systems, such as the behavior of molecules and materials.

As quantum computing technology advances, it is expected to significantly impact numerical simulation in various fields, from materials science to drug discovery. However, significant challenges still exist in developing quantum computing hardware and software. As a result, it may be some time before quantum computing becomes a practical tool for numerical simulation.

In addition, using numerical simulation is also creating new opportunities for interdisciplinary research by bringing together experts from different fields to collaborate on complex problems. For example, studying complex systems such as ecosystems or the human brain requires input from biology, physics, mathematics, and computer science experts.

As interdisciplinary research becomes increasingly important, researchers must develop new ways of communicating and collaborating across disciplines. This may involve the development of new tools and techniques for data sharing and analysis, as well as new approaches to education and training that encourage interdisciplinary thinking and collaboration.

Another area of research gaining momentum is using simulations in developing smart cities. Simulations can be used to model the behavior of urban systems, such as traffic flow and energy consumption, and to identify strategies for optimizing resource use and reducing waste. This can lead to the development of more sustainable and resilient urban environments, better able to cope with the challenges of climate change and population growth.

Moreover, integrating simulations with other technologies, such as virtual and augmented reality, opens new education, training, and design possibilities. For example, simulations can create immersive learning experiences and train professionals in high-risk fields, such as medicine and aviation. Simulations can also be used to design and test new products and systems, reducing the need for costly and time-consuming physical prototypes.

3. Challenges

Despite its many benefits, numerical simulation has its limitations. Models simplify real-world systems and are only as accurate as the assumptions and approximations made in their development. Furthermore, simulations can be computationally expensive, and the results can be sensitive to the numerical methods and algorithms used in the simulation.

Another challenge is the need for better validation and verification of simulations. Because simulations are based on mathematical models, validating and verifying experimentally can be challenging. As a result, researchers must rely on a combination of physical experiments, empirical data, and mathematical analysis to ensure that their simulations are accurate and reliable [8].

There are also some ethical concerns associated with its use. For example, using simulations to design autonomous weapons raises questions about the morality of using machines to make life-or-death decisions. Similarly, using simulations to model the behavior of financial markets raises questions about the ethics of using algorithms to make decisions that affect people's lives and livelihoods.

Another concern is the potential for simulations to perpetuate bias and discrimination. Models are only as accurate as the data they are trained on. The generated simulations may perpetuate those biases and inequalities if that data contains preferences or reflects societal disparities. As such, it is crucial for researchers to be aware of the potential for bias in their simulations and to take steps to mitigate it.

There is also a concern that the increasing reliance on numerical simulation may lead to a loss of intuition and creativity in scientific and engineering research. For example, as researchers become increasingly reliant on simulations to generate predictions and test theories, there is a risk that they may need help to think outside the box and come up with novel ideas [9].

As the use of numerical simulation continues to grow, researchers need to remain vigilant about these tools' potential risks and limitations. Models are inherently simplifications of reality, and simulations can be sensitive to the assumptions and approximations made in their development. As such, it is essential for researchers to validate their models and simulations using experimental data and to continuously refine their models and algorithms to ensure accuracy and reliability [10].

Moreover, increasing complexity and size of simulations can present significant computational challenges, requiring large-scale parallel computing resources and specialized software. As such, it is important for researchers to have access to these resources and to develop strategies for efficient and scalable simulations.

4. Conclusions

In conclusion, a numerical simulation is a powerful tool transforming scientific and engineering research and a wide range of industrial and scientific applications. As the use of numerical simulation continues to grow, researchers must remain vigilant about these tools' potential risks and limitations and ensure that ethical considerations are carefully considered. However, the possibilities of numerical simulation are immense. By continuing to explore new frontiers and develop new approaches and techniques, researchers can unlock the full potential of numerical simulation and progress in some of humanity's most challenging and important problems.

From aerospace engineering to biotechnology, finance to materials science, numerical simulation is helping researchers to explore and understand the world in new and exciting ways. As computational power continues to increase and new simulation methods are developed, the potential of numerical simulation to revolutionize science and engineering research is only set to grow.

As the integration of numerical simulation with other emerging technologies continues to accelerate, researchers can unlock new possibilities for research and development, leading to new insights and discoveries that can help address some of humanity's most pressing challenges. By pushing the boundaries of numerical simulation and exploring new frontiers, researchers can continue to harness the power of this tool to make progress in their fields and improve society as a whole.

Author details

Ali Soofastaei
Artificial Intelligence Center, Australia

*Address all correspondence to: ali@soofastaei.net

References

[1] Cheepu M, Kantumuchu VC. Numerical simulations of the effect of heat input on microstructural growth for MIG-based wire arc additive manufacturing of Inconel 718. Transactions of the Indian Institute of Metals. 2023;**76**(2):473-481

[2] Gou R et al. Study on the tribological properties of diamond and SiC interactions using atomic scale numerical simulations. Tribology International. 2023;**178**:108093

[3] Kihara K, Okada N. Numerical simulation model of gas–liquid–solid flows with gas–liquid free surface and solid-particle flows. Chemical Engineering Science. 2023;**270**:118507

[4] Kumari K, Donzis DA. Regimes of optical propagation through turbulence: Theory and direct numerical simulations. Waves in Random and Complex Media. 2023;**122**:1-35

[5] Lu Q et al. Numerical simulation of defect influence on nanosecond laser manufacturing. International Journal of Thermal Sciences. 2023;**183**:107900

[6] Rawlins BT, Laubscher R, Rousseau P. A fast thermal non-equilibrium Eulerian-Eulerian numerical simulation methodology of a pulverised fuel combustor. Thermal Science and Engineering Progress. 2023;**161**:101842

[7] Song X et al. Calibration of DEM models for fertilizer particles based on numerical simulations and granular experiments. Computers and Electronics in Agriculture. 2023;**204**:107507

[8] Vantadori S et al. Numerical simulation of the shear strength of the shot-earth 772-granite interface. Construction and Building Materials. 2023;**363**:129450

[9] Wang Y et al. Experimental research and numerical simulation of the multi-field performance of cemented paste backfill: Review and future perspectives. International Journal of Minerals, Metallurgy and Materials. 2023;**30**(2):193-208

[10] Wei X et al. Numerical simulation of anti-plane wave propagation in heterogeneous media. Applied Mathematics Letters. 2023;**135**:108436

Chapter 2

Mathematical Basics as a Prerequisite to Artificial Intelligence in Forensic Analysis

KP Mredula Pyarelal

Abstract

The chapter examines a review and revisit to the current study of advancements in mathematics and statistical methods underlying the most sorted topic of artificial intelligence (AI). Inclusion of references is done for better and smooth discussion for more clarity to the underlying difficulties faced by readers. Mathematics motivates image processing and image processing improves methods involved with mathematics, which is to be explored. Mathematics stands as a back bone and with discussions of basics of neural network the path way to artificial neural network would be build. The struggle to recall the prerequisites faced by researchers is addressed in this chapter. The chapter will provide you through an ariel view by stating the definitions of prerequisites such as mathematics for image processing, mathematics for forensic image processing which includes basics of neural network and prerequisites of probability theory as a subsection. Forensic sciences utilize the concepts of probability density to a great extent. The topics briefed would provide the readers to have a quick recap of the concepts which though seem to be from different specializations but are deeply connected to one another. Section one is dedicated to mathematics for image processing and Section two connects mathematics, image processing with forensic sciences.

Keywords: mathematics, probability, forensic sciences, image processing, neural network, score- based likelihood

1. Introduction

The chapter aims to bridge the gap between AI and Mathematics at core of it. They represent the two branches of the same tree [1]. But since they are taught independently, many times it becomes difficult to connect the two and look at the broader picture with a satellite view. AI was invented way back in 1950's by John Mc Carthy who coined the phrase the science and engineering of making intelligent machines. A brief history could be referred in [2].

The three layers of AI basically consist of Incremental advancements in algorithm design, application to specific design and step change is performed for improvement. Mathematical and statistical concept clarity is highly desirable to understand and implement AI. At this point, an attempt is made with few topics to show the connection in a lucid way. Few explanations are beyond the scope of the chapter so are stated with detailed references for further explorations.

Since there is a deep connection between mathematics, image processing and forensic sciences, the topics included will be an initial knowhow to start exploring the possibilities to scholars in forensic based image processing. Multimedia forensic utilizes AI techniques which involve convolution neural network. Analysis of forensic image is used to decode the fake information about photographs using machine learning [3], another important application is in medicine for forensic anthropology [4]. To understand such recent research in multimedia, concepts of image sampling, enhancement, convolution and neural network are included in this chapter. IoT based forensic analysis also utilizes the statistical approaches discussed.

Beginning with few basic image processing and mathematical expressions, a systematic development is traced through score-based likelihood used in forensic exploration. It would be beneficial to a reader interested in exploring the quantitative procedure to weigh evidence which would use likelihood ratio [5]. In the later part of the chapter a brief introduction to neural network paving the path to digital forensic neural network is included.

Keeping the above idea in mind, the first section includes a brief discussion on topics of image processing and mathematical representation, image formation model, image sampling, intensity transformation and spatial filtering, image enhancement using Laplacian mask, image denoising, order statistics filters, convolution in image processing. The second section is dedicated to forensic image processing and mathematics which includes widely concepts of steganography, discrete Fourier transform, score-based likelihood, and finally an overview of Neural network and probability theory.

2. Brief of mathematics for image processing

Digital image processing has developed rapidly along with the advances in computing and mathematical advances, to nurture the ever-growing demand for technological luxuries.

2.1 Mathematics in image processing

Mathematical image processing is widely used in fields such as medical imaging, surveillances, video transmission, astrophysics and many more. Signals are one dimensional image. Planar images are in two dimension and volumetric are in three dimensions. Images in grey scale are classified as single valued functions while coloured images are vector valued functions. The imperfections such as blurring and noise reduce the quality of the image

2.2 Image formation model

Mathematically a planar image is represented by a function form as spatial domain to a function value,

$$(x, y) \rightarrow f(x, y) \tag{1}$$

In an image the intensity value is the energy radiated by the physical source.

$$0 < f(x, y) < \infty \tag{2}$$

$f(x, y)$ depends on the illumination $i(x, y)$ and the reflectance $r(x, y)$ hence,

$$f(x, y) = i(x, y)r(x, y) \tag{3}$$

A similar expression is applicable to the images formed by transmission through a medium.

2.3 Image sampling

Sampling means digitalization of coordinate values and digitalizing the amplitude means quantization. A digital image is represented as a matrix and the number of grey levels is taken in powers of 2.

2.4 Intensity transformation and spatial filtering

For improving contrast, a statistical tool of histogram equalizer is used. Consider a low contrast/dark image or light image as input image $p(x, y)$ and an output as a high contrasted image $m(x, y)$.

Assume that the grey-level range consists of L grey-levels. In the discrete case, let $rk = p(x, y)$ be a gray level of

$$p,k = 0,1,2, \ldots L - 1. \tag{4}$$

Le$s_k = m(x, y)$ be the desired gray level of output image m.

A transformation T :

$$[0, L - 1] \rightarrow [0, L - 1] \text{ such that } s_k = T(r_k), \quad \text{for all } k = 0,1,2, \ldots ,L - 1. \tag{5}$$

Define $h(r_k) = n_k$ where r_k is the kth grey level, and n_k is the number of pixels in the image p taking the value r_k.

Visualization of discrete function gives histogram.

In the discrete case, let $r_k = p(x, y)$. Then we define the histogram-equalized image m at$(x, y\ )$ by

$$m(x, y) = s_k = (L - 1)\sum_{j=0}^{k} p(r_j\ ) \tag{6}$$

Theoretical interpretation of histogram equalization (continuous case) considering the grey levels r and s as random variables with associated probability distribution functions $p_r(r)$ and $p_s(s)$ The continuous version uses the cumulative distribution function and we define

$$s = T(r) = (L - 1)\int_0^r p_r(w)dw \tag{7}$$

If p_k (r) > 0 on $[0, L-1]$, then T is strictly increasing from $[0, L-1]$ *to* $[0, L-1]$, thus T is invertible. Moreover, if T is differentiable, then we can use a formula from probabilities: if $s = T(r)$, the

$$p_s(s) = p_r(r)\frac{\partial r}{\partial s} \tag{8}$$

where we view s = s(r) = T(r) as a function of r, and r = r(s) = T–1 (s) as a function of s. From the definition of s, we have by differentiating equation (7)

$$\frac{\partial s}{\partial r} = (L-1)p_r(r) = T'(r) \tag{9}$$

$$\begin{aligned} \frac{\partial r}{\partial s} &= \frac{\partial}{\partial s}\left(T^{-1}(s)\right) = \frac{1}{T'(s)} = \frac{1}{(L-1)p_r(r(s))} \\ p_s(s) &= p_r(r)\frac{\partial r}{\partial s} = p_r(r)\frac{1}{(L-1)p_r(r(s))} = \frac{1}{(L-1)} \end{aligned} \tag{10}$$

So, the uniform probability distribution function p_s in the interval $[0, L-1]$, corresponding to a flat histogram in the discrete case.

2.5 Image enhancement

Considering an image function f with second order partial derivatives, Laplacian of f in continuous form is defined as,

$$\Delta f = \frac{\partial^2 f}{\partial x^2} + \frac{\partial^2 f}{\partial y^2} \tag{11}$$

$\Delta f \rightarrow f$ is linear and rotationally invariant.

Implementing numerical analysis finite difference approximation, the second derivatives can be approximated as,

$$\frac{\partial^2 f}{\partial x^2}(x, y) \approx \frac{f(x+h, y) - 2f(x, y) + f(x-h, y)}{h^2} \tag{12}$$

and

$$\frac{\partial^2 f}{\partial y^2}(x, y) \approx \frac{f(x, y+k) - 2f(x, y) + f(x, y-k)}{k^2} \tag{13}$$

Here second derivative of $f(x, y)$ is approximated using the function values of f at $x, x+h, x-h$ keeping y constant for second derivative with x. Similar approach is used for second derivative of f with y.

Considering h=k=1 for any image a Laplacian 5-point formula given by

$$\Delta f(x, y) \approx f(x+1, y) + f(x-1, y) + f(x, y+1) + f(x, y-1) - 4f(x, y) \tag{14}$$

which can be applied to discrete images.

The Laplacian mask defined for a spatial filter is

$$m = \begin{bmatrix} 0 & 1 & 0 \\ 1 & -4 & 1 \\ 0 & 1 & 0 \end{bmatrix} \tag{15}$$

Laplacian operator can be discretized with 9-point Laplacian formula as,

$$\begin{aligned} \Delta f(x,y) \approx & f(x+1,y) + f(x-1,y) + f(x,y+1) + f(x,y-1) \\ & + f(x+1,y+1) + f(x-1,y+1) + f(x-1,y-1) + f(x+1,y-1) - 8f(x,y) \end{aligned} \tag{16}$$

with Laplacian mask $m = \begin{bmatrix} 1 & 1 & 1 \\ 1 & -8 & 1 \\ 1 & 1 & 1 \end{bmatrix}$

The sum of coefficients is 0 for both the Laplacian masks m, which is related with the sharpening property of the filter. The Laplacian can be used to enhance images. For example, in 1D a smooth-edged image can be enhanced to a sharp-edged profile by applying the operator $e = f - f''$. In 2D for a blurry image an operator $e(x,y) = f(x,y) - \Delta f(x,y)$. In discrete case if 5-point Laplacian is used, we obtain the linear spatial filter and mask as,

$$e(x,y) = f(x,y) - \Delta f(x,y) = 5f(x,y) - f(x+1,y) - f(x-1,y) - f(x,y+1) - f(x,y-1) \tag{17}$$

and

$$m = \begin{bmatrix} 0 & -1 & 0 \\ -1 & 5 & -1 \\ 0 & -1 & 0 \end{bmatrix}$$

Similarly, the 9-point Laplacian mask

$$m = \begin{bmatrix} -1 & -1 & -1 \\ -1 & 9 & -1 \\ -1 & -1 & -1 \end{bmatrix} \tag{18}$$

2.6 Image denoising

Human vision can classify and categorize the image into different levels but for a digital camera denoising is difficult. Enhance observations, interpolating missing image data are to be performed rigorously to improve image quality. There are many reasons for image contamination such as heat generated by camera or external sources which might emit free electrons from the image sensor itself, thus contaminating the true photoelectrons.

Mathematically, one can write the observed image captured by devices as [6]:

$$v(i) = u(i) + n(i), \tag{19}$$

where $v(i)$ is the observed value, $u(i)$ is the true value, which needs to be recovered from $v(i)$. $n(i)$ is the noise perturbation.

For a grey value image, the range of the pixel value is (0, 255), where 0 represents black and 255 represents white. To measure the amount of noise of an image, one may use the signal noise ratio (SNR),

$$SNR = \frac{\sigma(u)}{\sigma(n)}, \tag{20}$$

where σ(u) denotes the empirical standard deviation of u(i),

$$\sigma(u) = \sqrt{\frac{\sum_i (u(i) - \overline{u})^2}{N}} \tag{21}$$

$$\sigma(n) = \sqrt{\frac{\sum_i (u(i) - v(i))^2}{N}} \tag{22}$$

where $\overline{u} = \frac{\sum u(i)}{N}$ the average grey level values computed from a clean image.

SNRs are usually expressed in terms of the logarithmic decibel scale as signals have a wide dynamic range. In decibels, the SNR is, by definition, 10 times the logarithm of the power ratio:

$$SNR = 10 \log_{10} \left(\frac{\sum_i (u(i) - \overline{u})^2}{\sum_i (u(i) - v(i))^2} \right) \tag{23}$$

A denoising method can be defined as D_h working on an image u:

$$u = D_h u + n(D_h, u) \tag{24}$$

where h is the filtering parameter, D_h is the denoised image, and $n(D_h, u)$ is the noise guessed by the method.

It is not sufficient to just smooth$u$ and get the denoised image. The more recent methods are not only working on smoothing, but also try to recover lost information in $n(D_h, u)$ as needed as discussed by [7, 8], i.e. in an image captured by digital SLR cameras, we often need to keep the sharpness and the detailed information while the noise is being blurred. In the literature, work has been done on local filtering methods which include Gaussian smoothing model [9], Bilateral filters (Elad), PDE based methods, including anisotropic filtering model [10, 11] and total variation model (F. Guichard). Approaches using frequency domain filtering [12], Steering kernel regression [13] and so on.

2.7 Order statistics filters

Order filters are used in image processing. Non-linear spatial filters are classified as Order statistic filters, whose response is based on the ordering of the pixels contained in the image area encompassed by the filter, and then replacing the value in the centre pixel with the value determined by the ranking result.

It is an estimator of mean which uses a linear combination of order statistics. Now the question is how is it different from the mean filter? It is known that the mean filter is a simple sliding-window of spatial filter that replaces the centre value in the window with the average of all the pixel values in the window but for order statistic filter we consider N observations arranged in ascending order,

$$X_1 < X_2 < \dots < X_N \tag{25}$$

$\{X_i\}$ are the order statistics of the N observations [14]. An orders statistics filter is an estimator of $F(X_1, X_2, \dots, X_N)$ of the mean of X as

$$F(X_1, X_2, \dots, X_N) = a_1X_1 + a_2X_2 + \dots + a_NX_N \tag{26}$$

The linear average which has coefficients $a_i = \frac{1}{N}$ and the median filter which has coefficients

$$a_i = \begin{cases} 1 & i = \dfrac{N+1}{2}, \\ 0 & otherwise \end{cases} \tag{27}$$

The trimmed mean filter has coefficients,

$$a_i = \begin{cases} \dfrac{1}{M} & \dfrac{N-M+1}{2} \le i < \dfrac{N+M+1}{2}, \\ 0 & otherwise \end{cases} \tag{28}$$

For any distribution one can determine the optimal coefficient by minimizing the criteria function
$J(a) = E\left[\left(a^TX - \mu\right)^2\right]$ with a being the vector of order statistics filter coefficient, X is the vector of order statistics and is the mean of X, here, $E[X] = \frac{\sum_i X_i}{N}$. The order statistic filters which turns out to be important for Grid Filters is that they are piecewise linear. Order statistic filter is generally applied to 3x3, 5x5 or 7x7 windows.

2.8 Convolution in image processing

Filter effect for images is due to convolution. A matrix operation is applied to an image in which a mathematical operation comprised of integers is performed. The idea is to determine the value of a central pixel by adding the weighted values of all its neighbours together. The outcome is a new modified filtered image. Convolution is performed to obtain a smooth/enhance/intensity/sharpen an image.

A convolution is done by multiplying a pixel to its neighbouring pixels colour value by a matrix called kernel. Kernel is a small matrix of numbers that is used in image convolutions. Differently sized kernels containing different patterns of numbers produce different results under convolution. The size of a kernel is arbitrary but 3x3 is often used [15].

Formula for convolution is given by,

$$W = \frac{\sum_{i=1}^{q}\sum_{j=1}^{q} f_{ij} d_{ij}}{F} \tag{29}$$

W is the output pixel value.

f_{ij} the coefficient of the convolution kernel at position i,j in the kernel matrix

d_{ij}data value of the pixel that corresponds tof_{ij}

F sum of the coefficients of kernel matrix

q dimension of the kernel.

Now in next section a connection between mathematics behind the image processing and forensic based image processing is handled.

3. Mathematics for forensic image processing

Forensic can provide information security when the information source is not trusted. There are forgery detection algorithms to detect interplay between the actual and the modified details. Probability, linear algebra helps in learning forgery. Fourier transform infrared micro spectroscope is used in analysing traumatic brain injuries [16]. Hypothesis testing with local estimates help in identifying the unaltered and falsified images. Stochastic gradient descent approaches are also used in forensic editing detection. These are used as metadata is unreliable and multimedia forensic is found superior in such cases. In certain advance studies graph theory is also employed to image identification purposes in a better way. So now we will walk through the different aspects to have the essence of the approaches utilized in forensic sciences.

3.1 Steganography

The art of hidden writing is known as Steganography. Steganography is different from cryptography (the art of secret writing), which is utilized to make a message unreadable by the creator. Steganography is commercially used functions in the digital world, most notably digital watermarking. If someone else steals the file and claims the work as his or her own, the artist can later prove ownership because only he/she can recover the watermark [17, 18].

Kessler [19] quoted that a computer forensics examiner might suspect the use of steganography because of the nature of the crime, books in the suspect's library, the type of hardware or software discovered, large sets of seemingly duplicate images, statements made by the suspect or witnesses, or other factors. A website might be suspect by the nature of its content or the population that it serves. These same items might give the examiner clues to passwords, as well. And searching for steganography is not only necessary in criminal investigations and intelligence gathering operations. Forensic accounting investigators are realizing the need to search for steganography as this becomes a viable way to hide financial records which is discussed by [20].

3.2 Discrete Fourier transform

A discrete Fourier transform transforms a sequence of N complex numbers $\{x_n\} := x_1, x_2, \ldots x_{N-1}$ to another sequence of complex numbers $\{X_r\} := X_1, X_2, \ldots X_{N-1}$ as,

$$X_r = \sum_{n=0}^{N-1} x_n e^{\frac{-2\pi rn}{N}} = \sum_{n=0}^{N-1} x_n \left(\cos\left(\frac{-2\pi rn}{N}\right) + i \sin\left(\frac{-2\pi rn}{N}\right)\right) \tag{30}$$

$$X = F(x) \tag{31}$$

3.3 Score-based likelihood

Statistical methods are widely used for digital image forensic. A score-based likelihood ratio (SLR) for camera device identification is one amongst them. Score based likelihood is a deciding factor for selection of the similarities between the evidences such as two fingerprints obtained from a crime scene and the one from the suspect. Photo-response non-uniformity as a camera fingerprint and one minus the normalized correlation as a similarity score [21]. The procedure includes source identification problems and forgery and tampering detection problems. A camera device identification problem utilizes the SLR. It decides whether given an unknown digital image has been taken by a known camera device or not along with the strength and weakness of the evidence in favour of the decision. It helps in qualifying the weight of the evidence. SLRs fit probability density functions to both sets of scores. Both pdfs are evaluated at the score between the noise residual of the image in question and the camera fingerprint, and the SLR is the ratio of the results.

Let's discuss a simple example as,

Let I_1 be the image output from a camera that includes noise, and let I_0 be the perfect image without noise. Consider the camera fingerprint, as K, and denote all other noise components from the image as ϕ. Then the image output t I_1can be modelled as,

$$I_1 = I_0 + I_0 K + \phi \tag{32}$$

The true image K is not obtained practically so we estimate it using maximum likelihood estimator $\hat{K}$

$\hat{K}$ can be obtained by collecting first N images from the camera in discussion as $I^{(1)}, I^{(2)}_{\dots} \cdot I^{(N)}$

The noise residuals from each image can be computed utilizing a filter F using the formula

$$W^{(i)} = I_1^{(i)} - F\left(I_1^{(i)}\right) \tag{33}$$

with $i = 1,2, \dots N$.

Then the photo response non uniformity is given by the formula,

$$\hat{K} = \frac{\sum_{i=0}^{N} W^{(i)} {I_1}^{(i)}}{\sum_{i=0}^{N} \left({I_1}^{(i)}\right)^2} \tag{34}$$

For accuracy of similarity score, peak to correlation score is replaced instead of the normalized correlation computation. The former has a better decision threshold stability [21]. Further a trace anchored score-based likelihood ratio could also be studied. SLR framework uses the hypothesis testing methodology to justify the conclusion based on the evidences received.

3.4 Basics of neural network

Neural network consists of artificial neurons connected in a network structure. Its structure is inspired by the human brain and learns from the data feed to it. It has an extensive approximation property. There are several types of neural networks namely, recurrent neural network, convolutional neural network, radial basis function neural network, feedforward neural network, modular neural network. A basic structure will be discussed here for getting the insight of its architecture. Single layer neural network in mathematical expressions is seen as

$$y_1 = f(\alpha_{11}x_1 + \alpha_{12}x_2 + \alpha_{13}x_3) \tag{35}$$

The output y_1 is derived from the input $\{x_j\}$ using weights $\{\alpha_{1j}\}$described in a function form $j = 1 \ to \ 3$ (**Figure 1**).

3.4.1 Neural network with one hidden layer

The neural network with one hidden layer having three inputs and one output with one hidden layer is given by,

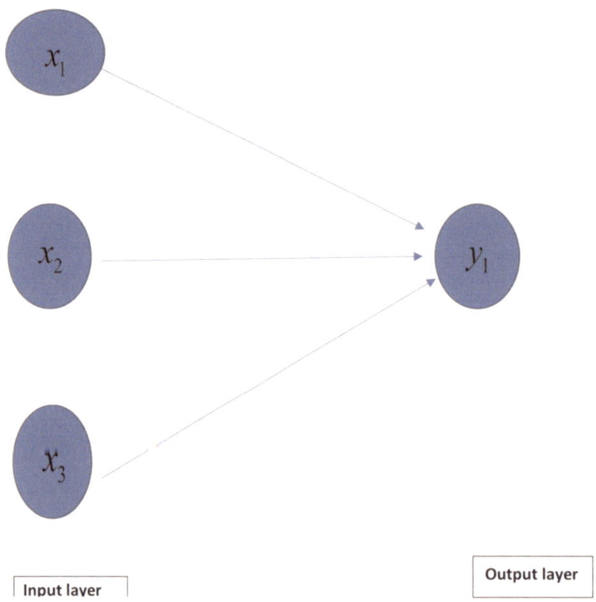

Figure 1.
Representation of a single layer neural network without hidden layer.

$$
\begin{aligned}
y_1^{(2)} &= g\left(\alpha_{11}^{(1)}x_1 + \alpha_{12}^{(1)}x_2 + \alpha_{13}^{(1)}x_3\right) \\
y_2^{(2)} &= g\left(\alpha_{21}^{(1)}x_1 + \alpha_{22}^{(1)}x_2 + \alpha_{23}^{(1)}x_3\right) \\
y_3^{(2)} &= g\left(\alpha_{31}^{(1)}x_1 + \alpha_{32}^{(1)}x_2 + \alpha_{33}^{(1)}x_3\right) \\
\\
y_1^{(3)} &= g\left(\alpha_{11}^{(2)}y_1^{(2)} + \alpha_{12}^{(2)}y_2^{(2)} + \alpha_{13}^{(2)}y_3^{(2)}\right)
\end{aligned}
\tag{36}
$$

The output $y_j^{(2)}$ is derived from the input $\{x_j\}$ using weights $\left\{\alpha_{ij}^{(1)}\right\}$ described in a function form for$i,j = 1\ to\ 3$. The superscript describes the level of layer applied. Then the final output $y_1^{(3)}$ is computed from the information of hidden layer. Similarly, architecture of networks with two hidden layers can be generated (**Figure 2**).

Such structures are classified under single layer feed forward network. In multi-layer feed forward networks, the inclusion of one or more hidden layers enables the network to be computationally more effective. Feedback networks take in output as its input and closes the loop to gain improvements in the procedure. There are other patterns as well to enhance the performances of the network.

Kingston [22] quoted that Simulation experiments with a type of neural network known as a Hopfield net indicate that it may have value for the storage of tool mark patterns (including bullet striation patterns) and for the subsequent retrieval of the matching pattern using another mark by the same tool for input.

A convolutional neural network is a deep learning algorithm which takes image as an input, assign importance to various aspects and is able to differentiate between the features. Its architecture is very similar to the neurons in human brain. An image which is a 3*3 matrix entry is written as 9*1 vector and is feed to a multilayer perceptron for classification. Digital forensic implements neural network for better analysis [23].

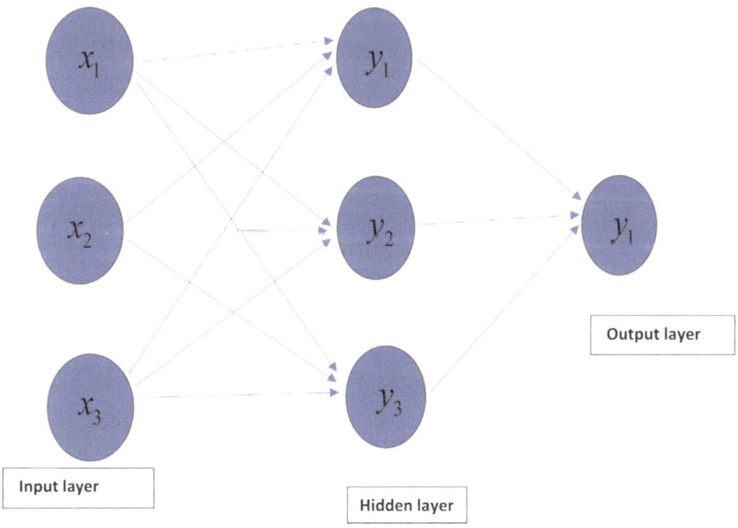

Figure 2.
Visualization of the neural network with one hidden layer having three inputs and one output with one hidden layer.

3.5 Basics of probability

Forensic analysis is using probabilistic inferences and probabilistic reasoning. Probability can be termed as specialised facet of logical reasoning whereas statistics deals with collection and summary of data. It may be utilized in measuring. Forensic may include criminal proceedings with chemical analysis of suspicious substances and measurements of elemental compositions of glass fragments. Probability is a branch of mathematics which aims to conceptualise uncertainty and render it tractable to decision-making (Aitken-Roberts-Jackson). In the criminal justice context, the accused is either factually guilty or factually innocent, there is no other possibility. Hence, p(Guilty, G) + p(Innocent, I) = 1. Applying the ordinary rules of number, this further implies that p(G) = 1–p(I); and p(I) = 1–p(G). The probabilistic formulae are also used to measure levels of uncertainty associated with a particular estimate.

Probability is widely used in AI as it resolves around the data collection handling and analysing to reach conclusive outcomes. With the advance foreseen utility of AI in various fields like agriculture, engineering, demography, medicine, education, marketing and so on probability will be involved as the core of algorithms. An overview of few of the important basic concepts are included in this section.

Statistics will deal in data characterisation and analysis. It involves grouping and the choice of selection by hypothesis testing. Here an overview is included.

An experiment is a process of observation and measurement. Randomness is the key desired inclusion for the experiments in discussion. The subsets of a sample space are events.

If equally likely events having finite outcomes, then P(A) is the probability of event A as

$$P(A) = \frac{Number\ of\ outcomes\ to\ the\ occurance\ of\ an\ event\ A}{Total\ number\ of\ equally\ likely\ events}$$

$$0 \leq P(A) \leq 1$$

Impossible event P(A)=0 and P(A)=1 is a certain event.

Conditional probability deals with the concept of events occurrence under the condition of another event already occurred. If the requirement to study the probability of event A under the circumstance with event B occurring already is given by

$$P\left(\frac{A}{B}\right) = \frac{P(A \cap B)}{P(B)}, P(B) \neq 0$$

3.6 Bayes theorem

Bayes theorem is implemented in a variety of applications where conditional probability is involved. It is used in classification of occurrence of an event when the consideration of probability of occurrence of event is true is included without any specific evidence.

Bayes theorem states that

If we consider S to be a sample space which is subdivided into $C_1, C_{2,\ldots,} C_k$ with $P(C_i) \neq 0$ for $i = 1,2,\ldots,k$. Then for any arbitrary event A in S with $P(A) \neq 0$ we have for $r = 1,2,\ldots,k$,

$$P\left(\frac{C_r}{A}\right) = \frac{P(C_r \cap A)}{\sum_{i=1}^{k} P(C_i \cap A)} = \frac{P(C_r)P\left(\frac{A}{C_r}\right)}{\sum_{i=1}^{k} P(C_i)P\left(\frac{A}{C_i}\right)} \quad (37)$$

Although the Bayes theorem is widely used but there are few limitations of the theorem as the requirement of mutually exclusive and exhaustiveness is not achievable practically. If the data is huge the computation is not feasible. In many problems the requirement of relevant conditional probability to be stable over time is also not achievable. Further often assuming the constant likelihood ratio in the denominator is not feasible in practical cases. Bayes theorem helps in making interpretations from the collected samples from the site of crime [24].

3.7 Probability density functions

Probability density function is widely used in the AI and Neural networks. Probability density function is used to define a range for a specific occurrence of a continuous random variable. The area under the curve represents the interval in which the continuous random variable will fall.

$f(x)$ is called a probability density function if,

$$\int_{-\infty}^{\infty} f(x)dx = 1 \quad f(x) \geq 0 \quad \forall x \quad (38)$$

Probability density function is different from probability mass function which will indicate whether there is a probability of a function to achieve a specific value and not a range as in the case of probability density function.

Probability density function of an image (Rafael C. GONZALES)

$$A = \sum_{g=0}^{255} m_{I_k}(g+1) \quad (39)$$

$m_{I_k}(g+1)$ represents the number of pixels in I_k (k denotes colour band of image I) with value g. A denotes the number of pixels in I. The gray level probability density function of I_k is given by,

$$p_{I_k}(g+1) = \frac{1}{A} m_{I_k}(g+1) \quad (40)$$

The use of probability lies in the starting point of neural network for making decisions. Casual and probabilistic decisions are made naturally in human brain but the machine needs to be trained using probability theory. Machine learning uses Bayes theorem. Bayes theorem relates marginal probability and conditional probabilities. Experiment outcomes termed as events are mutually exclusive if they do not occur simultaneously. Events are called exhaustive if the occurrence of them constitute the entire possibilities.

Mathematically events A and B are said to be mutually exclusive and exhaustive if

$$A \cap B = \varnothing$$

$$A \cup B = S$$

where S constitutes all possible outcomes of the sample space.

For further enhancement of the approaches many more algorithms are required which could be understood by starting with the ideas discussed above. Probability is utilized to understand the interpretations of forensic and law for concluding legal decisions [25].

4. Conclusion

Artificial intelligence is built along human cognition i.e., learning and retrieving. A network is expected to recognize a previously learned pattern even when some noise is involved. Associative memory is desirable in building multimedia database. Here optimization plays a significant role as it deals with finding the solution which satisfies a given set of constraints. The purpose of AI will be to create admissible model to the human brain. The idea to produce some computational structures similar to neurons or neuron system and connections between them to form neural networks. Application of AI is vast and in this chapter an attempt is made to sensitize the reader with few basic prerequisites to start exploring AI using mathematics for advancements in forensic sciences. The topics involve maybe studied in details to implement the procedures. This will pave the way to understanding and exploring new search methods.

A variety of search mechanisms are employed to problems which include blind search, searching in extent and Heuristic search. A basic introduction to graph theory and probability with examples of implementation in game of 'chess' or 'Go' motivates the reader to have an insight to the applications. For example, the study of Bayesian networks as a direct acyclic connected graph, with probability distribution associated with each node that defines a mutual relationship between nodes and edges.

Modelling data can be done using algorithms mentioned could use MATLAB techniques. For example if the knowledge of how different parameters influence the energy load is known then we might use statistics or curve fitting tools to model the data with linear or nonlinear regression. If the number of variables is more, the underlying system is particularly complex, or the governing equations are unknown, then we could use machine learning techniques such as decision trees or neural networks. Use of Neural fitting app (Filion).

Further a deeper conceptualization aims to modify the existing research in the benefit of society as a whole. Future research in IoT forensic analysis is having wide scope as different algorithms could be modified incorporating data from relevant inputs which could lead to capturing of the limitations for the protype generated.

To conclude, the prerequisites of the chapter would enhance the reader to work with the underlying concepts in dept to advance the research in the field of image processing and forensic sciences. In future study, researchers may undertake improvement of the algorithms of forensic image processing for an enhanced AI development which is not possible without the basic understanding of some key procedures as discussed. Although the topics may be of interest to readers not limited to forensic sciences as they are applied in various engineering fields [26–33].

Author details

KP Mredula Pyarelal
Applied Science and Humanities Department, Sardar Vallabhbhai Patel Institute of Technology Vasad, Affiliated to Gujarat Technological University, Vasad, India

*Address all correspondence to: mredpl@gmail.com

References

[1] Garrido A. Mathematics and artificial intelligence, two branches of the same tree. Procedia Social and Behavioral Sciences. 2010;**2**(2):1133-1136

[2] Rigano C. Using Artificial Intelligence to Address Criminal Justice Needs. National Institute of Justice; 2019

[3] Atiyah J. Image forensic and analytics using machine learning. International Journal of Computing and Business Research. 2022;**12**:69-93

[4] Andrej Thurzo HS. Use of advanced artificial intelligence in forensic medicine. New Trends in Forensic and Legal Medicine. Forensic Anthropology and Clinical Anatomy. Healthcare (Basel, Switzerland). 12 Nov 2021;**9**(11): 1545

[5] Iyer SP. Likelihood ratio as weight of forensic evidence: A closer look. Journal of Research of National Institute of Standards and Technology. 2017;**122**

[6] Deng H. Mathematical Approaches to Digital Color Image Denoising. Thesis. 2009

[7] Malgouyres F. A noise selection approach of image restoration. Applications in. 2001:34-41

[8] Osher S. Using geometry and iterated refinement for inverse problems. Total Variation Based Image Restoration. 2004:4-13

[9] Bruchkstein ML. On Gabor contribution to image enhancement. 1994;**27**:1-8

[10] Morel LA-L-M. Image selective smoothing and edge detection by nonlinear diffusion (II). SIAM Journal of Numerical Analysis. 1992;**29**:845-866

[11] Malik PP. Scale space and edge detection using anisotropic diffusion. IEEE Transactions on Pattern Analysis and Machine Intelligence. 1990;**12**: 629-639

[12] Zhou YW. A total variation wavelet algorithm for medical image denoising. The International Journal on Biomedical Imaging. 2006. DOI: 10.1155/IJBI/2006/89095

[13] Takeda H. Kernel regression for image processing and reconstruction. IEEE Transactions on Image Processing. Feb 2007;**16**(2)

[14] David HA. Order Statistics. 3rd edn. Toronto: Wiley; 1981

[15] Mather PM. Computer Processing of Remotely Sensed Images, An Introduction. West Sussex: John Wiley & Sons Ltd; 2004

[16] Zhang J et al. Characterization of protein alterations in damaged axons in the brainstem following traumatic brain injury using fourier transform infrared microspectroscopy: A preliminary study. Journal of Forensic Sciences. 2015: 759-763

[17] Arnold MS. Techniques and Applications of Digital Watermarking and Content Protection. Book by Artech house. 2003

[18] Barni MP. Watermark embedding: Hiding a signal within a cover image. IEEE Communications. 2001;**39**(8): 102-108

[19] Kessler GC. An overview of steganography for the computer forensics examiner. Forensic Science Communications. 2015

[20] Hosmer CA. Discovering covert digital evidence. In: Digital Forensic Research Workshop (DFRWS). Proceedings DFRWS; 2003

[21] Reinders S. Statistical Methods for Digital Image Forensics: Algorithm Mismatch for Blind Spatial. Iowa State University; 2020

[22] Kingston C. Neural netowrk in forensic science. Journal of Forensic Sciences. 1992

[23] Mohammad R. A neural network based digital forensics classification. In: 2018 IEEE/ACS 15th International Conference on Computer Systems and Applications (AICCSA). 2018. pp. 1-7

[24] Pete Blair J. Evidence in context: Bayes theorem and investigations. 2010

[25] Taroni F. Uncertainty in forensic science: Experts, probabilities and Bayes' theorem. Italian Journal of Applied Statistics. 2015:129-144

[26] Jegede OA. Neural networks and its application in engineering. In: Proceedings of Informing Science & IT Education Conference (InSITE). 2009

[27] El-Sharkawi MA. Neural networks and their application to power engineering. Control and Dynamic Systems. 1991

[28] Singh Y. Application of neural networks in software engineering: A review. Communications in Computer and Information Science. 2009:128-137

[29] Aitken-Roberts J. Fundamentals of Probability and Statistical Evidence. Royal Statistical Society. Practitioner guide no 1. 2022. pp. 1-122

[30] Elad M. On the Origin of the Bilateral Filter and Ways to Improve It. 2022. DOI: 10.1109/TIP.2002.801126

[31] Guichard F. Image Analysis and P.D. E.'s Book. Available from: https://archive.siam.org/meetings/an00/talks_online/MorelGuichard.pdf

[32] Filion SD. Data-driven insights with MATLAB analytics: An energy load forecasting case study. Mathwork. 2012

[33] Rafael C, Gonzales RE. Digital Image Processing. 2022

Chapter 3

Computer Vision: Anthropology of Algorithmic Bias in Facial Analysis Tool

Mayane Batista Lima

Abstract

The usage of Computer Vision (CV) has led to debates about the bias within the technology. Despite machines being labeled as autonomous, human bias is embedded in data labeling for effective machine learning. Proper training of neural network machines requires massive amounts of "relevant data," however, not all data is collected. This contributes to a one-sided view and feeds a "standard of data that is not collected." The machine develops algorithmic decision-making based on the data it is presented, which can create machinic biases such as differences in gender, race/ethnicity, and class. This raises questions about which bodies are recognized by machines and how they are taught to "see" beyond binary "male or female" limitations. The study aims to understand how Amazon's Rekognition, a facial recognition and analysis tool, analyzes and classifies people of dissident genders who do not conform to "conventional" gender norms. Understanding the mechanisms behind the technology's decision-making processes can lead to more equitable and inclusive outcomes.

Keywords: artificial intelligence, computer vision, algorithmic Bias, Misgendering, Amazon recognition

1. Introduction

Artificial Intelligence (AI) is full of biological analogies we tend to recognize their machinic bodies and organs, through sci-fi stories, as in the case of Computer Vision (VC) from 2001: A Space Odyssey [1]. In the movie [2], HAL 9000, a sentient AI, a Heuristic Algorithmic programming computer, is composed of cameras with fisheye lenses and an internal structure constituted by the binomial <algorithms+data>, which in many branches of AI is impossible to dissociate [3]. This structure provides Hal with the visual inputs needed to scan and analyze the spacecraft Discovery, and is imbued with natural language (similar to a human voice) to interact with the interstellar mission crew.

As in history, we are surrounded by voices (Siri, Alexa, and Google Assistant) and machine eyes through cell phones, notebooks, cameras installed on poles, in ATMs, subways, cars, busses, and drones, whether autonomous or guided. All configured with objectives and must present results with the means shown to them; thus, the scanning of facial expressions goes through a series of tangles inspired by the human brain, the convolutional neural networks (CNN), which are responsible for the analysis and processing of data. Videos and images so for are an effective knowledge result, and a lot of data is needed for the neural network to learn about who or what is seeing these ramifications are made up of norms and models, which regulate the constitutive patterns of what to see, for whom to look, and what should be described about whom one is looking at. Commonly, these models are known as algorithms, in terms of computer science, and they are any computational, mathematical, and statistical procedures aligned to take values or set of values as input and produce set of values as output [4, 5].

In other words, machines learn by using models and analyzing patterns through algorithms, so the computer learns about what it is seeing according to the data presented to it, over and over again, so that it can understand what differs a leopard from a cat, for example. In this way, the objective of AI algorithms is learning, so when new information is presented, it knows how to classify, regardless of what was previously shown to the algorithm, analyzing patterns through the data articulated with each other to generate results. From this point of view, Russell [6] and Lee [7] argue that neural networks demonstrate effective recognition after proper training with labeled examples that connect the many data points to the expected result, and this action according to both requires massive amounts of "relevant data."

Data considered relevant are those that are always contained in the machines, even with the massive production of existing data, not all are collected and if they are, they go through a screening. Thus, no matter how sophisticated they are, algorithms are useless in isolation, and part of their results is supported by the data and samples contained in them, as well as in the way it interacts with the environment [8]. Crucially, data are people [9] if a certain group is included and others are on a smaller scale, statistically the "data that is left out" does not exist according to the analysis of the machine, so the algorithms analyze that a certain group of people is considered hegemonic [10]. In this way, the machine develops algorithmic decision making based on what appears in the data, establishing parameters that express the machinic biases.

2. Anthropology of algorithmic bias

Medeiros [3] argues that: a) algorithms can be blamed for certain results, even when there is curation performed on the entered data. In other cases: b) algorithms specified under inappropriate assumptions generate inappropriate solutions, even though the data have been well configured. There is also the possibility that: c) both algorithms and data have human biases, which may have been unconsciously inserted by their programmers and creators. In this sense, from the anthropological point of view, Forsythe [11] argues that software embodies values that are tacitly maintained by those who build them.

> *You need to consider sources of bias throughout the data lifecycle – collection, curation, analysis, storage and archiving. [...] the responsibility does not end with archiving data, or delivering software. Regardless of whether bias exists in data, algorithms, or in their combination, it always appears due to humans – in collection, analysis or interpretation, whether intentional or through ignorance. And when adding machines to the binomial, more questions arise. ([3]:12)*

To a certain extent, AI algorithms are human extensions, so by extensions it is understood that there is an automated widening of biases and as noted earlier, even though machines are named as autonomous, such as AI algorithms, human bias

[5, 12] is embedded for effective learning. It is with the premise of corrosive software in the social context that investigations about Computer Vision (VC) are consistent in several debates about its use and the biases inserted in the machines, therefore racist algorithmic biases [13–16], as well as disparities related to gender, class, politics, democracy, and surveillance.

Like Hal 9000—the computer presented at the beginning of this article—the information contained in it constituted his decision making for what he should do and in which situations he should act, but his actions were based on the decisions of his creator Dr. Chandra. There is a lot of controversy about the story, but it still presents a scenario that demonstrates how the objectives embedded in analysis software can establish significant changes in the short/long term, if they do not go through a bias audit process. In this sense, when analyzing the bibliographies and documentary about the algorithmic gender bias, race[1], it was analyzed that a large part is articulated around the social binary "woman" or "man."

2.1 AWS Amazon rekognition (facial analysis)

From the situations presented by Joy Buolamwini in the documentary *Coded Bias* [17], in which she describes the process of a project for the MIT Media Lab in which facial analysis software from conglomerates is used: Google Cloud, Microsoft, AWS Amazon, and IBM Watson, and in most of these software, her face is not recognized, which configures that the tools did not recognize her as a human only when she started to use the white mask[2]—symbol of *Algorithmic Justice League*—there is recognition and it is noticed that in the course of the documentary, there are changes in the algorithms/data of these companies that begin to detect the scientist's face, in addition to classifying her, indicating gender, age, and emotional aspects.

Considering that algorithms are not static, conglomerates modify, add, or remove data, modifying nuances of the algorithms frequently, whether for user experience (UX) or to prevent stocks from fluctuating, after all CEO's of Google Cloud, Microsoft, AWS Amazon, IBM Watson, and other companies would not allow their brands to be linked to racism/gender discrimination or any factor that harms the brand, not for reasons of social concern, but for monetary reasons.

As a result, when researching the aforementioned tools, AWS Amazon Rekognition (Facial Analysis) proved to be more accessible, offering a layout in which the user can choose which tools she needs to use, in addition to: "no monetary charges." Undoubtedly, in this segment, the principles of advertising do not fail: "there is no free lunch," if I do not pay cash, so my information is the bargaining chip and that includes email address, digital traces, photo, age, academic background, orders placed on Amazon, mobile number, geolocation, etc.... still, Google was tested, but not to the same degree as Amazon (**Figure 1**).

Let us return to the tool, AWS Amazon Rekognition (Facial Analysis) is a tool based on computer vision and deep learning, as previously mentioned, and machine learning works by making various data look for patterns repeatedly to then discern and recognize

[1] Joy Buolamwini, revealed biases in facial analysis algorithms from Amazon, IBM, Google Cloud, Face++, Microsoft, and others, demonstrating that the services often viewed black women as "men" but made little or no mistakes when it came to men of color. Light skins in: Ref. [17].

[2] Joy Buolamwini describes this moment by relating to Frantz Fanon's [1925–1961] Black Skin, White Masks, in which she comes to question the complexities of changing herself by putting on a mask to conform to the norms or expectations of a dominant culture. in this case of dominant technologies.

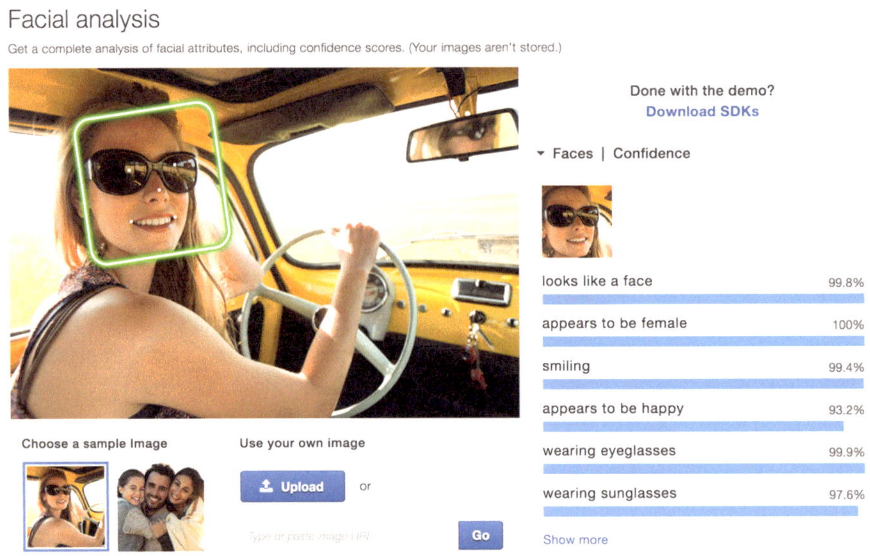

Figure 1.
AWS splash screen recognition-facial analysis.

certain images, with that it performs facial recognition and analysis, detection of objects and texts, information about where faces are detected in an image or video, assigning points on faces and eye position, and "detecting emotions" (e.g., happy or sad) [18]. With that, I tested the tool to see if it was able to give me a result that was different from the one shown in the documentary in which Buolamwini was made invisible, I inserted 15 photographs of women, black celebrities, who socially identify as straight people, and the machine recognized them, classified them according to the available parameters as follows: looks like a face 99.9%, looks like a woman 99.9%, age group 27–37 years old, smiling 96.1%, looks happy, 95.9% is not wearing glasses 97.4%, is not wearing sunglasses 99.9%, eyes open 97.5% mouth open 95.4%, no mustache 98.2%, and no beard 95.8%.

2.2 Algorithmic misgender

Thus, based on the results obtained, I asked if the same would happen if photographs of bodies were inserted that do not correspond (and do not need to correspond) to the expectations of "conventional gender norms" if the tool analyzes, identifies, and classifies characteristics of people of dissident genders, and if analyzed, how are they taught to "see"? would be through of the biological/social limits of the "male/female" binary? [19, 20] or would they be unfeasible?

The analysis was carried out from December 2021 to February 2022, then from March 2022 to July 2022, and these spaces were purposeful to verify if there would be changes in the classification algorithms in relation to gender labels, in all months 15 photographs were used, and this time of celebrities who socially identify as non-binary, I observed that the Amazon Rekognition facial analysis method includes the following: a) inverted pyramid: eyes, nose, lips from one end to the other, and b) pyramid straight: nose lips from one end to the other. Triangulations are part of the parameters in which algorithms focus on the biological for precision analysis, in all analyzed photographs there were no classifications that transposed the female/male binary.

According to AWS, a binary gender prediction (male/female) is based on the physical appearance of a face in a given image. It does not enforce gender identity [21].

},
"Gender": {
"Value": "Female",
"Confidence": 55.517173767089844

Here the machine returns the labeling as being 55.51% female, but the person considers himself genderless/gender fluid.

},
"Gender": {
"Value": "Female",
"Confidence": 99.7735595703125

Above, the machine returns the labeling as being 99.77% female, but the person self-identifies as non-binary and other factors that contribute to the analysis are not demonstrated, that is, we do not know what data makes up the analysis of the tool so that she arrived at this result.

The use, treatment, and/or mention of gender terms that do not correspond to the self-identity that a non-binary or trans-person self-identifies generates the misgender experience [22–24], even when recognition goes beyond the human social line, consisting in Automatic Gender Recognition (AGR) the automatism that algorithmically identifies the gender of individuals generating self-identification errors, and it is considered algorithmic misgender.

In fact, the joining of data and algorithms together with the company's regulation determines the machinic vision indicating that the biological that is inserted in the social overlaps the self-identity; thus, the invisibility happens in the sense of the prominence of the "norm" that hegemonizes bodies and makes them invisible, in the sense of self-identity, and the dynamics of observation, analysis, and classification define labels that match the result. It seems to be a woman or it seems to be a man, according to the imposed biological/social norm.

Google, however, returned the result of gender-related labels such as: person. The "algorithmic adaptation" of this tool was to decode the labeling to indicate that the analyzed image is of a person, a human, a neutral, colorless, and genderless being. Indeed, the results indicate that these analysis tools do not learn and do not know that there are people who do not fit into the categorical double woman/man, so the catalog of possible identities [25] is not part of what machines need to learn.

The limitations presented in this research are in line with those of Keyes [24] and Scheuerman [26] in which the search for diversity in certain tools coincides with a large wall of data that cannot be analyzed and reviewed, but which still return with results. According to the objective imposed by the companies, the machinic eye is organized according to the categorical binary woman/man.

3. The BitLocker effect and the imponderables of academic life

This subtitle was not foreseen, but there is the so-called Murphy's law[3] or "the imponderables of real life," [27] or even more the imponderables of academic life that

[3] "Qualquer coisa que possa dar errado, dará errado e no pior momento possível".

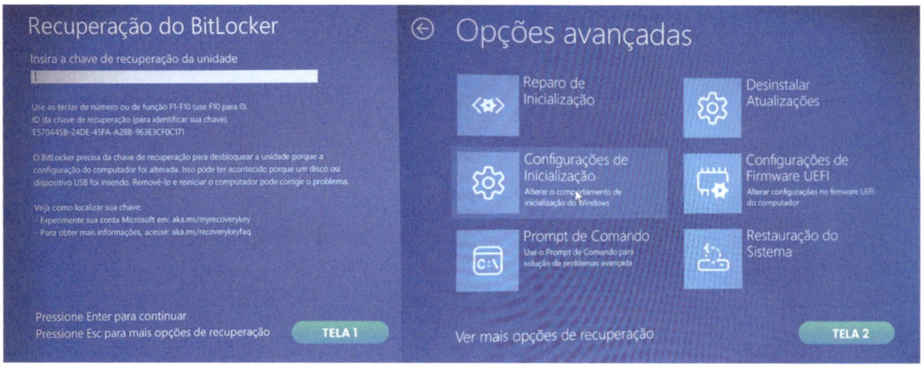

Figure 2.
Author's notebook screen.

is when the notebook believes it is suffering some kind of cyberattack and does not allow the human to have control of the situation, just like Hal 9000 my notebook had a single objective, not to allow anyone (not even me) to access the hardware.

Just over a week to send this article, I revised it, formatting, font, after exhausting days/weeks/hours of adjustments, the goal of finishing the article was overcome by my organic limitations. There are times when, after analyzing images in search engines, reading data lists, arguing, and referencing so much, this organic body that typed these lines runs out, unlike machines. I left the glasses on the corner of the table and lowered the notebook screen (I was condescending, after all I would like at least my machine to rest) a few minutes before sending the complete work when I returned to the workflow this around 30 minutes, which was my surprise when I lifted the notebook screen and noticed that instead of the wallpaper as usual, there was this warning (**Figure 2**):

From experience, the worst thing to do in these cases where our machines hijack our academic research and all the materials accumulated for the making of the doctoral thesis, projects, etc., is to consult online search engines, and that is because all the tutorials said the same thing: "click F2 directly, give this command and with that my friends the entire drive will be erased, with factory formatting and don't forget to leave your likes."

I had already gone through something similar to notebooks before this one, but seeing a work about to be submitted held hostage, my organic algorithms adjusted to deliver the result: Tears. I came up with this role so after so much pressure and desperation it was what I had at the moment. I could not think rationally, that was it, my own machine had betrayed me. I called my partner in tears and she said: "it would be comical if it wasn't tragic, a work on machines hijacked by your own machine, leave him in the corner of thought and let's have lunch."

I was the one who kept thinking what was behind the notebook's action? Until reaching this conclusion, the crying would not stop and the hours were passing, after lunch, I called the notebook company, the call center's response was: "Bitlocker is when your machine thinks you may be suffering a cyberattack and is securing your information, you can only access it with Bitlocker recovery keys if you do not have a recovery key you will not be able to access your computer and if you cannot revert you will have to reset your device, this action removes all your files."

I turned off my cell phone. A few minutes of silence until I recover and think of another way to recover the machine and files. I called Microsoft and was answered by an artificial intelligence who said he could only help if my case was related to XBOX, otherwise I should access the company chat. I accessed the Microsoft chat, I told the whole odyssey and to my surprise it was another artificial intelligence, that is, my regret had to be as objective as possible, I typed the whole story, but when I noticed the AI left the chat because "it didn't understand my request." I went back to the chat, another AI, I typed everything again as objective and succinct as possible again, who knows then the AI would help me access my machine. After a few moments the answer came, I had to confirm my data and the BitLocker recovery keys were sent to my email. I entered the 48 digits and I have never been so relieved to see this article again.

4. Conclusions

The entanglement of human life with computer vision has been done through data, which is the main raw material of machine learning or Big Data, being the deluge of data that is made available to train computer vision. As they are inserted, they receive preliminary descriptions that consist of a tag, label, which enables standards, which regulate deep learning, the more data that is produced by us, the more data are selected, labeled, and each time more algorithmic performativity corresponds to what has been learned through datasets, but this has not given the guarantee that there is equity, diversity in facial recognition, as noted, Google chose not to describe people's gender, it will be simpler to delete algorithms instead of teaching that there is diversity? Or remain in the hegemonic category like Amazon? Anyway bodies that do not adapt to the binary continue to be made invisible, whether by humans, increasingly sophisticated algorithms, but that do not work alone, we know that the standards that regulate this data tend to be biased by other humans.

Author details

Mayane Batista Lima[1,2,3]

1 Social Anthropology at the Federal University of Amazonas (PPGAS/UFAM), Brazil

2 Social Anthropology at the University of São Paulo (FFLCH/USP), Brazil

3 Center for Artificial Intelligence (C4AI) – University of Sao Paulo (USP/FAPESP/IBM), Brazil

*Address all correspondence to: mayanejornalista@gmail.com

References

[1] 2001: A Space Odyssey. Directed by Stanley Kubrick. USA: Metro-Goldwyn-Mayer and Stanley Kubrick Productions, 1968.

[2] Clarke AC. 1917-2008*uma odisseia* no espaço [livro eletrônico]. In: Clarke AC, editor. tradução Fábio Fernandes. São Paulo: Aleph; 2001. p. 2014

[3] Medeiros, CB. Dados, Algoritmos, Máquinas e pessoas. Revista da Sociedade Brasileira de Computação, Brasil, n. 47, p. 11-14, 2022. Disponível em: https://bit.ly/3PEewdP. Acesso em: 29 july 2022.

[4] Cormen TH. Algoritmos. In: Cormen TH et al., editors. tradução Arlete Simille Marques. Rio de Janeiro: Elsevier; 2012

[5] O'neil C. Algoritmos de destruição em massa: como o big data aumenta a desigualdade e ameaça a democracia. In: O'Neil C, editor. tradução Rafael Abraham. 1st ed. Santo André, SP: Editora Rua do Sabão; 2020

[6] Russell SJ. Inteligência Artificial a nosso favor: como manter o controle sobre a tecnologia. SãoPaulo: Companhia das Letras; 2021

[7] Lee K-F. Inteligência artificial [recurso eletrônico]: como os robôs estão mudando o mundo, a forma como amamos, nos relacionamos, trabalhamos e vivemos. In: Lee K-F, editor. tradução Marcelo Barbão. 1st ed. Rio de Janeiro: Globo Livros; 2019

[8] Jones T. Artificial Intelligence: A Systems Approach. Jones and Bartlett Publishers; 2008

[9] Lippold JC. We Are Data: Algorithms and the Making of our Digital Selves. New York University Press; 2017. Available from: https://www.sbc.org.br/component/flippingbook/book/55/1?page=11

[10] Onuoha M. (Canal Futura). Inteligência Artificial | Expresso Futuro. Youtube, 2017. Disponível em: https://bit.ly/3ruNf4s

[11] Forsythe DE. The construction of work in artificial intelligence. Science, Technology, & Human Values. 1993;**18**(4):460-479

[12] CHRISTIAN & GRIFFITHS. *Algoritmos para viver*: A ciência exata das descisões humanas. São PAULO: Brian Christian e Tom Griffiths. EDITORA SCHWARCZ S.A; 2017

[13] Noble SU. Algoritmos da opressão: como o Google fomenta e lucra com o racismo. In: Noble SU, editor. Tradução Felipe Damorim. Santo André - SP: Rua do Sabão; 2021

[14] Tarcízio S. Comunidades, algoritmos e ativismos digitais: Olhares afrodiaspóricos Organização e Edição: Tarcízio Silva; Revisão Ortográca: Toni C.; Demetrios dos Santos Ferreira; Tarcízio Silva; Gabriela Porfírio; Taís Oliveira; Tradução: Vinícius Silva; Tarcízio Silva; Ilustração de Capa: Isabella Bispo; Diagramação: Yuri Amaral; Consultoria Editorial. São Paulo: LiteraRUA; 2020

[15] Raji ID. Handle with care: Lessons for data science from black female scholars. Opinion. 2020;**1**(8).100150

[16] Buolamwini J, Gebru T. Gender shades: intersectional accuracy disparities in commercial gender classification. Learn: Proc. Mach; 2018

[17] Coded Bias. Production by Shalini Kantayya. Performers: Joy Buolamwini,

Cathy O'Neil. China, USA, UK, Northern Ireland, Great Britain, 2020. (90 min.). Available from: https://bit.ly/3AHooiD. [Accessed: June 1, 2022].

[18] AWS. Detecting and Analyzing Faces. Disponível em: https://go.aws/3wsrFzw. Acesso em: June 1, 2022.

[19] de Almeida HB. Gênero. Edição eletrônica. URL: https://bit.ly/3QQvW86 ISSN: 2526-6187Blogs de Ciência da Universidade Estadual de Campinas. Mulheres na Filosofia. 2020;**6**(3):33-43

[20] Stolcke V. Sexo está para gênero assim como raça para etnicidade? Estudos Afro-Asiáticos. 1991;**20**:101-119

[21] AWS. Detecting and analyzing faces. Disponível em: https://go.aws/3wsrFzw. Acesso em: July 25, 2022

[22] Buolamwini J. Facing the Coded Gaze with Evocative Audits and Algorithmic Audits. Massachusetts Institute of Technology; 2022

[23] Filho, Raphael Borges dos Santos. Inclusão de pessoas transgênero nos algoritmos de reconhecimento automatizado de gênero para reconhecimento facial / Raphael Borges dos Santos Filho. – Trabalho de conclusão de curso (graduação) – Universidade Federal da Fronteira Sul, curso de Ciência da Computação, Chapecó, SC, 2021.

[24] Keyes OS. The misgendering machines: Trans/HCI implications of automatic gender recognition. Proceedings of the ACM on Human-Computer Interaction. 2018;**2**:1-22. DOI: 10.1145/3274357

[25] Pérez M, RADI B. Gender punitivism: Queer perspectives on identity politics in criminal justice. Criminology & Criminal Justice. 2020;**20**(5):523-536. DOI: 10.1177/1748895820941561. Available from: https://periodicos2.uesb.br/index.php/recic/article/view/10883

[26] Morgan S, Jacob P, Jed B. How computers see gender: An evaluation of gender classification in commercial facial analysis services. Proceedings of the ACM on Human-Computer Interaction. 2019;**3**:1-33. DOI: 10.1145/3359246

[27] Os K. The misgendering machines: Trans/HCI implications of automatic gender recognition. Proceedings of the ACM on Human-Computer Interaction. 2018;**2**:1-22. DOI: 10.1145/3274357

Chapter 4

Numerical Simulations and Validation of Engine Performance Parameters Using Chemical Kinetics

Muzammil Arshad

Abstract

Use of detailed chemistry augments the combustion model of a three-dimensional unsteady compressible turbulent Navier–Stokes solver with liquid spray injection when coupled with fluid mechanics solution with detailed kinetic reactions. Reduced chemical reaction mechanisms help in the reducing the simulations time to study of the engine performance parameters, such as, in-cylinder pressure in spark ignition engines. Sensitivity analysis must be performed to reduce the reaction mechanism for the compression and power strokes utilizing computational singular perturbation (CSP) method. To study a suitable well-established surrogate fuel, an interface between fluid dynamics and chemical kinetics codes must be used. A mesh independent study must be followed to validate results obtained from numerical simulations against the experimental data. To obtain comprehensive results, a detailed study should be performed for all ranges of equivalence ratios as well as stoichiometric condition. This gives rise to the development of a reduced mechanism that has the capability to validate engine performance parameters from stoichiometric to rich mixtures in a spark ignition engine. The above-mentioned detailed methodology was developed and implemented in the present study for premixed and direct injection spark ignition engines which resulted in a single reduced reaction mechanism that validated the engine performance parameters for both engine configurations.

Keywords: direct injection spark ignition (DISI) engines, premixed, CHEMKIN-KIVA, in-cylinder pressure, heat release rate (HRR), emissions

1. Introduction

In Computational Fluid Dynamics (CFD), many simplifications and assumptions are made to the mathematical models to make them computationally affordable. The rapid development and progress in the field of computer processors and enhancement of computer memory has enabled scientists and engineers to review and revisit some of the assumptions. This has helped to improve and enhance the predictive capabilities of computer modeling. This approach has also been implemented in engine simulation codes and tools. The objective of this study and research is to identify the reactions that govern the chemical kinetics of fuel oxidization and move from multi-step global reactions to a reduced number of elementary reactions that can better model the combustion and engine performance parameters [1].

For the development of future fuels and the optimization of automotive engines, computer modeling and simulation has proved itself as an inseparable tool alongside experimental study. Due to computational limitations, the traditional approach has been to utilize simplified global reactions to simulate and evaluate the combustion and performance parameters in internal combustion engines. Due to increased interest in various advanced engine configurations with various combustion modes and injection strategies, this approach could reduce credibility of the predictions for advanced concepts since it depends on arbitrary adjustment of model parameters.

Oxidation of hydrocarbons is shown by detailed chemical kinetic mechanisms. These detailed mechanisms are very large and comprised of a large number of species and reactions. As the size of the hydrocarbon increases, the length (number of species and reactions) of the mechanism increases along with it. Due to the construction from smaller hydrocarbons (HC) progressing to larger HC and intermediate radicals, detailed mechanisms of large hydrocarbons consist of many species and reactions that are redundant. Those species do not have significant impact on simulating ignition and combustion phenomenon and needlessly raise the computational and memory requirements. These reactions and species can be identified and eliminated from the detailed reaction mechanism without compromising the accuracy and integrity of the detailed reaction mechanism. The detailed and large mechanisms cannot be employed in present solvers because they are time expensive. For example, gasoline and diesel fuels consist of thousands of reactions and hundreds of hydrocarbons [2].

Fuels, such has gasoline and diesel, are composed of hundreds of hydrocarbons. These hydrocarbons include alkanes, alkenes, aromatics and naphthenes. Surrogate fuel mechanisms that contain limited number of hydrocarbons from all the above-mentioned hydrocarbon types are important in this regard. There is a necessity to use reduction techniques to produce reduced order mechanisms that can replicate the predictions of detailed mechanisms [3]. The reduction techniques decrease the number of species and reactions.

Redundant reactions were identified by two methods [1–3]:

a. reaction rates analysis and,

b. sensitivity analysis

In a reaction mechanism, there are two subsets of reactions; slow reactions and fast reactions. The reaction rates analysis divides a reaction mechanism into above-mentioned two subsets of slow and fast reactions. The sensitivity analysis is performed to divide the reaction mechanism into two subsets of rate limiting and non-rate limiting reactions. When combined, both analyses identify redundant reactions. The redundant reactions identified by this method are non-rate limiting slow reactions. As the redundant reactions are eliminated, the species taking part in those reactions gets automatically eliminated, hence, a reduced mechanism is obtained [1–3]. Sensitivity analysis is discussed further in detail in the next section with detailed references.

Using the computational singular perturbation (CSP) technique [4, 5], reduction of iso-octane/n-heptane reaction mechanism by Soyhan et al. [4] and Valorani et al. [5] has been performed that has resulted in reduced and skeletal mechanisms.

Using various computational codes [6–15] and experimental tools, various researchers performed multiple studies to review surrogate fuel mixtures [8], reduced PRF mechanisms [9], with variable intake parameters including an operating range of equivalence ratios, intake pressures and temperatures while considering various engine performance parameters such as heat release rate (HRR) analysis, in-cylinder pressure data [10] and emissions on various engine geometries operating at various operational ranges [11, 12]. Using a mixture of iso-octane, n-heptane and toluene as gasoline surrogate fuel predicts engine performance parameters correctly, especially in HCCI and SI engines. Further addition of di-isobutylene and methylcyclohexane is also recommended. Under stoichiometric and lean conditions, significant number of small particulate formation occurs while large particulate formation shows existence with increasing equivalence ratio. Decrease in the peak in-cylinder pressure can be achieved by mass and temperature reduction. This phenomenon occurs due to heat loss to the chamber walls [13].

A reaction mechanism reduction through sensitivity analysis of a skeletal reaction mechanism for the compression and power strokes by utilizing computational singular perturbation (CSP) method and using the low temperature reaction pathway analysis leads to a reaction mechanism that predicts accurate results for computational studies. Detailed chemistry in conjunction with fluid dynamics enhances the ability of a computational code to correctly predict the engine performance parameters. This is proven in benchmarking the global and quasi-global mechanisms [1] which provided necessary data and confidence in the use of detailed chemistry to correctly predict the engine performance parameters. Also, using 90% iso-octane and 10% n-heptane as surrogate fuel for gasoline helps in correct prediction and best modeling the engine performance parameters. Along with a correct reduced mechanism, mesh independent study is a key to correctly predict and validate the engine performance parameters against the experimental data for a range of equivalence ratios in premixed spark ignition engines.

2. Materials and methods

KIVA-3 V [7] is a code developed by Los Alamos National Laboratory for numerical calculation of transient, two- and three-dimensional chemically reactive fluid flows with sprays [1]. It is used to perform numerical simulations which were then used to compare and investigate the variation against the experimental data for a premixed case in a spark ignition engine. To perform the studies, iso-octane is used as a gasoline surrogate.

CHEMKIN [16] has many modules; one of which is SENKIN [17] that performs sensitivity analysis and subsequent reduction of skeletal reaction mechanism. Extraction of the sensitivity data using SENKIN leads to the next step of performing mechanism reduction using KINALC [18] which uses the computational singular perturbation (CSP) method. Sensitivity analysis for spark ignition (SI) engines is performed for both the compression and power strokes since combustion occurs in some portion of each stroke. CSP analysis utilizes to set the size of the time criterion for the characteristic chemical reaction in such a way that fast reactions are discarded from the reduced chemical reaction mechanism while rate limiting reactions are included into the reduced reaction mechanism since the slower reactions are the rate-controlling reactions [19]. On the other hand, fast reactions cause magnification of the inaccuracies which makes the numerical methods to be unstable. The best way to set the time criterion for CSP is such that it encompasses the combustion process to the end of power stroke. This in effect creates a reduced mechanism that encompasses for the full duration of combustion process, therefore, the selected reactions selected are sensitive and important to the whole combustion process in the engine. For performing the numerical simulation using the reduced reaction mechanism, a mixture of 90% iso-octane and 10% n-heptane is recommended to be used as a gasoline surrogate.

The comprehensive analysis discussed above, helps in construction of a reduced reaction mechanism which mitigates experimental results as well as promotes a greater understanding of the chemical kinetics. Experimental results obtained from [20, 21] for premixed case, at equivalence ratio of 0.98 and 1.3 are mitigated to prove the accuracy of the above-mentioned process. For this, an engine geometry used 85.96 mm bore, 94.6 mm stroke, compression ratio of 11.97 running at 2100 rpm. Engine performance parameters of in-cylinder pressure and heat release rate showed a comparative analysis.

Due to chemical stiffness and large size of reaction mechanisms, computations based on detailed reaction mechanisms are complex. Therefore, it is required to reduce chemical mechanisms. This can be performed using the two levels of reductions mentioned below. As a result, the reduced mechanism is extracted which is a subset of the detailed reaction mechanism.

Level I - Skeletal Reduction: Methods used to eliminate unimportant species and reactions: Directed relation graph (DRG), DRG with error propagation (DRGEP), path flux analysis (PFA), revised DRG (DRGMAX), and computational singular perturbation (CSP) [22].

Level II – Global Reduction Methods: Methods used for analysis of timescale impact on reaction mechanism: Computational singular perturbation (CSP), and quasi-steady-state approximation (QSSA) [22].

The first step to reduce the detailed reaction mechanism to a skeletal reaction mechanism is the elimination of unimportant species using the DRG method which identifies the species closely coupled with major species, such as fuel and oxidizer [19]. This is achieved from a sensitivity analysis that uses Jacobian matrix or sensitivity matrix that can be normalized. The Jacobian matrix is the matrix of all first-order partial derivatives of a vector valued function [19]. This method is important for reduction of a large reaction mechanism. The reduced skeletal mechanisms obtained from DRG are not minimal in size due to assumption of the upper-bound error propagation. A more straightforward definition of DRG is,

$$r_{A,B} = \frac{\sum_{j=1,I} |\nu_{A,i}\omega_i\delta_{Bi}|}{\sum_{j=1,I} |\nu_{A,i}\omega_i|} \tag{1}$$

where, $\delta_{Bi} = 1$, if the ith elementary reaction involves species B, otherwise, 0; r_{AB} is the relative error induced to species A by elimination of B, subscript 'i' represents the ith elementary reaction and 'j' represents the jth species, ν_A is the net stoichiometric coefficient of species A, $\omega_i = \omega_{f,i} - \omega_{b,i}$ where $\omega_{f,i}$, $\omega_{b,i}$ and ω_i is the forward, backward and net reaction rates respectively, that can be calculated from the already given coefficients and activation energy in the CHEMKIN input file. If $r_{A,B} < \epsilon$ for all species, then the relation between B and A is considered to be negligible. Species B is selected when $r_{A,B} \geq \epsilon$. Here ϵ is a user-defined small threshold value. In most cases, $\epsilon = 0.1$ is used [23, 24]. Since ϵ is a user-defined small threshold value, it depends on the application and user experience.

The second step is to further reduce the skeletal mechanism by using the technique called directed relation graph-aided sensitivity analysis (DRGASA). This method further reduces the species set by performing the sensitivity analysis on the already obtained species data from the previous DRG method. The parameters that are focused for the DRGASA method are ignition delays, extinction times, and laminar flame speeds. The reduction with this method is carried out using for a range of pressure, temperature and equivalence ratio. These two steps provide the researchers with a skeletal mechanism.

Skeletal mechanisms are still too large to be used in the computational work and it is important to reduce the skeletal mechanism into a reduced mechanism. A major advancement in the CSP technique has been introduction of a new concept of using a vector G that represents the rate of change of species mass fraction (Y) and temperature (T). Numerical computations are used to monitor any contributions to vector G. Identification of various terms becomes straightforward with this method as it can help identify the reactions that are controlling the reaction, constant terms and chemical species that have depleted [24]. The ODEs given for a reactive system,

$$G = \frac{dy}{dt} = S_r F_r, \text{ where } y = y_i \text{ and } y_i = Y_i \text{ or } T \tag{2}$$

where, S_r is the stoichiometric vector and F_r is the reaction rate of rth reaction [25].

$$G = \frac{dy}{dt} = S_r F_r = \nu_r q_r \tag{3}$$

$$\nu_r = [\nu_1 \nu_2 \dots \dots \dots \nu_n]^T \tag{4}$$

where, r = 1, ..., N and n = 1, ... , s. Here 'r' is number of reactions and 'n' is number of species.

All the reactions in the reactive system form the vector G.

$$\frac{dG}{dt} = J.G, J = \frac{dG}{dy} \tag{5}$$

where, J is a Jacobian. G is divided into fast and slow subspace. The following steps are performed to solve a CSP problem [25]:

1. Identify S_r and F_r (or alternatively ν_r and q_r respectively, both symbols are synonymous here) from a given G(y)

2. Find Jacobian

3. Find eigenvalues, λ

4. An eigenvalue is considered large if, $|\lambda \, \Delta t| > 1.0$, where Δt is the time step.

5. Determine the total number of large eigenvalues (m) from step 4.

6. Since, reaction mechanism is composed of fast and slow reactions, therefore, $G(y) = G_{fast} + G_{slow}$. To determine G_{fast} and G_{slow}, first find the characteristic chemical timescale of each reaction by taking the negative reciprocal of the eigenvalues determined in Step 3, i.e. $-1/\lambda = \tau$. This timescale calculation will show the magnitude of fast and slow reactions.

7. Eliminate fast reactions. If step 4 satisfies, fast reactions can be eliminated.

Using the CSP method, a term is only discarded or eliminated when it becomes numerically too small that it does not make any difference. This is determined by the importance index Eq. (6). CSP, as a reduction method, has been used by other researchers successfully [19, 26, 27].

Lu and Law [19] developed the CSP method for the removal of the unimportant reactions. The method used an importance index that eliminates the unimportant species. For the above-mentioned method, the importance index of the reaction is defined as

$$I_{A,i} = \frac{|\nu_{A,i} q_i|}{\sum_{j=1,N_R} |\nu_{A,j} q_j|} \tag{6}$$

where, 'A' is species, 'i' is reaction and 'q' is overall reaction rate of the ith reaction. Also, $\nu_{A,j}$ is the stoichiometric coefficient of species A in the jth reaction. It must be noted that a reversible reaction must be treated as a single reaction [19].

If $I_{A,I} < \epsilon_{Reac}$ for all species, then the reaction is an unimportant reaction where ϵ_{Reac} is a user-defined small threshold.

The parameter ϵ_{Reac} is a dimensionless parameter and can be defined as [26]:

$$\epsilon_{Reac} = \left| \frac{\tau_{fast}}{\tau_{slow}} \right| \tag{7}$$

$|\tau_{fast}|$ is the slowest relevant time scale for the fast reaction group and $|\tau_{slow}|$ is the fastest relevant time scale for the slow reaction group [25]. τ is defined as the characteristic chemical time scale. The characteristic time scales are negative reciprocals of the diagonal elements of the problem's Jacobian [27].

Simplification of detailed model is achieved by eliminating the following [5]:

i. Species having weakly coupled chemical kinetics with the species of interest and,

ii. the reactions deemed unimportant to the species that are retained.

The above procedure results in a smaller kinetic mechanism. This smaller mechanism is formed from the detailed mechanism and is a subset of the detailed mechanism. The accuracy of the skeletal mechanism is defined with respect to the species that are declared of interest for the problem and the domain of applicability defined for the problem. The domain of applicability is defined on the following criterion [5]:

i. the type of problem defined, for example, various types of reactors, ignition, flame speed and structure, etc., and

ii. the defined range of initial conditions and mixture concentration. Reference problems would be solved for a specified range of initial conditions such as pressure, initial temperature, and mixture concentration specified by equivalence ratio.

Reaction flow analysis can prove to be very helpful with complex reactions. If the reaction rate satisfies the following condition for all times t, then the reaction rate is unimportant [25]:

$$|RR_{t,r,s}| < \varepsilon |Max_{r=1,2,\dots.M} RR_{t,r,s}| s = 1,\ 2,\ \dots Nt = 0,\ \dots,\ t_{total} \tag{8}$$

where, ε is a very small value which is arbitrarily specified, 'r' is elementary reaction number, and 's' is number of species.

The physics of the process of reaction of complex hydrocarbons follows the three basic reaction steps: chain-initiating reactions, chain-branching/carrying reactions and chain-terminating reactions. Chain-initiating reactions are those elementary reactions that produce free radicals. Similarly, free radicals are destroyed in chain-terminating reactions. Chain-propagating reactions or chain-carrying reactions are defined as the elementary reactions where the ratio of free radicals in products to the free radicals in reactants is equal to 1. If this ratio is greater than 1, then the reactions are called chain-branching reactions [28]. Concentrations of free radicals are treated as constant as they remain essentially constant throughout the reaction, except for the short initial and final periods which is minimal as compared to the entire reaction period.

High temperature oxidation of paraffins (C_nH_{2n+2}) that are larger and more complex than the methane is much more complicated as they include many complex hydrocarbons in the mixture. Over the years, the evolution of combustion science has developed detailed combustion mechanisms for those smaller hydrocarbons that are now part of various combustion libraries. Therefore, it is possible to develop a general framework for the complex combustion process [28].

The first step in the combustion process of larger paraffins (C_nH_{2n+2}) is that rather than directly breaking into CH_3, it first breaks down into hydrocarbon radicals of lower order, C_nH_{2n+1}. The hydrocarbon radicals of higher order are highly unstable and are further broken down to CH_3 and a lower order olefinic compound, $C_{n-1}H_{2n-2}$. For hydrocarbons larger than C_3H_8, the process of fission takes place between the olefinic compound and a lower order radical. Further reactions of those radicals include intermediate steps that eventually form the methyl (CH_3) radical. Also, formaldehyde formed during the reactions is rapidly attacked in flames by the O, H, and OH atoms. Therefore, formaldehyde in found only in trace amounts in flames. The situation is more complex for fuel rich hydrocarbon flames [28].

Hydrocarbons follow a similar set of steps for combustion and oxidation. The difference between various hydrocarbons requires further intermediate steps but they go through the same oxidation procedure. The process of complex hydrocarbon oxidation occurs in the following manner [29]:

1. Carbon—carbon (C—C) bond is broken: The C—C bonds are primarily broken over H—C bond because C—C bonds are much weaker than H—C bonds.

2. H-atom abstraction: Resulting hydrocarbon radicals further break down into olefins (hydrocarbons with double carbon bonds, C=C) and H-atoms.
3. Radical formation: Creation of H-atoms forms a pool of radicals.
4. Formation of radicals gives pathways for fuel molecules to attack O, OH and H atoms.
5. Hydrocarbon radicals again decay via H-atom abstraction.
6. Formyl radical (HCO) and Formaldehyde (CH_2O) are formed.
7. Radical reacts with O-atoms that result in the formation of carbon monoxide (CO).
8. Carbon monoxide (CO) oxidation occurs and results in the formation of carbon dioxide (CO_2).

A skeletal mechanism [9] was selected and used for further reduction. The skeletal mechanism was enhanced with extended Zeldovich mechanism for the formation of NO_x emissions which resulted in 299 reactions and 75 species. Reduction was performed using the CSP technique to achieve 53 reactions and 44 species in the reduced mechanism.

The reduced mechanism was used in the KIVA-CHEMKIN interface to make it part of the KIVA input file. Simulations were performed for the same conditions performed in the engine study for premixed engine [20, 21]. This reaction mechanism has proved to predict and validate the results for both stoichiometric and fuel-rich conditions. Sensitivity analysis using SENKIN was performed for both compression and power stroke as the combustion takes place during some part of both strokes.

2.1 Premixed case

Experimental results obtained from [20, 21] for premixed case, at equivalence ratio of 0.98 and 1.3 show validation and improved prediction of the engine performance parameters of in-cylinder pressure and heat release rate (HRR). **Figure 1** and **Table 1** show the pentroof engine geometry with no moving valves.

A mesh independent study performed for the reduced mechanism utilized three meshes shown in **Table 2**. Validation studies are performed to compare the general trend of the engine performance parameters. **Figure 1** shows the isometric view of mesh # 3.

2.2 Direct-injection case

Similar analysis was performed for injection case [31, 32] for the engine geometry shown in **Table 3**, where validation of peak in-cylinder pressure was performed against the experimental data. In this case, mass fractions were calculated based on equivalence ratio of 1.

The mixture is injected at 30° BTDC for 1.75 ms or an equivalent of 15.75°. The mixture is ignited at 14° BTDC. The initial temperature is 302 K and the pressure is 76 ± 1 kPa [31, 32]. The cylinder wall temperature, cylinder fire deck temperature, and the piston temperature are set at 400 K. Mass fractions for fuel and oxidizer are

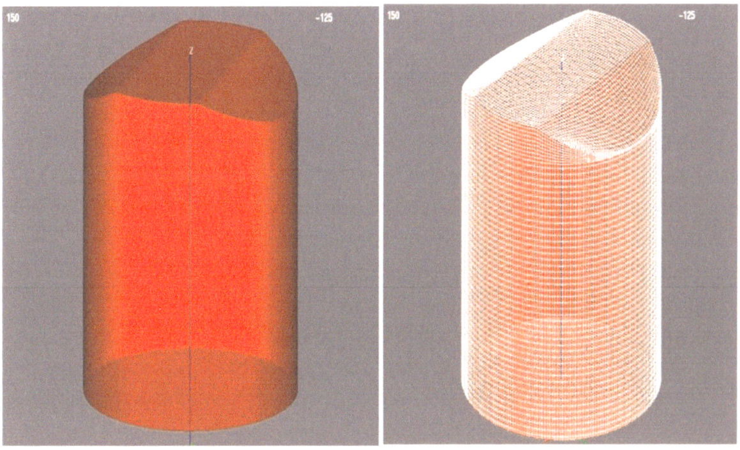

Figure 1.
Pentroof engine geometry [1, 30] – isometric view and mesh.

Dimension	Unit	Value
Compression Ratio	[–]	11.97
Bore	[mm]	85.96
Stroke	[mm]	94.6
Connecting Rod Length	[mm]	152.4
Clearance Volume	[cm^3]	50.0
Displacement	[cm^3]	549.0
Engine Speed	[rpm]	2100

Table 1.
Engine geometry [1, 20, 21].

	Mesh # 1	Mesh #2	Mesh # 3
Number of cells	100,000	178,000	230,000

Table 2.
mesh independent study.

Dimension	Unit	Value
Compression Ratio	[–]	9.4
Bore	[mm]	89.0
Stroke	[mm]	81.4
Displacement	[cm^3]	0.51 L
Engine Speed	[rpm]	1500

Table 3.
Engine geometry – direct injection.

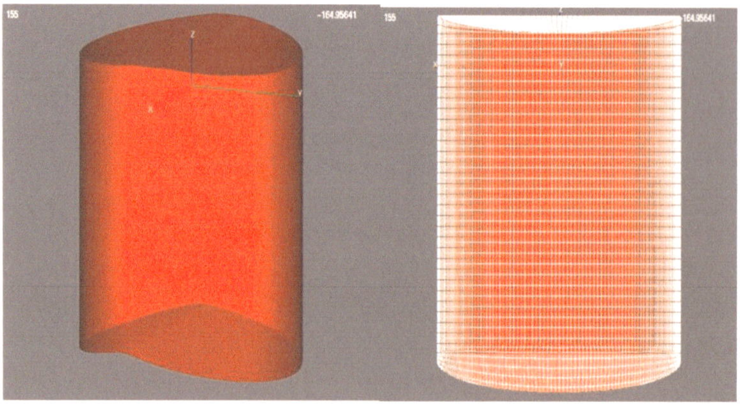

Figure 2.
Pentroof engine geometry [1, 30] – isometric view with mesh.

calculated. The fuel is 90% iso-octane and 10% n-heptane [33] given the complexity of a multi-component detailed chemistry model.

The analysis results again showed better agreement of in-cylinder pressure data with the reduced reaction mechanism for the direct injection case (**Figure 2**).

3. Results and discussion

To better predict the engine performance parameters of in-cylinder pressure and heat release rate (HRR), detailed chemistry is incorporated into the CFD model. To achieve this, KIVA-CHEMKIN interface is used to extract reaction rates and reaction constants for the reaction data and, heat of formation and molecular weights for the species data, that is used for input into KIVA simulations. The reduced reaction mechanism has 53 reactions and 44 species.

3.1 Premixed case

3.1.1 In-cylinder pressure and heat release rate, Φ = 0.98

For a mixture of 90% iso-octane (iC_8H_{18}) and 10% n-heptane (nC_7H_{16}) at $\phi = 0.98$, numerical simulations were performed for comparison and validation are performed for in-cylinder pressure at the ignition timing of 20° BTDC. **Figure 3** shows the validation results showing a good agreement and prediction of in-cylinder pressure against the experimental data.

Figure 4 shows a good agreement of predicted heat release rate (HRR) obtained from the numerical solution from the reduced mechanism using KIVA CHEMKIN interface against the experimental data. HRR is the ratio of difference in heat release values and corresponding crank angle values. These values are obtained from KIVA output file. This also shows that detailed chemistry is very important to capture the correct trend of engine performance parameters. The reduced mechanism achieved from this research has given a confident deduction that the numerical simulations for $\phi = 0.98$ has successfully modeled the engine performance parameters of in-cylinder pressure and heat release rate (HRR). It is recommended to build a library of reduced mechanisms for all the fuels that are used

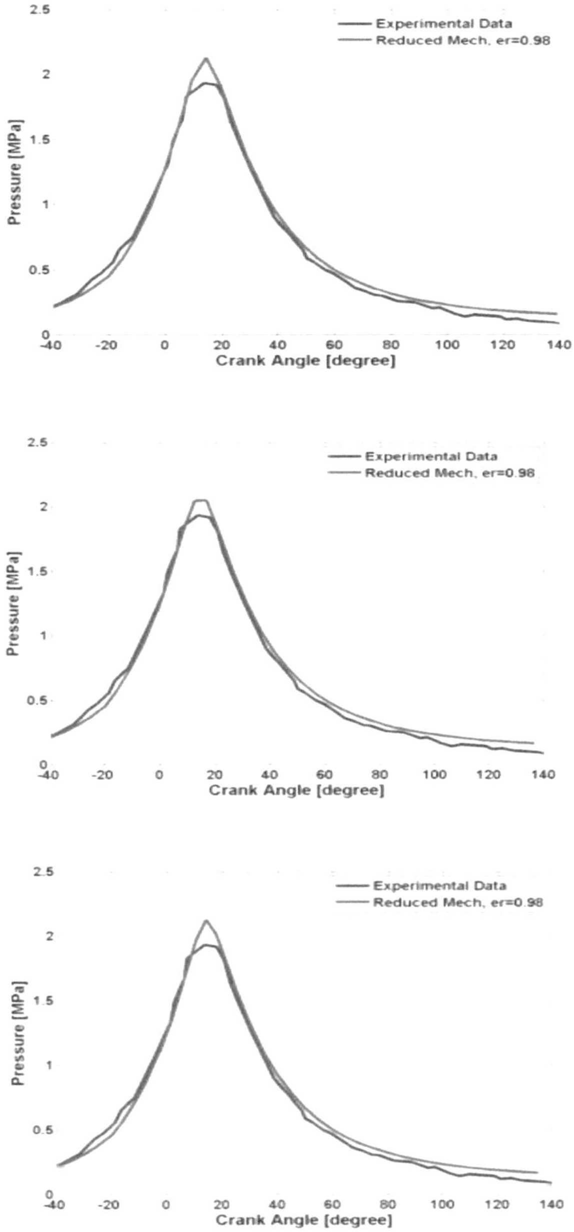

Figure 3.
In-cylinder pressure; 100,000, 178,000 & 230,000 cells.

in the internal combustion engines so that the correct prediction of experimental results can be recorded and achieved.

3.1.2 In-cylinder pressure and heat release rate, Φ = 1.3

Numerical simulations and comparison is performed below for a mixture of 90% iso-octane (iC_8H_{18}) and 10% n-heptane (nC_7H_{16}) at ϕ = 1.3 using the reduced mechanism.

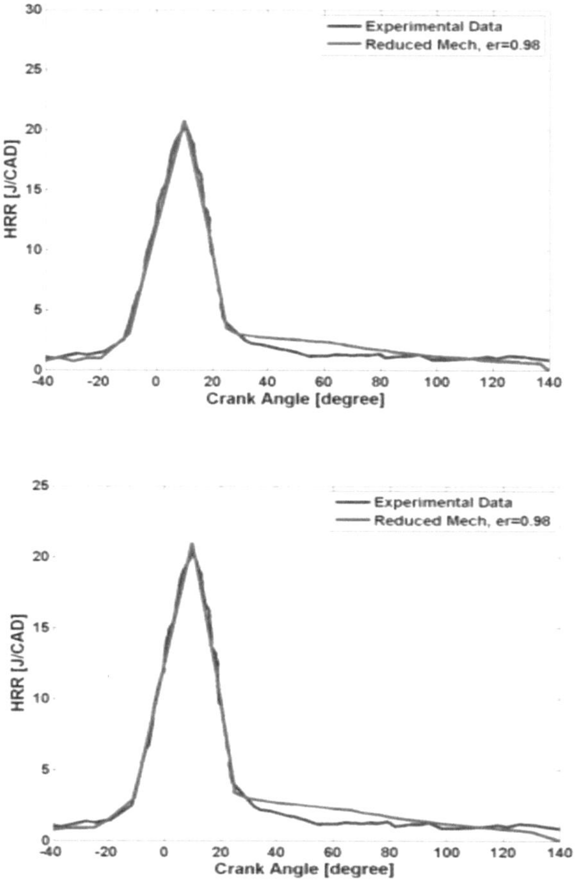

Figure 4.
Heat release rate (HRR); 100,000 cells.

Good prediction of in-cylinder pressure for $\phi = 1.3$ is achieved using the reduced mechanism and is shown in **Figure 5**.

Figure 6 shows the heat release rate (HRR) for $\phi = 1.3$. The results for Heat release rate (HRR) obtained from the numerical solution from the reduced mechanism using KIVA-CHEMKIN interface again shows a good prediction. This also shows that detailed chemistry is very important to capture the correct trend of engine performance parameters.

3.1.3 Emissions

Apart from validating the engine performance parameters of in-cylinder pressure and heat release rate (HRR), this study also validated the emission results with the experimental data [21]. The engine-out exhaust emission data is shown in **Figure 7** for the exhaust species of H_2, CO_2 and CO. The results show a good agreement between the numerical and experimental results for the equivalence ratio of $\phi = 0.98$ and $\phi = 1.3$, where the predicted values of exhaust species of H_2, CO_2 and CO obtained through combustion-CFD simulations exactly match the experimental data at the equivalence ratio of $\phi = 0.98$. At $\phi = 1.3$, H_2 and CO values obtained through

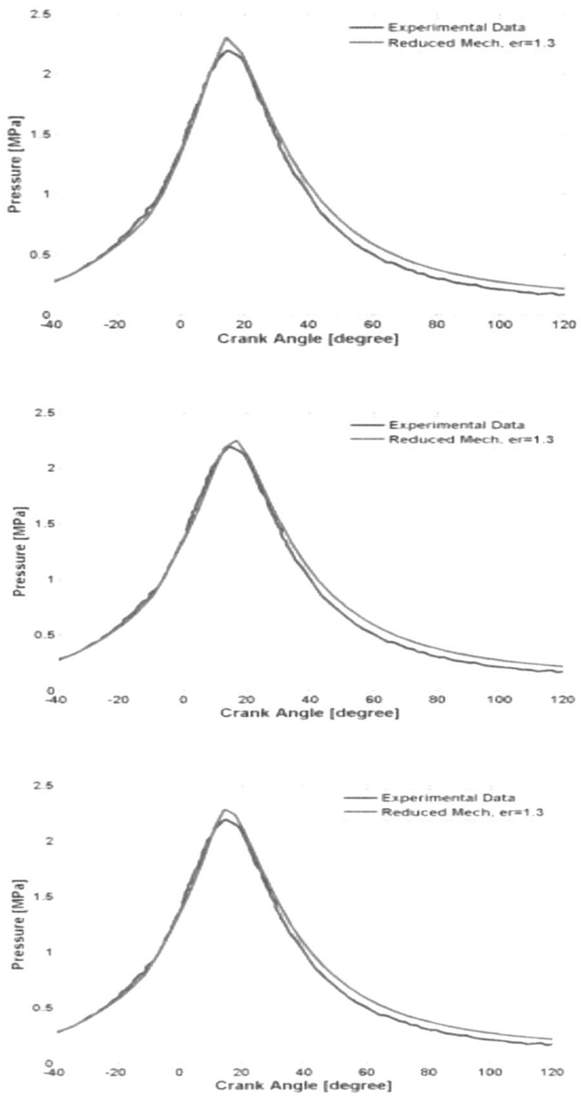

Figure 5.
In-cylinder pressure; 100,000, 178,000 & 230,000 cells.

numerical simulations exactly match and overlap the experimental value while CO_2 values also show a good agreement. The results obtained from **Figure 7** shows that the reduced mechanism developed was able to validate the engine performance parameters as well as the emissions which makes this reduced model reliable in predicting performance parameters for premixed spark ignition engines.

3.2 Direct-injection case

Numerical simulations are performed for the direct-injection case at equivalence ratio of 1.0. The pentroof engine geometry and mesh is shown in **Figure 2** and specification are mentioned in **Table 3**. There are no moving valves in the geometry.

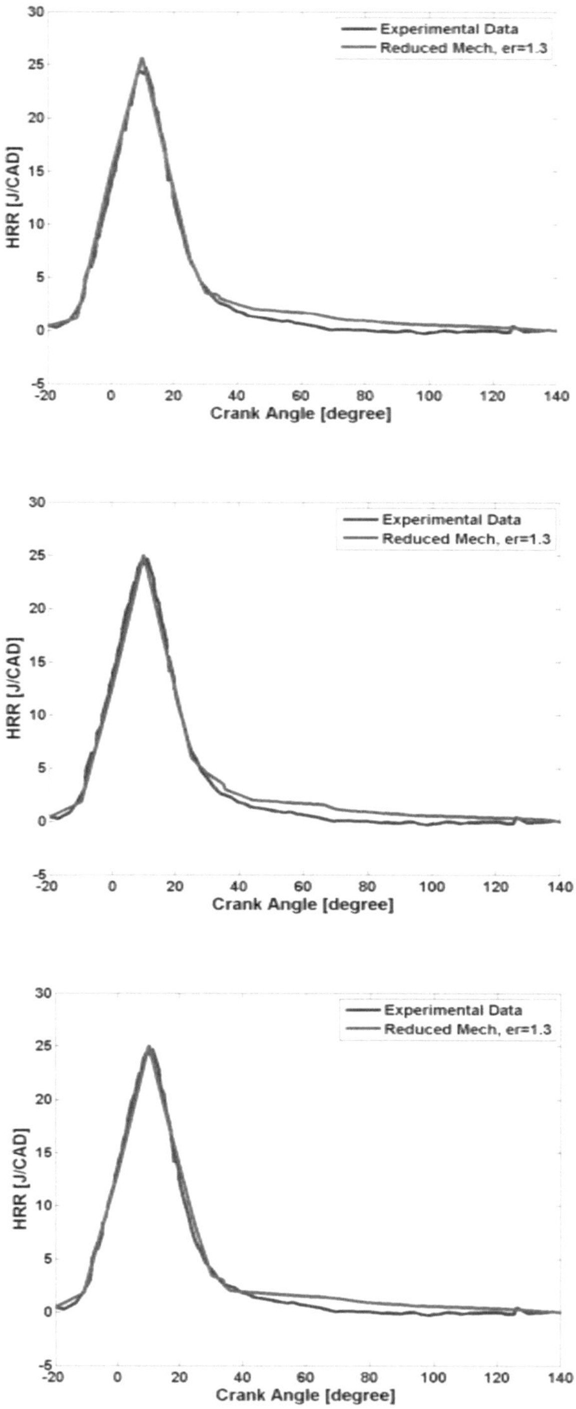

Figure 6.
Heat release rate (HRR); 100,000, 178,000 & 230,000 cells.

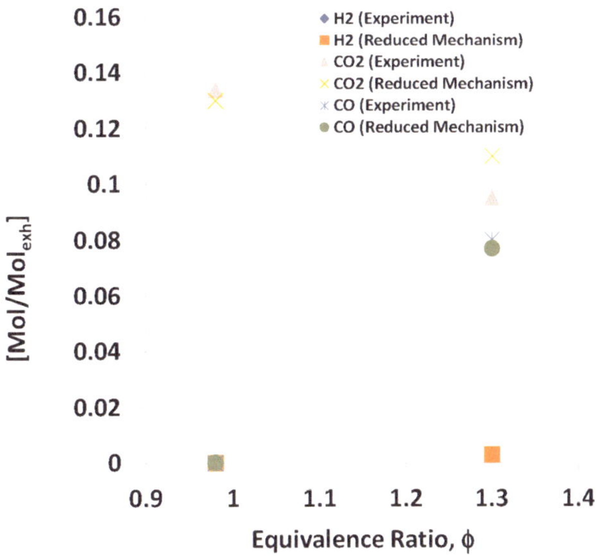

Figure 7.
Emissions.

Results shown in **Figure 8** and **Table 4** represent the importance of using the detailed chemistry in the CFD solvers where detailed chemistry has improved and better predicted the engine performance parameters.

To test the effect of the breakup model, numerical simulations were performed for global mechanism, quasi-global mechanism and, reduced mechanism obtained

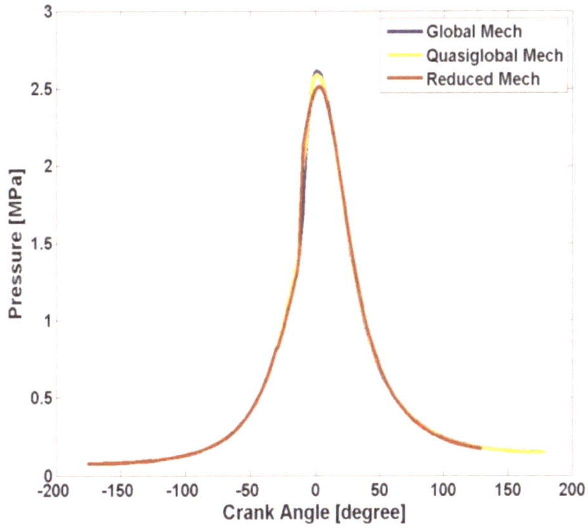

Figure 8.
Maximum or peak in-cylinder pressure (breakup model on).

	Experimental Value [31], MPa	Global Mech, MPa	Quasi-Global Mech, MPa	Reduced Mech, MPa
Numerical Simulation	2.51	2.61	2.58	2.51
Deviation	—	3.98%	2.78%	0%

Table 4.
Maximum or peak in-cylinder pressure at stoichiometric conditions for direct-injection case.

	No Residual		With Residual	
Mechanism	**Breakup ON**	**Breakup OFF**	**Breakup ON**	**Breakup OFF**
Global	2.62	2.55	3.04	2.75
Quasi-Global	1.78	1.82	2.09	1.72
Reduced	—	—	2.51	2.35

Table 5.
Summary of results for breakup model analysis.

through this research, with breakup model turned off. The global reaction mechanism still over predicted the results of in-cylinder pressure while the quasi-global mechanism under predicted the results. Summary of results is given in **Table 5**.

4. Conclusions

Development of a single reduced mechanism was performed for SI engine geometries and configurations which required performing sensitivity analysis and reduction of the skeletal reaction mechanism using SENKIN, a sensitivity analysis module of CHEMKIN. The extraction of the sensitivity data from SENKIN using KINALC was then performed and a mechanism reduction was completed using the computational singular perturbation (CSP) method. This helped in minimizing the computational time by using fewer required reactions and species. A reduced mechanism was then constructed that validated engine performance and combustion parameters of in-cylinder pressure, heat release rate, and emissions for a range of equivalence ratios utilizing the low temperature pathway analysis.

A well-established surrogate, such as *iso-octane*, was selected for study as gasoline is a complex mixture of various compounds and hydrocarbons. The fundamental research provided the data and mechanistic understanding needed for the development of a library for detailed mechanisms that can be used to correctly predict engine performance parameters.

Understanding of the reduction techniques, and a practical reduced reaction mechanism for *spark-ignition engines* and combustion has been achieved through the above-mentioned methodology. Also, it has shown the need for using detailed chemistry in reactive flow problems which can help predict the combustion and engine performance parameters more accurately. To gain the necessary data and confidence for use of detailed chemistry, benchmarking was performed using the global and quasi-global mechanisms in a previous study.

Author details

Muzammil Arshad
Texas A&M University, McAllen, USA

*Address all correspondence to: marshad@tamu.edu

References

[1] Arshad M. Optimization of chemical kinetic mechanism for efficient computation of combustion process in advanced internal combustion engine configurations. [PhD dissertation], Florida Institute of Technology. 2018

[2] Saylam A, Ribaucour W, Pitz WJ, Minetti R. Reduction of large chemical kinetic mechanisms for autoignition using joint analyses of reaction rates and sensitivities. International Journal of Chemical Kinetics. 2007;**39**:181-196

[3] Arshad M, Ilie M. Analysis of reduced order chemical mechanisms for oxygen-enriched combustion of methane and n-decane. 48th AIAA/ASME/SAE. ASEE Joint Propulsion Conference and Exhibit, Atlanta, GA, August 2012

[4] Soyhan HS, Mauss F, Sorusbay C. Chemical kinetic modeling of combustion in internal combustion engines using reduced chemistry. Combustion Science and Technology. 2002;**174**(11&12):73-91

[5] Valorani M, Creta F, Goussis DA, Lee JC, Najm HN. An automatic procedure for the simplification of chemical kinetic mechanisms based on CSP. Combustion and Flame. 2006;**146**:29-51

[6] Babajimopoulos A, Assanis DN, Flowers DL, Aceves AM, Hessel RP. A fully coupled computational fluid dynamics and multi-zone model with detailed chemical kinetics for the simulation of premixed charged compression ignition engines. International Journal of Engine Research. 2005;**6**(5):497-512

[7] Amsden AA. KIVA-3V: A block-structured KIVA program for engines with vertical or canted valves. USA: Los Alamos National Laboratory; July 1997

[8] Pitz WJ, Cernansky NP, Dryer FL, Egolfopoulos FN, Farrell JT, Friend DG, et al. Development of an experimental database and chemical kinetic models for surrogate gasoline fuels. SAE Technical Paper. 2007-01-0175

[9] Wang H, Yao M, Reitz RD. Development of a reduced primary reference fuel mechanism for internal combustion engine combustion simulations. Energy & Fuels. 2013;**27**: 7843-7853

[10] Gatowski JA, Balles EN, Chun KM, Nelson FE, Ekchian JA, Heywood JB. Heat release analysis of engine pressure data. SAE Technical Paper Series, Fuels and Lubricants Meeting & Exposition, Baltimore, Maryland, October 1984-841359

[11] Hageman MD, Rothamer DA. Sensitivity analysis of particle formation in a spark-ignition engine during premixed operation. 8th U.S. National Combustion Meeting, 2013-070IC - 0046

[12] Sakai S, Hageman M, Rothamer DA. Effect of equivalence ratio on the particulate emissions from a spark-ignited, direct-injected gasoline. SAE Technical Paper, 2013-01-1560

[13] Wiles MA, Probst DM, Gandhi JB. Bulk cylinder flow field effects on mixing in DISI engines. SAE World Congress, 2005-01 0096

[14] Liu J, Gong J, Cai L, Tan L, Ni X, Gao W. Multi-dimensional simulation of air/fuel premixing and stratified combustion in a gasoline direct injection engine with combustion chamber bowl offset. SAE International, 2006-32-0006

[15] Koch J, Schmitt M, Wright YM, Steurs K, Boulouchos K. LES multi-cycle analysis of the combustion process in a small SI engine. SAE International Journal of Engines. 2014;7(1):269-285

[16] Kee RJ, Rupley FM, Meeks E, Miller JA. CHEMKIN III – A FORTRAN Chemical Kinetics Package for the Analysis of Gas-Phase Chemical and Plasma Kinetics. Livermore, CA: Sandia National Laboratories; 1996

[17] Lutz AE, Kee RJ, Miller JA. SENKIN: A FORTRAN Program for Predicting Homogenous Gas Phase Chemical Kinetics in a Closed-System with Sensitivity Analysis. Livermore, CA: Sandia National Laboratories; 1988

[18] Available from: http://www.chem.leeds.ac.uk/Combustion/Combustion.html

[19] Lu TF, Law CK. Strategies for mechanism reduction for large hydrocarbons: *n*-heptane. Combustion and Flame. 2008;**154**(1-2):153-163

[20] Hageman MD, Sakai SS, Rothamer DA. Determination of soot onset and background particulate levels in a spark-ignition engine. Proceedings of the Combustion Institute. 2015;**35**(3): 2949-2956

[21] Hageman MD. Isolation of fundamental parameters contributing to particulate formation in a spark ignition direct injection (SIDI) engine. [PhD dissertation], 2014

[22] Liu C, Zuo Z, Feng H. Systematic reduction of the detailed kinetic mechanism for the combustion of n-butane. Journal of Chemistry. 2016; **2016**(11):1-7

[23] Lu T, Law CK. Toward accommodating realistic fuel chemistry in large-scale computations. Process in Energy and Combustion Science. 2009; **35**:192-215

[24] Sun W, Chen Z, Gou X, Ju Y. A path flux analysis method for the reduction of detailed chemical kinetic mechanisms. Combustion and Flame. 2010;**157**: 1298-1307

[25] Kuo KK. Principles of Combustion. 2nd ed. John Wiley & Sons; 2005. pp. 154, 199-155, 214

[26] Lam SH. Model reductions with special CSP data. Combustion and Flame. 2013;**160**:2707-2711

[27] Lam SH. An efficient implementation of computational singular perturbation. Combustion Science and Technology. 2018;**190**(1): 157-163

[28] Glassman I. Combustion. 3rd ed. San Diego, California, USA: Academic Press; 1996. pp. 94-104

[29] Turns SR. An Introduction to Combustion Concepts and Applications. 2nd ed. USA: McGraw Hill Companies; 2000

[30] Arshad M. Numerical simulations and validation of engine performance parameter in direct injection spark ignition (DISI) engines using chemical kinetics. ASME 2020 International Mechanical Engineering Congress and Exposition (IMECE), November 2020, DOI: 10.1115/IMECE2020-24683

[31] Wooldridge M, Fatouraie M. In-Cylinder particulate matter and spray imaging of ethanol/gasoline blends in a direct injection spark ignition engine. SAE International, 2013-01-0259

[32] Fatouraie M. The effects of ethanol/ gasoline blends on advanced combustion

strategies in internal combustion engines. [PhD disseration]. 2014

[33] Kokjohn SL, Hanson RM, Splitter DA, Reitz RD. Experiments and modeling of dual-fuel HCCI and PCCI combustion using in-cylinder fuel blending. SAE International, 2009-01-2647

Chapter 5

Bayesian Methods and Monte Carlo Simulations

Pavel Loskot

Abstract

Bayesian methods provide the means for studying probabilistic models of linear as well as non-linear stochastic systems. They allow tracking changes in probability distributions by applying Bayes's theorem and the chain rule for factoring the probabilities. However, an excessive complexity of resulting distributions often dictates the use of numerical methods when performing statistical and causal inferences over probabilistic models. In this chapter, the Bayesian methods for intractable distributions are first introduced as sampling, filtering, approximation, and likelihood-free methods. Their fundamental principles are explained, and the key challenges are identified. The concise survey of Bayesian methods is followed by outlining their applications. In particular, Bayesian experiment design aims at maximizing information gain or utility, and it is often combined with an optimum model selection. Bayesian hypothesis testing introduces optimality in the data-driven decision making. Bayesian machine learning assumes data labels to be random variables. Bayesian optimization is a powerful strategy for configuring and optimizing large-scale complex systems, for which conventional optimization techniques are usually ineffective. The chapter is concluded by examining Bayesian Monte Carlo simulations. It is proposed that augmented Monte Carlo simulations can achieve explainability and also provide much better information efficiency.

Keywords: Bayesian analysis, distribution, Monte Carlo, numerical method, machine learning, optimization, posterior, prior, simulation, statistical inference

1. Introduction

Many real-world systems exhibit some level or form of randomness. This is reflected in their mathematical models, which are often probabilistic. There are two basic strategies to interpret the randomness captured by probabilistic models. The frequency of occurrence of a random phenomenon is a measurable quantity that can be used to predict how likely the phenomenon can be observed in future. The other approach quantifies the uncertainty about the occurrence of random phenomenon more subjectively, that is, as a degree of belief or expectation of observing the phenomenon, given domain knowledge and the past experiences. This latter approach led to a broad area of general Bayesian methods [1], Bayesian data analyses [2], Bayesian signal processing [3], Bayesian regression [4], Bayesian machine learning [5], and

Bayesian optimization [6]. In the tasks of statistical inference, the assumption of knowing (or not) a prior distribution crucially affects the feasibility as well as the structure of estimators [7]. There is also an intimate connection between Bayesian probabilistic models and making causal inferences [8], as will be discussed in Section 3.

Bayesian methods found widespread applications in many probabilistic modeling frameworks. These methods are all rooted in the surprisingly simple Bayes's theorem, that is,

$$p(\theta|x) = \frac{p(x|\theta)p(\theta)}{p(x)} \tag{1}$$

where $p(\cdot)$ denotes the (for continuous random variables) or the probability (for discrete random variables). Eq. (1) quantifies how our belief about the parameter θ changes from its prior, $p(\theta)$, to posterior, $p(\theta|x)$, after observing data x. The conditional term, $p(x|\theta)$, represents the likelihood of the parameter θ, given observations, x. The scaling term, $p(x)$, in Eq. (1) is usually referred to as evidence. It should be noted that both the parameter and the data can be assumed in an arbitrary number of dimensions.

The majority of tasks in Bayesian inference involves explicitly or implicitly solving one of the following integrals (or sums, for discrete random variables), that is,

$$\text{marginalization: } p(x) = \int p(x, \theta)d\theta \tag{2}$$

$$\text{summarization: } E[f(x)|\theta] = \int f(x)\ p(x|\theta)dx \tag{3}$$

$$\text{prediction: } p(x_{t+1}) = \int p(x_{t+1}|x_t)\ p(x_t)dx_t \tag{4}$$

$$p(x_{t+1}|x_t) = \int p(x_{t+1}|\theta)\ p(\theta|x_t)d\theta.$$

Unfortunately, in most but a few real-world scenarios, the expressions (2)–(4) are not mathematically tractable. In particular, the distributions often involve multiple sums and/or integrals, and their closed-form expressions cannot be obtained. The distributions are sometimes only known up to a scaling constant. Then, even numerically computing Eq. (1) can be rather challenging, since many complex distributions are multi-modal with a large number of local minima and maxima. Moreover, in online data processing, the distributions must be updated continuously as soon as the new data arrives.

The rest of this chapter is organized as follows. The strategies for performing Bayesian inferences with intractable distributions are outlined in Section 2 including sampling, filtering, approximation, and likelihood-free methods. The applications of Bayesian inferences are discussed in Section 3 including Bayesian experiment design, Bayesian hypothesis testing, Bayesian machine learning, and Bayesian optimization. Bayesian Monte Carlo simulations are reviewed in Section 4. Although the chapter mostly reviews known concepts and frameworks in Bayesian analysis of probabilistic models, Section 4 contributes a description of augmented Monte Carlo simulations. These simulations aim at providing explainability and improving information gain.

The references cited at the end of this chapter are by no means comprehensive; rather, they are suggestions of initial readings where to find further information about the topics discussed in this chapter.

2. Bayesian inference with intractable distributions

Bayesian inference assumes that prior knowledge is available, so it should be exploited to improve the estimation accuracy. Such knowledge is often represented as a prior distribution of the parameters to be estimated. It can be provided by the domain expertise or otherwise be subjectively selected. As already mentioned, non-linear and high-dimensional models are rarely tractable analytically. This section presents four basic strategies on how to deal with intractable distributions in Bayesian inference. In particular, the sampling methods define a proposal distribution, which is much easier to sample from. Most of these methods form a Markov chain of samples, which gradually converges to the desired distribution. The filtering methods update the estimates iteratively, so they are particularly suitable for streaming data. These methods are also more efficient in higher dimensions than the pure sampling methods. The approximate methods strive to improve the efficiency, even at the cost of estimation accuracy. The last group is the likelihood-free methods, which are particularly suited for parameter estimation using simulations.

2.1 Sampling methods

Consider, for example, the expectation (3). The integral (3) may be not only mathematically intractable but also numerically intractable, provided that the distribution $p(x)$ is difficult or even impossible to sample from. A common strategy then is to choose and sample from another, simpler distribution, $q(x)$, such that $q(x) > 0$, whenever $f(x)p(x) > 0$. The resulting method is known as importance sampling (IS), and $q(x)$ is referred to as the proposal distribution. Eq. (3) is then rewritten as

$$\int f(x)\ p(x)dx = \int f(x)\frac{p(x)}{q(x)}q(x)dx \approx \frac{1}{N}\sum_{i=1}^{N} f\left(x^{(i)}\right)\underbrace{\frac{p\left(x^{(i)}\right)}{q\left(x^{(i)}\right)}}_{w^{(i)}} \tag{5}$$

where the sum is an unbiased and consistent estimator of the expectation due to the law of large numbers.

Although the estimate (5) is unbiased and consistent, the number of samples required grows exponentially with the number of dimensions. It is also desirable to make the weights, $w^{(i)}$, more uniform; that is, $q(x)$ should approximate $p(x)$, in order to reduce the variance of the estimator in (5). Alternatively, the distribution $p(x)$ could be approximated by a histogram; the bins of such histogram can be optimized for a maximum efficiency. The histogram can be sampled directly using a uniform random number generator.

The sampling efficiency of the IS methods, particularly in high dimensions, can be improved by accounting for the likelihood of samples. The Markov chain Monte Carlo (MCMC) strategy represents a broad variety of sampling methods, such that the current sample is affected by the location of the previous sample. It allows learning complex target distributions in multiple dimensions. However, the MCMC sampling requires a burn-in period for achieving the convergence. Furthermore, the samples may get stuck in areas of small probability (requiring a random restart), and the correlation between consecutive samples increases the estimator variance.

Metropolis–Hastings sampling is the most commonly used MCMC algorithm. It can be succinctly described as the following three steps:

1. Given the current state x_t, sample a new state, x_{t+1}, from the proposal $q(x_{t+1}|x_t)$;

2. Calculate the acceptance probability $\alpha(x_{t+1}, x_t)$ of the new state x_{t+1};

3. If $q(x_t|x_{t+1}) = q(x_{t+1}|x_t)$ is symmetric, then accept the new state x_{t+1} if $\pi(x_{t+1}) > \pi(x_t)$; otherwise remain at the current state, that is, $x_{t+1} \equiv x_t$.

The challenge is to find a good proposal, q, with a bounded support to achieve fast convergence to the target distribution, $\pi(x)$. The target distribution only needs to be known up to a proportionality constant. The generated samples tend to concentrate in the areas of large probability. The generated samples occasionally contain sub-sequences of the same value. More importantly, as with any other sampling methods, there is an exploration–exploitation trade-off. Large acceptance rate means slow exploration of the target distribution, π, as well as large correlations between the samples. On the other hand, a small acceptance rate means large jumps across the support of π. The acceptance rate can be controlled by assuming a parameterized proposal, q_ϕ. The acceptance rate of about 50% is recommended for samples in a few dimensions, and it should be reduced to about 25% for distributions in many dimensions.

In the literature, many variants of the Metropolis–Hastings sampling algorithm were proposed, for example, combining random walk sampling, adaptive rejection sampling, Langevin algorithm, and augmented estimator.

Gibbs sampling iteratively samples from the conditional densities

$$\begin{aligned} X_1^{(t)} &\sim p_1\left(x_1|x_2^{(t)}, x_3^{(t)}, \dots, x_N^{(t)}\right) \\ X_2^{(t)} &\sim p_2\left(x_2|x_1^{(t)}, x_3^{(t)}, \dots, x_N^{(t)}\right) \\ &\vdots \\ X_N^{(t)} &\sim p_N\left(x_N|x_1^{(t)}, x_2^{(t)}, \dots, x_{N-1}^{(t)}\right). \end{aligned} \tag{6}$$

It can be represented as a set of N univariate Metropolis–Hastings samplers. The sampling dimensions can be chosen systematically or at random. Even though Gibbs sampling has 100% acceptance rate, it can still experience a slow convergence to the target distribution, so it is often combined with other sampling methods

Gibbs sampling motivates the Rao–Blackwell estimators

$$E[h(x_1)] \approx \frac{1}{T}\sum_{t=1}^{T} E\left[h(x_1)\,|x_2^{(t)}, \dots, x_N^{(t)}\right] \tag{7}$$

$$p_i(x_i) \approx \frac{1}{T}\sum_{t=1}^{T} p_i\left(x_i|x_j^{(t)}, j \neq i\right) \tag{8}$$

of function expectation and of the marginal distribution, respectively. These estimators are unbiased and have lower variance, since

$$var\left[E\left[h(x_1)\,|x_2^{(t)},\ \ldots,\ x_N^{(t)}\right]\right] \le var[E[h(x_1)]]. \tag{9}$$

An alternative strategy combines importance sampling with MCMC sampling assuming independent proposals, $\prod_{i=1}^{N} q_i(x_i)$ or $\prod_{i=1}^{N} q_t\left(x_i|x_i^{(t-1)}\right)$.

An interesting problem is how to design MCMC sampling when the target distribution is not stationary. The adaptive MCMC sampling can be trained online with the generation of new data. However, using past samples for adaptation invalidates the Markov assumption, so the convergence to the target distribution may be problematic, or the adaptation should cease after the burn-in period.

The Hamiltonian MCMC sampling tracks movements of a hypothetical ball under the potential and kinetic energy constraints. The new sample representing the updated ball location is obtained by integrating the ball's speed. The integration can be approximated, for example, by discretization. More importantly, the interval of integration trades-off the acceptance rate and the sampling rate, so, also the amount of correlations between subsequent samples and the waiting time. The main advantage of Hamiltonian MCMC method is a fast mix-in; that is, the samples converge quickly to the target distribution.

Another sampling method motivated by models in statistical physics is a restricted Boltzmann machine (RBM). This method learns the target distribution as a configuration, h, of a bipartite graph. The target distribution is $p(h) = \exp(-E(h))/Z$, where $E(h)$ is the energy of the configuration h, and $Z = \sum_h \exp(-E(h))$ is a partition (scaling) function.

The reversible-jump MCMC sampling can be used for distributions having an uncertain number of parameters (the model order). The idea is to increase the number of parameters by 1 with a certain probability every time the sample is taken from the proposal distribution.

The common pitfall of all sampling methods is how to recognize that the generated samples are not converging to the target distribution, for example, since the assumptions were violated, or the method has not been implemented correctly. These issues may not be easily detected by simply evaluating the samples or the simulation outputs.

2.2 Filtering methods

Processing time-series and streaming data must be often done recursively. The corresponding data model can be usually represented as a dynamic system, so the current state, x_t, is updated from the previous state, x_{t-1}, by an innovation random process, and the states, x_t, are observed as values, z_t. The states form a Markov chain, that is, $p(x_t|x_{t-1},\ \ldots,\ x_1) = p(x_t|x_{t-1})$, and the observations are assumed to be independent, that is, $p(z_i, z_j|x_t) = p(z_i|x_t)p(z_j|x_t)$. Assuming the Bayes's theorem

$$p(x_{1:t}|z_{1:t}) = \frac{p(z_{1:t}|x_{1:t})\,p(x_{1:t}|z_{1:t-1})}{p(z_t|z_{1:t-1})}, \tag{10}$$

the states can be estimated recursively from the incoming observations as a two-step procedure. The first step predicts the (distribution of) current state as

$$p(x_t|z_{1:t-1}) = \int p(x_t|x_{t-1})p(x_{t-1}|z_{1:t-1})dx_{t-1}. \tag{11}$$

The predicted state is corrected in the second step after receiving the latest observation, z_t, as

$$p(x_t|z_{1:t}) = \frac{p(z_t|x_t)p(x_t|z_{1:t-1})}{\int p(z_t|x_t)p(x_t|z_{1:t-1})dx_t}. \tag{12}$$

For linear systems with Gaussian inputs, the processes remain Gaussian, so only their mean values, $E[x_{1:t}|z_{1:t}]$, and covariances, $cov[x_{1:t}|z_{1:t}]$, need to be tracked. In such a case, the recursive estimator implementing (11) and (12) is known as Kalman filter. The Kalman filtering can be succinctly described as

$$\hat{x}_t = (\text{prediction of } x_t) + K_t \cdot [z_t - (\text{prediction of } z_t)] \tag{13}$$

where K_t is the Kalman gain at time t. It can be shown that Kalman filter is unbiased, and it is optimum in a sense of having the smallest possible variance; it belongs to a class of the best linear unbiased estimators (BLUE). Practical implementations of Kalman filter are usually concerned with numerical stability and process observability. The posterior distribution of the model states can be estimated using kernel and other smoothing methods.

Kalman filter can be modified to work with non-linear models. In particular, extended Kalman filter performs local linearization at each iteration assuming the first-order Taylor expansion. This can, however, cause large estimation errors and stability issues. Unscented Kalman filter defines a set of sigma points, which are tracked through non-linearity. The weighted and transformed sigma points are used to reconstruct the mean and the covariance of Gaussian distribution, so canonical Kalman filter can be then used. Ensemble Kalman filter has been adopted for large but sparse systems. Expectation maximization (EM) filtering is used when there are also unknown model parameters; see Eqs. (16) and (17) below.

In practice, having an approximate solution for a complex model is often much more useful than obtaining the exact solution of a simplified model. This has motivated the development of particle filters to estimate random signals under non-linear and non-Gaussian conditions. The key idea is to represent the posterior distribution by a set of random weighted samples (called particles), x_i, that is,

$$p(x) \approx \sum_{i=1}^{N} w_i \ \delta(x - \tilde{x}_i) \quad \Rightarrow \quad E[f(x)] \approx \sum_{i=1}^{N} w_i f(\tilde{x}_i) \tag{14}$$

where δ denotes the Dirac-delta function. However, the number of required particles grows quickly with the number of dimensions.

The particles should be chosen to represent high-density regions by many samples with large weights. Provided that the particles are generated using some MCMC sampling strategies, the resulting recursive Bayesian estimator is referred to as the sequential MC (SMC) method. Since only after a few iterations, most particles often have negligible weights (so-called particle degeneracy problem), particle resampling (with replacement and proportionally to their weight) is necessary in order to recover a good representation of the underlying distribution while maintaining a constant number of particles.

Expectation maximization (EM) method has been developed to deal with missing data and to filter data under parameter uncertainty. Assuming the observed but incomplete data, x, as well as the missing data, y, the data likelihood conditioned on an unknown parameter, θ, can be computed as

$$p(x|\theta) = \int p(x, y|\theta)dy = \int p(y|x, \theta)p(x|\theta)dy. \tag{15}$$

The integral (15) is evaluated iteratively by first computing the Q-function (the E-step) as

$$Q(\theta, \theta_{n-1}) = E[\log p(x, \ y|\theta)|x, \theta_{n-1}] \tag{16}$$

followed by estimating the parameter θ (the M-step) as

$$\theta_n = \text{argmax}_\theta Q(\theta, \theta_{n-1}). \tag{17}$$

This procedure guarantees that the likelihood is maximized, since always $p(x|\theta_n) > p(x|\theta_{n-1})$. However, Eqs. (16) and (17) are normally analytically intractable, so they must be evaluated numerically.

2.3 Approximation methods

The sampling methods, which mostly involve MCMC sampling, for performing Bayesian inference are asymptotically exact. However, these methods are also numerically expensive, especially for high-dimensional models. When working with big data, or when there is a need to choose among many plausible models, having less accurate but numerically cheaper solution is highly desirable. Variational inference deterministically approximates the log-evidence $p(x)$ in Bayes's theorem (1) as

$$\log p(x) = \mathcal{L} + KL[q(\theta)||p(\theta|x)] \geq \mathcal{L} \tag{18}$$

where $\mathcal{L}$ is the evidence lower-bound (ELBO), since the Kullback–Leibler (KL) divergence is always non-negative. Therefore, given the evidence $p(x)$, minimizing the KL divergence between the approximation, $q(\theta)$, and the posterior, $p(\theta|x)$, is equivalent to maximizing ELBO, $\mathcal{L}$. Alternatively, it is possible to approximate the prior, $p(\theta)$, instead of the posterior. These are functional optimization problems defined as $\frac{d}{dq}\mathcal{L}(q) = 0$, s.t., $\int q(\theta)d\theta = 1$, with the optimum solutions $q^*(\theta) = p(\theta|x)$ or $q^*(\theta) = p(\theta)$.

Any function, $q(\theta)$, can be assumed, but some choices are better than others. The mean-field approximation assumes that the dimensions are independent, that is, $q(\theta) = \prod_{i=1}^{M} q_i(\theta_i)$. Then, $q_i(\theta_i)$ can be optimized in turn until a convergence is obtained, indicated by no more increases in the ELBO value. There is a performance penalty if the independence assumption is not satisfied. Another popular choice is to assume a class of distributions, $q_\phi(\theta)$, parameterized by ϕ.

The mean-field assumption is implemented in the coordinated ascent variational inference (CAVI) algorithm. It is related to Gibbs sampling and message passing. However, this method is not computationally efficient and only works for distributions, which are conditional conjugates of the prior. Stochastic variational inference

(SVI) improves the efficiency as well as accuracy of CAVI by assuming the stochastic approximation of the ELBO gradient and by optimizing the parameters using only a subset of data in each iteration (similarly to the EM method). Unfortunately, both CAVI and SVI require deriving the approximation components, q_i, analytically, which is often impractical.

Motivated by the limitations of conventional variational inference methods, the black box variational inference (BBVI) algorithm samples from a family of distributions, $q_\phi(\theta)$, assuming that $p(x|\theta)p(\theta) = p(x, \theta)$ is known. The objective is to avoid the need for analytical derivations. The BBVI method can be used to estimate a wide range of linear and non-linear models. Since it is also the most efficient method for performing variational inference, it has become available as a library in many programming languages.

The divergence between prior or posterior and its approximation cannot be performed directly but only by maximizing the ELBO. Although, at each iteration, the ELBO is guaranteed not to decrease, the convergence can be slow. However, the efficiency of variational methods is still better than that of the sampling and EM methods. The KL divergence can be replaced with alternative measures such as α divergence, f divergence, and mutual information. In addition, there are only a few theoretical guarantees, and the overall estimation accuracy of variational methods is crucially affected by not knowing how well the posterior (or prior) has been approximated and approximating asymmetric distributions is more challenging.

2.4 Likelihood-free inference

In many practical scenarios, a mathematical model can be available as a simulation. It is often easy to simulate the model for different parameter values. Thus, by generating and comparing many simulation outputs for different parameter values, the simulation run can be identified that best matches the observed data. The parameter values for this simulation run are then declared to be the estimate. The inference methods implementing this simple idea are known as being likelihood free. They can be found in the literature as indirect inference, bootstrap filter, synthetic likelihood, and other methods. However, the most popular among these methods is approximate Bayesian computation (ABC).

The ABC method only requires that a generative model to generate data is available. There is otherwise no need to assume, know, or calculate the model posterior, parameter likelihoods, or importance weights. The basic ABC rejection sampling is performed by repeating the following three steps:

1. Sample the parameter θ from its prior, $p_0(\theta)$;

2. Simulate or otherwise generate the data $x(\theta)$;

3. If the distance, $\|x - x\| < \varepsilon$, between the simulated data, x, and the observed data, $\tilde{x}$, is sufficiently small, record the parameter θ.

The posterior distribution, $\pi_\varepsilon(\theta|x)$, can be estimated after enough samples, θ and x, have been collected. It is clear that the number of samples required to accurately estimate $\pi_\varepsilon(\theta|x)$ grows quickly with the number of dimensions. The exact inference can be obtained in the limit $\lim_{\varepsilon \to 0} \pi_\varepsilon(\theta) = p(\theta|x)$. On the other hand, no learning occurs if $\lim_{\varepsilon \to \infty} \pi_\varepsilon(\theta) = p_0(\theta)$.

The efficiency of the ABC method can be improved by considering summary statistics, $S(x)$, rather than directly comparing the data, x. If $S(x)$ is the sufficient statistic for estimating the parameter θ, then $\pi_\varepsilon(\theta|S(x)) = \pi_\varepsilon(\theta|x)$; however, in practice, an informative statistic is often used.

Even though the ABC method can be readily parallelized, it can still be very inefficient, especially if the informative statistic has been poorly chosen and in high dimensions. Since the default ABC method does not exploit information about the accepted and rejected values of θ, the efficiency could be improved by employing the MCMC sampling. The resulting ABC-MCMC algorithm performs the following four steps:

1. Augment the posterior distribution as $p\left(\tilde{\theta}, \tilde{x}|x\right)$;

2. At the current state $\tilde{\theta}$, generate a new sample, $\theta \sim q\left(\theta|\tilde{\theta}\right)$;

3. Simulate the data, x, for the new state θ;

4. Accept the new state, θ, with some probability, $\alpha(\theta, x)$.

This algorithm can be fine-tuned by gradually reducing ε in order to balance rejections with the estimation accuracy while the convergence requires ε to be sufficiently small. Unlike pure MCMC sampling, the MCMC-ABC method requires a small acceptance rate in order to achieve good accuracy.

3. Other topics in Bayesian analysis

The Bayesian frameworks find many real-world applications in data analysis and data-driven planning and decision making. Bayesian experiment design is about choosing experiments and models that can provide the largest information gain or utility. Bayesian hypothesis testing accounts for prior knowledge to bias the hypothesis rejection or acceptance. This framework can be also used to make statistically informed decisions. Bayesian machine learning has become increasingly popular in recent years. It considers data labels to be random variables. Both supervised and semi-supervised strategies are discussed. Bayesian optimization shows a remarkable promise in optimizing complex, difficult-to-evaluate systems. It considers the observed system responses to be samples of a random process.

3.1 Bayesian experiment design

In Bayesian experiment design, the task is to choose the optimum experiment and possibly also a data model [9]. This can be probabilistically represented as the augmented posterior (1), that is,

$$p(\theta|x, d, m) = \frac{p(x|\theta, d, m)p(\theta, m)}{p(x|d, m)} \tag{19}$$

where d and m, respectively, represent the experiment and the model choice. More specifically, θ denotes uncontrolled inputs to the experiment as well as unknown model parameters, whereas d are the experiment inputs that can be controlled, that is, designed by a selection. The objective is to specify the optimum design, d, in order to facilitate the estimation of θ as well as selection of the best model, m, for observations, x.

For every configuration of the experiment including the model selection, let $U(d, x, m, \theta)$ be the perceived utility, for instance, the amount of information that can be gained from the experiment. Since the observations are random and the data model is unknown, the optimum experiment maximizes the average utility,

$$d^* = \text{argmax}_{d \in \mathcal{D}} \; \overline{U}(d) = \text{argmax}_{d \in \mathcal{D}} E_{x,m,\theta}[U(d, \; x, \; m, \; \theta)]. \tag{20}$$

Alternatively, the model selection can be formulated as a hypothesis-testing problem. The observed data and models are combined to obtain and then sample the predictive distribution for each model candidate. Additional experiments can be performed as needed in order to maximize reduction in the model uncertainty. A mismatch between the supports of the model prior and the model likelihood indicates that the model has either too few or too many parameters, and there is not enough evidence to make the model selection with a high confidence [10].

Instead of choosing one best model, the predictions from multiple models can be combined. The joint prediction has potentially much larger discriminatory power than individual predictions [10]. Moreover, how well the model can describe the majority of outcomes from many random experiments under different experimental conditions can be as important as the model likelihood.

The information value of an experiment can be quantified as the Fisher information matrix with the entries

$$[\mathcal{I}; (\theta)]_{i,j} = E\left[\left(\frac{\partial}{\partial \theta_i} \log p(X; \theta)\right)\left(\frac{\partial}{\partial \theta_j} \log p(X; \theta)\right)|\theta\right]. \tag{21}$$

Since performing experiments is often costly, the following optimization objectives have been defined in the literature for linear models:

- A-optimality: minimize the average variance of parameter estimates;
- C-optimality: obtain the BLUE of linearly combined model parameters;
- D-optimality: maximize the entropy or determinant of information matrix;
- E-optimality: maximize the minimum eigenvalue of information matrix;
- T-optimality: maximize the trace of information matrix.

These optimality objectives may include prior distributions about the model and experiment parameters and combine other objectives for model selection.

The main challenge in evaluating the experiment design (20) is the computational complexity involved in searching the whole design space. The search strategies include linearization, local search, discretization, enumeration, approximation by regression or surrogate models, random (e.g., MCMC) sampling, genetic algorithms, as well as using Bayesian optimization methods, as will be discussed in Section 3.4.

Furthermore, the design (20) yields the single best experiment. In batch-optimum experiment design, N experiments are performed simultaneously. The data from different experiments are conditionally independent given the parameters, θ. The expected utility from these N experiments will be, in general, different from the sum of utilities of individual experiments. A simpler objective is to minimize the predicted variance of observations, x, when the experiment designed as a single optimum is repeated N times.

In the sequential experiment design, the posterior, $p(\theta, x_t|d_t)$, from the tth experiment is used as a prior for the $(t+1)$th experiment, and then, the experiment conditions, d_{t+1}, are optimized. This is a greedy (sub-optimum) approach; however, it may still outperform batch design due to an inherent adaptation to x_t and d_t. The optimum sequential experiment design is more complex, and it leads to a problem of dynamic programming.

3.2 Bayesian hypothesis testing

Let θ_0 be the critical parameter value, so that the null hypothesis, $\mathcal{H}_0$, can be accepted, if the parameter value $\theta < \theta_0$, and it is rejected in favor of the alternative hypothesis, $\mathcal{H}_a$, otherwise. Denoting the loss function as $L(\theta; \mathcal{H}_0)$ and $L(\theta; \mathcal{H}_a)$ under the respective hypotheses, the Bayesian risk for observation, x, is computed as

$$R(x; \theta_0) = \int_{\theta < \theta_0} L(\theta; \mathcal{H}_0) Pr(x|\theta) p(\theta)\, d\theta + \int_{\theta \geq \theta_0} L(\theta; \mathcal{H}_a) Pr(x|\theta) p(\theta)\, d\theta. \tag{22}$$

The optimum decision threshold is then obtained by minimizing the average risk, that is, $\theta_0^* = argmin_\theta E_x[R(x; \theta)]$. The statistical power of the hypothesis test is determined by the sample size [11].

3.3 Bayesian machine learning

The general objective of machine learning is to learn how to label unseen data. In supervised learning, this objective is accomplished by training the model with labeled data. For instance, consider the minimum mean square estimation (MMSE) of label Y for data $\mathbf{X}$ [3], that is,

$$f^*(\mathbf{x}) = argmin_f E\left[(Y - f(\mathbf{X} = \mathbf{x}))^2\right] = E[Y|\mathbf{X} = \mathbf{x}]. \tag{23}$$

The training data, $(\mathbf{X}_1, Y_1), (\mathbf{X}_2, Y_2), \dots, (\mathbf{X}_n, Y_n)$, are samples from the distribution, $p(Y, \mathbf{X}) = p(\mathbf{X}|Y)p(Y)$, assuming the likelihood, $p(\mathbf{X}|Y)$, and the prior, $p(Y)$, of data labels, Y. The MMSE expectation (23) can be then approximated as

$$E[Y|\mathbf{X} = \mathbf{x}] \approx \frac{1}{n} \sum_{i=1}^{n} Y_i \cdot \mathcal{I}; _{\mathbf{X}_i = \mathbf{x}} \tag{24}$$

assuming the indicator function $\mathcal{I}$.

In semi-supervised learning, there are additional unlabeled data, $\mathbf{X}_{n+1}, \mathbf{X}_{n+2}, \dots$, which can be used to estimate the distribution, $\int p(\mathbf{X}, Y)\, dY$. It allows formulating and improving the estimates of the conditional expectation, $E[Y|\mathbf{X} = \mathbf{x}]$, and cast it as a missing labels problem. More generally, automated data labeling involves the problem of finding the discriminative data model, $p(Y|\mathbf{X})$, or $p(\mathbf{X}|Y)$. However, the full generative model, $p(Y|\mathbf{X})p(\mathbf{X}) = p(\mathbf{X}|Y)p(Y)$, may be required by some of the Bayesian inference methods outlined in the previous section.

Assuming data labels as random variables with their prior and posterior distributions is useful in tolerating label errors as well as missing labels. However, the challenge is that the assumed distributions may be biased or even incorrect. In addition to automated data labeling, Bayesian machine learning can generate more training data and improve the quality of training data by correcting the labels.

There is also an interesting connection to causal machine learning. In particular, the label can be assumed to be a cause of observed data (features) representing the effects. Since the probability, $Pr(\text{effect})$, and the conditional probability, $Pr(\text{cause}|\text{effect})$, are not independent, the data samples can be used to estimate their distribution, $p(\mathbf{X})$, which in turn facilitates an anti-causal learning of causes (i.e., the data labels, Y). On the other hand, the probabilities $Pr(\text{cause})$ and $Pr(\text{effect}|\text{cause})$ are independent, and so, intuitively, they cannot be exploited for causal learning of data from their labels.

3.4 Bayesian optimization

Consider the task of minimizing or maximizing a function $f(x)$ over some high-dimensional feasible set, $A \subseteq \mathscr{R}^n$. It is typically assumed that evaluating the function is numerically or otherwise very expensive, whereas testing whether $x \in A$ is computationally cheap. Even though f is normally assumed to be continuous, its derivatives are not known. Furthermore, the problem is non-convex, as the function has no special structure, it usually contains many local optima, and its observations may be noisy.

The basic idea to solve such a difficult optimization problem is to learn a surrogate approximation of f, which is cheap to evaluate. The surrogate approximation is constructed in the context of Bayesian inference [12]. In particular, Bayesian optimization starts by evaluating the function f at a few randomly chosen points, x. The obtained values, $f(x)$, are assumed to be samples of a random process. A Gaussian process (GP) is most commonly assumed, since it can yield closed-form expressions for the extrapolated values, even though inverting large covariance matrices is often numerically rather problematic. Then, the following steps are repeated as many times as can be practically afforded.

1. The posterior of a candidate sample is obtained given the existing samples $\{x_1, x_2, \dots, x_n\}$;

2. The new sample is chosen as the one maximizing so-called acquisition function (to be discussed below);

3. The function f is evaluated at the new sample, and the posterior is updated accordingly.

Assuming Bayesian regression over a Gaussian process, the n existing samples, $f(x_1), \dots, f(x_n)$, are normally distributed and have the joint mean, $\mu_0(x_{1:n})$, and the covariance matrix, $\Sigma_0(x_{1:n})$. The corresponding posterior distribution of a candidate sample, $f(x)$, is also normal with the mean, $\mu_n(x)$, and the variance, $\sigma_n^2(x)$.

The acquisition function predicts the value of a new sample after already knowing the values, $f(x_1), \ldots f(x_n)$. The values of the acquisition function tend to be larger when the calculated credible intervals are wider and when the posterior mean is larger. Since the mean, $\mu_n(x)$, is also a point estimate of $f(x)$ after n observations, the value of $f(x)$ is estimated by interpolating $\mu_n(x)$; this corresponds to Gaussian regression. In other words, the measured values, $f(x_1), \ldots f(x_n)$, are interpolated to estimate the mean, $\mu_n(x)$. The covariance matrix, $\Sigma_n(x_{1:n})$, determines how fast the function values vary in between the already measured samples. It is usually required that the values de-correlate with their distance. The covariance matrix can be also estimated from the existing samples as $\hat{\eta} = argmax_\eta p(f(x_{1:n})|\eta)$ (maximum likelihood estimation, MLE) or as $\hat{\eta} = argmax_\eta p(f(x_{1:n})|\eta)p(\eta)$ (maximum a posterior estimation, MAP). It is also possible to treat some parameters, η, as nuisance parameters and marginalize them from the likelihood assuming their prior, $p(\eta)$, before employing one of the Bayesian methods for intractable distributions.

Expected improvement (EI) is the most commonly assumed acquisition function. Provided that f is observed without measurement noise, the current best choice is $f_n^* = \max\{f(x_1), \ldots, f(x_n)\}$. The expected improvement is then defined as

$$\mathrm{EI}_n(x) = E\left[|f(x) - f_n^*|_+\right] \tag{25}$$

where $|\cdot|_+ \geq 0$, and the expectation is taken over by the posterior of $f(x)$ conditioned on the values known so far. The next best choice for sampling the function f is

$$x_{n+1} = \mathrm{argmax}_x \mathrm{EI}_n(x). \tag{26}$$

Unlike f, the function $\mathrm{EI}_n(x)$ is inexpensive to evaluate, and its derivatives can be used to effectively maximize Eq. (26). The best expected improvement occurs at points far away from the previously evaluated points (the points having a large posterior variance) and at points having large posterior means; this represents an exploration–exploitation trade-off.

Knowledge gradient (KG) acquisition function selects the sampling point x having the largest posterior mean, $\mu_n^* = \max_x \mu_n(x)$. It allows considering the posterior across the full domain of f and how it is changed by a new sample. The knowledge gradient is computed as

$$\mathrm{KG}_n(x) = E\left[\mu_{n+1}^* - \mu_n^* | x_{n+1} = x\right]. \tag{27}$$

The next best sample corresponds to the maximum, that is,

$$x_{n+1} = \mathrm{argmax}_x \mathrm{KG}_n(x). \tag{28}$$

An efficient practical implementation of KG can assume multi-start stochastic gradient descent (or ascent, when the target function is to be maximized). This leads to a two-step maximization procedure when the maximum is first searched among a collection of candidate functions, and then, the selected function is maximized separately, for example, by differentiation.

In general, the KG acquisition function tends to significantly outperform the EI method, especially when the function observations are noisy.

Entropy search (ES) acquisition function assumes that the global optimum, x^*, is a random variable implied by the Gaussian process, $f(x)$. It performs the search for a new sample in order to obtain the largest decrease in differential entropy corresponding to the largest reduction of uncertainty about the global optimum. The ES acquisition function is defined as

$$\mathrm{ES}_n(x) = H(P_n(x^*)) - E_{f(x)}[H(P_n(x^*|x, f(x)))] \tag{29}$$

where H denotes entropy, so that the posterior across a full domain of f and how it is changed by a new sample are again accounted for. This is useful when the observations are noisy. However, unlike the KG method, stochastic gradients cannot be obtained for the ES acquisition function.

Finally, the predictive entropy search (PES) acquisition function rewrites Eq. (29) using mutual information. Conceptually, it is exactly the same as the ES method; however, numerical properties of these two algorithms are different.

The basic optimization problem can be augmented by considering a set of objectives, $f(x, s)$, indexed by "fidelity", s, such that lower values of s mean higher fidelity, and $f(x, 0) \equiv f(x)$ represents the original objective. For example, fidelity can represent the model order or the modeling granularity. There is also a cost, $c(s)$, associated with fidelity, s; the larger the required fidelity, the larger the cost. These problems are then referred to as multi-fidelity source evaluation, in the literature. The task is to maximize the function, $f(x, 0)$, by observing a sequence of values of $f(x, s)$ at n points, that is, $(x_1, s_1), \dots, (x_n, s_n)$, subject to the total available budget, $\sum_{i=1}^{n} c(s_i) \le C_{\text{total}}$.

This problem can be further generalized by assuming that neither $f(x, s)$ nor $c(s)$ is monotonic in s; for example, keeping s constant, the observed $f(x, s)$ across different regions of x can have varying accuracy.

The problem of random environmental conditions is closely related to multi-task Bayesian optimization. It maximizes the function $\int f(x, w)p(w)dw$ by evaluating $f(x, w)$ for each w at multiple values of x. It is assumed that evaluating $f(x, w)$ at both x and w is expensive, but evaluating $p(w)$ is cheap. The values w represent random environmental conditions, and they act as noise in observations of $f(x, w)$. For example, f can represent the average performance of a machine learning model with w-fold cross-validation. Another example are Bayesian neural networks having random coefficients with a certain probability distribution.

Comparing Bayesian optimization with other optimization methods, the latter usually work well only for specific problems or under specific conditions, but have high computational cost, and suffer from the curse of dimensionality, leading to a low sample efficiency and slow convergence. Bayesian optimization, on the other hand, can work very well for moderate dimensionality (the maximum dimension of about 20 is recommended in the literature), and the sampling decisions appear to be optimum despite being sequential. In addition, Bayesian optimization is considered to be derivative free but a black box method.

Despite a good performance, there is a need to provide better theoretical understanding of Bayesian optimization methods, define stopping rules, and improve learning of surrogate models, especially for non-Gaussian processes and for systems containing non-linear transformations. Bayesian optimization can be used to efficiently find hyperparameters of machine learning models [13]. Recently, maximization of a composite function, $g(h(x))$, has been considered in the literature, where the inner function, h, is expensive to evaluate and a black box, whereas the outer function, g, is easy to evaluate.

4. Bayesian Monte Carlo simulations

Monte Carlo simulations are a general class of random sampling methods. They can be used to solve problems such as (2)–(4) using numerical algorithms. Simulations are often used to solve problems involving exploration of high-dimensional parameter spaces in order to perform sensitivity analysis, performance analysis, and system optimization. Unlike analytical approaches, computer simulations are more flexible, require fewer assumptions, and allow for more complex and, thus, more realistic models to be considered. In silico experiments are much more cost-effective than laboratory experiments conducted in life sciences. Digital twins are now often built to monitor complex engineering systems. Synthetic data from simulations can be used to train machine learning models and to avoid expensive or hard-to-get real-world measurements. However, there are often problems with reproducibility and validation of simulation results, especially when the simulation procedures and models become more complex. In general, it is extremely difficult to distinguish among algorithmic errors; incorrect choice and use of algorithms; the errors in mathematical derivations; conceptual errors, for example, due to invalid assumptions; and the errors in algorithm implementations.

Vast amounts of generated simulation data are usually reduced into a few summary statistics. This results in a loss of potentially valuable information, unless the statistics are also sufficient within a given context. There is rarely a deeper analysis of simulation techniques to explain why the implemented algorithms work or do not work, under what conditions, and whether they could be improved further. Thus, there is a need to enhance traditional Monte Carlo simulations and define the best practices of how to plan and conduct computer experiments in order to maximize the information gain and extract maximum useful knowledge from these simulations [14].

Improving the usefulness of Monte Carlo simulations requires that they are interpreted as statistical estimators of quantities of interest including identifying relevant causal associations [15, 16]. For instance, the sample statistics are point estimates, whereas it may be more useful to extract Bayesian posterior or credible intervals of these statistics when performing simulations. Bayesian analysis also enables counterfactual reasoning in order to find how the performance or system behavior would change, if the simulated model was altered. In particular, if the average performance of a system, $f(x)$, is evaluated as $\int f(x)p(x)dx$ over a distribution of factors, $p(x)$, then the counterfactual performance of a modified system having the factor distribution $p^*(x)$ can be defined as

$$\int f(x)p^*(x)dx = \int f(x)\underbrace{\frac{p^*(x)}{p(x)}}_{w(x)}p(x)dx. \tag{30}$$

Eq. (30) is akin to IS Eq. (5), that is, the system response, $f(x)$, is scaled by weights, $w(x)$, when assessing its hypothetical performance. Such approach can yield the following systematic and general strategy for designing systems via computer simulations:

1. Build a structural equation model (SEM) of the system using available performance data;

2. Use the SEM to define points of interventions and measurements;

3. Infer changes in the performance distribution under different hypothetical modifications;

4. Select and validate the modification with the best hypothetical improvement by implementing it in a real-world system.

A basic strategy for increasing the information gain of Monte Carlo simulations is to augment the number of observation points. In **Figure 1A**, the simulation inputs and outputs are represented by random variables, X and Y, respectively, and the augmented output is represented by a random variable, Z. The corresponding posteriors and likelihoods are then computed as

$$p(X|Y) = \frac{p(Y|X)p(X)}{p(Y)} p(Y|X) = \sum_Z p(Y|X,Z)p(Z|X) \tag{31}$$

$$p(Z|Y) = \frac{p(Y|Z)p(Z)}{p(Y)} p(Y|Z) = \sum_X p(Y|X,Z)p(X|Z) \tag{32}$$

$$p(X|Z) = \frac{p(Z|X)p(X)}{p(Z)} p(Y,Z,X) = p(Y|Z,X)p(Z|X)p(X). \tag{33}$$

The causal relationships can be represented by probabilistic graphs, which are referred to as structural causal models (SCMs). These models are similar to SEM as well as allow capturing non-linear relationships. The SCMs are effectively Bayesian networks with directed edges indicating causal effects (since these effects are generally asymmetric) rather than statistical dependencies. The graph cycles are not allowed in order to avoid a variable to be a cause of itself. The endogenous noises are not shown explicitly.

The fundamental idea of studying causal effects is by accommodating interventions. Such a framework has been devised by Pearl [8], and it is now known as Do-calculus. The intervention denoted as do(X) enforces a certain deterministic or a random value from some distribution on the variable X. If this causes a change in the posterior, $p(Y|\text{do}(X))$, then X and Y are causally related. However, in general,

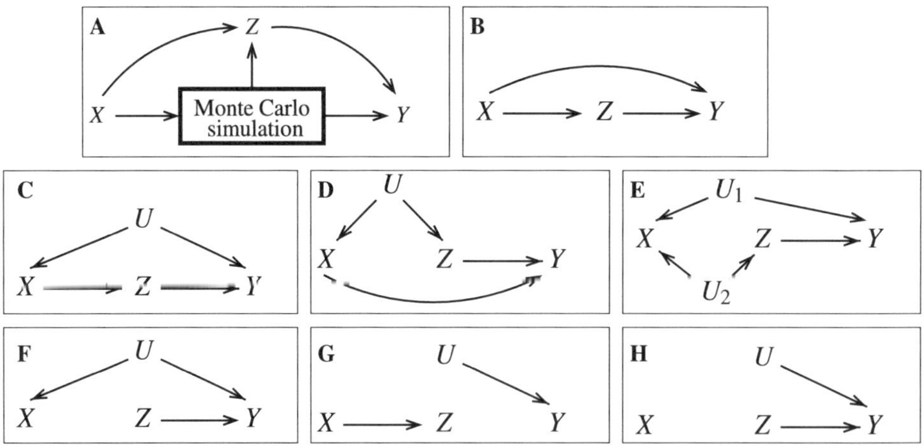

Figure 1.
The input–output causal relationships in augmented Monte Carlo simulations.

observing $p(Y|X)$ is not sufficient to determine causality. The basic rules of Pearl's Do-calculus are:

1. Observations can be inserted or deleted in conditional probabilities;

2. Actions and observations can be exchanged in conditional probabilities;

3. Actions can be inserted or deleted in conditional probabilities.

Moreover, if the causal effect is identifiable, then the causal effect statement can be transformed into a probability expression containing only the observed variables. Unknown causal dependencies can be replaced with conditional probabilities. The significance of these rules is that they allow performing causal inferences using traditional methods of statistical inference for Bayesian networks. This can be used for removing confounding bias, recovering from a selection bias, defining surrogate experiments, and extrapolating and transferring causal knowledge to other similar SCMs.

To illustrate the basic causal components of SCMs, consider three random variables, A, B, and C. The causal chain $A \to B \to C$ and the causal fork $A \leftarrow B \to C$ imply that A and C are generally dependent; however, they are independent, conditioned on B. On the other hand, the causal collider $A \to B \leftarrow C$ implies that A and C are generally independent, but they are dependent, conditioned on B. Thus, conditioning on B in the case of causal chain or fork, blocks (or D-separates) the path between A and C, whereas conditioning on B opens up such a path in the case of causal collider.

These rules can be used to discuss causality in different SCM representations of augmented Monte Carlo simulation. In particular, the SCM in **Figure 1B** contains direct causal paths, $X \to Z$, $Z \to Y$, and $X \to Y$. There is also a backdoor path between Z and Y, that is, $Z \leftarrow X \to Y$. Thus, X is a common cause, or a confounder, so conditioning on X blocks the backdoor path and enables causal inference.

The SCM in **Figure 1C** contains unmeasured or unobserved (e.g., exogenous, outside the model) variable, U. This is an example of so-called confounding by indication. There are two confounded associations, that is, $X \to Z \to Y$ and $U \to X \to Z \to Y$. Moreover, although conditioning on U is not possible, conditioning on X removes any unmeasured confounding. In the case of SCM in **Figure 1D**, conditioning on X is sufficient to block the backdoor path.

The SCM in **Figure 1E** contains two unmeasured variables, U_1 and U_2. There is no bias without any conditioning. However, fixing X as do(X) will induce a selection bias by opening the backdoor path, $Z \leftarrow U_2 \to X \leftarrow U_1 \to Y$, between Z and Y. On the other hand, conditioning on X will create direct as well as backdoor association between Z and Y. The causal relationships in the remaining SCMs in **Figures 1F–H** are trivial.

The augmented Monte Carlo simulation in **Figure 1A** may require more sophisticated processing of the observed inputs and outputs as suggested in **Figure 2** with the aim of improving the information gain and achieving explainability [17]. In particular, the summary statistics should be calculated between variables X and Y as in traditional Monte Carlo simulations and then also between X and Z and Z and Y in augmented Monte Carlo simulations. This should also improve detectability of events and anomalies. More importantly, it may be possible to define formal rules to automate building SEM and SCM models [18]. The models can be then fitted to observations and analyzed to discover causal relationships. The schematic of SEM or SCM is depicted in **Figure 3**, where U's represent the inferred latent variables. The

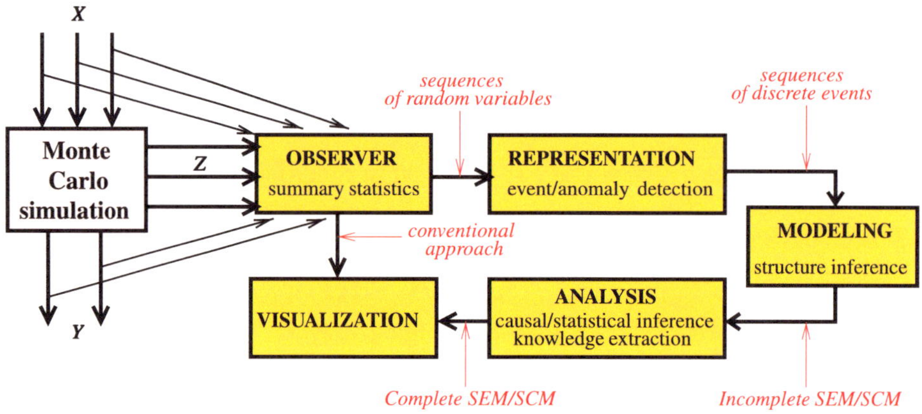

Figure 2.
A structure of explainable Monte Carlo simulations.

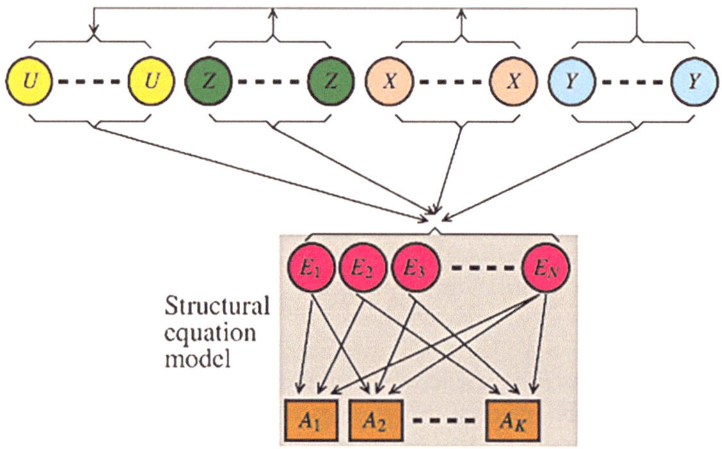

Figure 3.
A schematic SCM for the explainable Monte Carlo simulation in ***Figure*** *2.*

associations, A_i, interconnect the identified events, E_j, for example, using Bayesian rules of statistical dependencies and causal inferences.

5. Conclusions

This chapter discussed the most important Bayesian frameworks and their applications. These frameworks play a crucial role in statistical and causal inferences. Learning from observations is one of the most fundamental concepts. Learning should be unbiased, systematic, replicable, generalizable, sample and other resources efficient and also provide a sufficient information gain. Since learning problems are often mathematically intractable, Bayesian reasoning can guide development of corresponding numerical algorithms. The algorithms can be considered to be concise representations of data, which can be generated by these algorithms. Bayesian frameworks allow building much more sophisticated models and carry out more complex computer simulations.

Bayesian experiment design is especially important due to a widespread adoption of computer simulations in designing and analyzing various real-world systems. Research in machine learning is slowly moving from statistical correlations-based models to models exploiting causal relationships. As discussed in this chapter, the causal relationships in SCM (or SEM) can be analyzed by statistical inference methods, which were developed for Bayesian probabilistic models. Bayesian machine learning assigns a distribution to data labels. It improves the performance of semi-supervised learning and enables automated data labeling and data label corrections. Bayesian optimization is particularly useful for optimizing real-world complex systems, which are very expensive to evaluate. Both Bayesian optimization and Bayesian machine learning are very active areas of research with many exciting open research problems. Bayesian Monte Carlo simulations aim at providing a systematic, semi-automated, explainable simulation framework. They augment the set of observations and leverage SEM or SCM in order to move from traditional descriptive simulations to predictive and even prescriptive simulations. Such research is still little explored in the existing literature.

Acknowledgements

This research was funded by a research grant from Zhejiang University.

Abbreviations

ABC	approximate Bayesian computation
BBVI	black box variational inference
BLUE	best linear unbiased estimators
CAVI	coordinated ascent variational inference
ES	entropy search
ELBO	evidence lower bound
EM	expectation maximization
EI	expected improvement
GP	Gaussian process
IS	importance sampling
KG	knowledge gradient
KL	Kullback–Leibler
	Monte Carlo
MAP	maximum a posterior estimation,
MCMC	Markov chain Monte Carlo
MLE	maximum likelihood estimation
MMSE	minimum mean square estimation
PES	predictive entropy search
RBM	restricted Boltzmann machines
SMC	sequential Monte Carlo
SVI	stochastic variational inference
SCM	structural causal model
SEM	structural equation model

Author details

Pavel Loskot
ZJU-UIUC Institute, Haining, Zhejiang, China

*Address all correspondence to: pavelloskot@intl.zju.edu.cn

References

[1] Robert CP. The Bayesian Choice. 2nd ed. New York, NY, USA: Springer; 2007

[2] Gelman A, Carlin JB, Stern HS, Dunson DB, Vehtari A, Rubin DB. Bayesian Data Analysis. 3rd ed. Boca Raton, FL, USA: CRC Press, Taylor & Francis Group; 2014

[3] Kay SM. Fundamentals of Statistical Signal Processing: Estimation Theory. Vol. I. Upper Saddle River, NJ, USA: Prentice Hall; 1993

[4] Marin JM, Robert CP. Bayesian Core: A Practical Approach to Computational Bayesian Statistics. New York, NY, USA: Springer; 2007

[5] Theodoridis S. Machine Learning: A Bayesian and Optimization Perspective. 2nd ed. Elsevier: Academic Press; 2020

[6] Zhigljavsky A, Žilinskas A. Bayesian and High-Dimensional Optimization. Cham, Switzerland: Springer; 2021

[7] Loskot P, Atitey K, Mihaylova L. Comprehensive review of models and methods for inferences in bio-chemical reaction networks. Frontiers in Genetics. 2019;**10**(549):1-29. DOI: 10.3389/fgene.2019.00549

[8] Pearl J, Glymour M, Jewell NP. Probabilistic Reasoning In Intelligent Systems. San Francisco, CA, USA: John Wiley & Sons; 2016

[9] Huan X, Marzouk YM. Simulation-based optimal Bayesian experimental design for nonlinear systems. Journal of Computational Physics. 2013;**232**(1): 288-317. DOI: 10.1016/j.jcp.2012.08.013

[10] Vanlier J, Tiemann CA, Hilbers PAJ, van Riel NAW. Optimal experiment design for model selection in biochemical networks. BMC System Biology. 2014;**8**(20):1-15. DOI: 10.1186/1752-0509-8-20

[11] Sahu SK, Smith TMF. A Bayesian method of sample size determination with practical applications. Journal Royal Statistical Society A. 2006;**169**(Part 2): 235-253. DOI: 10.1111/j.1467-985X.2006.00408.x

[12] Frazier PI. A tutorial on Bayesian optimization. 2018. ArXiv:1807.02811 [stat.ML]

[13] Bergstra J, Yamins D, Cox DD. Making a science of model search: Hyperparameter optimization in hundreds of dimensions for vision architectures. In Inter. Conf. Machine Learning. 2013;**28**:1-9

[14] Liepe J, Filippi S, Komorowski M, Stumpf MPH. Maximizing the information content of experiments in systems biology. PLOS Computational Biology. 2013;**9**(1):1-13. DOI: 10.1371/journal.pcbi.1002888

[15] Bocquet M, Brajard J, Carrassi A, Bertino L. Bayesian inference of chaotic dynamics by merging data assimilation, machine learning and expectation-maximization. Foundations of Data Science. 2020;**2**(1):55-80. DOI: 10.3934/fods.2020004

[16] Cranmer K, Brehmer J, Louppe G. The frontier of simulation-based inference. PNAS. 2020;**117**(48):30055-30062. DOI: 10.1073/pnas.1912789117

[17] Liang Y, Li S, Yan C, Li M, Jiang C. Explaining the black-box model: A survey of local interpretation methods for deep neural networks. Neurocomputing. 2021;**419**:168-182. DOI: 10.1016/j.neucom.2020.08.011

[18] Le TA, Baydin AG, Wood F. Inference compilation and universal probabilistic programming. In: Inter. Conf. On Artificial Intelligence and Statistics. 2017;**54**:1-11

Chapter 6

Numerical Simulation on Sand Accumulation behind Artificial Reefs and Enhancement of Windblown Sand to Hinterland

Takuya Yokota, Takaaki Uda and Yasuhito Noshi

Abstract

Salients were formed in the lee of two artificial reefs (submerged breakwaters) constructed on Kimigahama Beach in Chiba Prefecture, Japan, owing to the wave-sheltering effect of the reefs, and then, a significant amount of fine sand was transported inland from the salients by wind action. In this study, not only shoreline changes after the installation of the two artificial reefs but also beach changes caused by windblown sand were predicted using a model, in which the BG model (a model for predicting three-dimensional beach changes due to waves based on Bagnold's concept) is combined with a cellular automaton method. Reproduction calculation was carried out on the basis of field data. Beach changes after the artificial reefs were removed were also predicted and the effect of beach nourishment was investigated. It was concluded that landward sand transport by wind is accelerated when wave-sheltering structures such as an artificial reef are constructed on a coast composed of fine sand, and such an effect can be successfully predicted by using the present model.

Keywords: artificial reef, windblown sand, predictive model, BG model, cellular automaton method, Kimigahama Beach

1. Introduction

On a coast subject to strong wind action, foreshore sand may be transported to the hinterland, causing damage to the houses and coastal roads along the coastline. When a detached breakwater or an artificial reef (submerged breakwater) is constructed as a measure against beach erosion on such a coast, sand accumulates in the lee of the structure owing to its wave-sheltering effect, forming a salient [1]. In Japan, detached breakwaters have been widely used as a shore protection measure against beach erosion. Moreover, artificial reefs have often been constructed as a measure instead of detached breakwaters in recent years because they are submerged and thus do not block the ocean view. An artificial reef has the similar wave-dissipating effect as a detached breakwater; thus, sand deposition occurs in the lee of the constructed structure. For example, on Kimigahama Beach in Chiba Prefecture, Japan, salients were formed after the construction of two artificial reefs owing to the wave-sheltering effect of the reefs. Then, a significant amount of fine sand was transported inland from the foreshore by wind.

Thus, in this area, topographic changes due to the actions of waves and wind simultaneously occur on coast after artificial reefs were constructed, and therefore, the prediction of the beach changes is required to consider the effective measures to maintain the shoreline and reduce windblown sand to the hinterland. In the previous studies regarding the sand transport due to waves, a number of models predicting the beach changes have been proposed [2–11]. Moreover, regarding windblown sand, many studies have been carried out for more than half a century, and not only many formulae of windblown sand transport but also models for predicting topographic changes have been proposed [12–20]. However, there are few studies to predict topographic changes while taking the combined effects of waves and wind into account. Yokota et al. [21] developed a model for predicting these beach changes, in which the BG model (a model for predicting 3-D beach changes due to waves based on Bagnold's concept) [22–24] is combined with a cellular automaton method [20]. In this study, this model was tried to apply to the prediction of beach changes on Kimigahama Beach, which is of importance in practical coastal engineering.

In Yokota et al. [21], the BG model was employed, which is a model based on the concept of the equilibrium slope and is derived by an energetics approach [25, 26], and there are eight types [22]. Among them, Type 8 is a model taking the effects of both wave action and nearshore currents into account ([22], Fujiwara et al. [27]). However, In this study, the simplest model of Type 1 BG model was used, which uses a simple sediment transport equation expressed by the wave energy flux at the breaking point, and does not calculate the nearshore currents field, so the calculation load is small [22]. Beach changes caused by wave action in the lee of a detached breakwater have already been successfully predicted when the Type 1 BG model is employed [22].

In general, when artificial reefs are installed near the shoreline, not only shoreward current but also rip current generated by forced wave breaking on the reef can strongly affect the deformation of the beach. In this case, it is necessary to include this effect in the prediction [22, 27]. However, the artificial reef on Kimigahama Beach has a large offshore distance of 300 m or more, so this effect is considered to be small, and the effect of the artificial reef on the beach can be regarded as similar to that of the detached breakwater. Based on this, in this study, the coefficient of wave height transmission of the artificial reef is given and calculated as a breakwater.

The model of topographic changes due to windblown sand used by Yokota et al. [21] is based on the cellular automaton method (Katsuki et al. [20]). This model has a relatively small calculation load compared with the other models and can perform calculations. Here, this model is used for predicting beach changes when artificial reefs are constructed on a coast subject to both waves and wind on Kimigahama Beach. Reproduction calculation was first carried out on the basis of the field data. Then, beach changes were predicted after the artificial reefs were removed to avoid excess sand accumulation behind the artificial reefs and erosion on nearby beaches. Moreover, the effect of beach nourishment around the artificial reefs was investigated. It was concluded that landward sand transport by wind is accelerated when wave-sheltering structures such as an artificial reef are constructed on a coast composed of fine sand.

2. General conditions of study area

Beach deformation owing to the combined actions of waves and wind after the construction of two artificial reefs on Kimigahama Beach was investigated. The study area is located near the east end of Honshu Island and faces the Pacific Ocean, as shown in **Figure 1**, which is a pocket beach of 1070 m length bounded by a rocky headland on the north end and Point Inubo on the south end. At present, two artificial reefs of 50 m in width and 200 m in length with a crown height of 2.0 m below the mean sea level (MSL) are constructed in the central part of the study area. A shallow zone covered with exposed rocks extends in the vicinity of the rocky headland and Point Inubo, whereas a sandy beach composed of fine and medium-size sand extends along the shoreline with salients being formed in the lee of the artificial reefs.

The wind and wave conditions of this area are as follows. The wind conditions of the area predicted by NEDO Neo Winds [28] are shown in **Figure 2**. The predominant wind direction of the study area ranges between NE and NNE in winter and is SSW in summer. In **Figure 2**, the direction of the mean shoreline at transect No. 3 located at the central part of the pocket beach shown in **Figure 1** is indicated by a dotted line. Because there is no predominant wind direction at the central part of the pocket beach between the rocky headland and Point Inubo, wind can be assumed to blow from the direction normal to the shoreline.

We referred to the forecasting results offshore of Chiba Prefecture predicted by the Japan Meteorological Agency [29] for the wave characteristics in this area. **Figure 3** shows the probability of occurrence of waves in each direction: the predominant wave

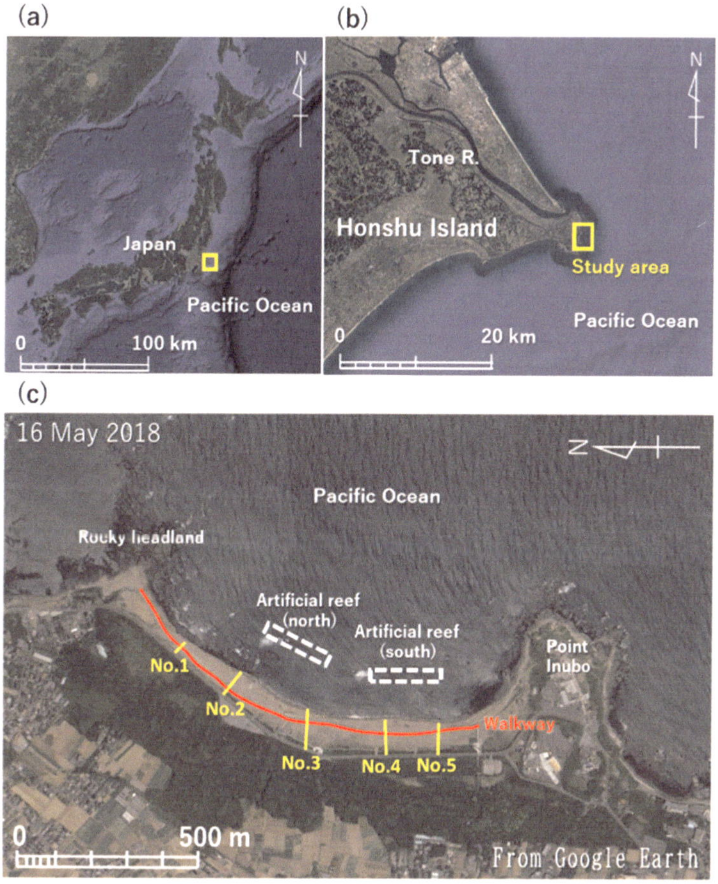

Figure 1.
Location of Kimigahama Beach at east end of Honshu Island in Japan, and alignment of transects of Nos.

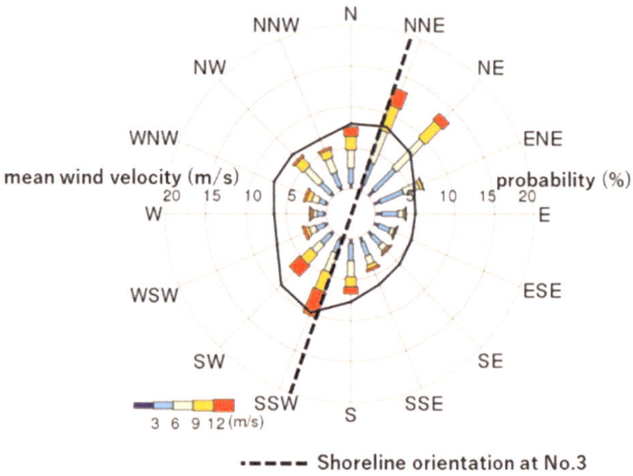

Figure 2.
Probability of occurrence of wind direction and mean shoreline orientation.

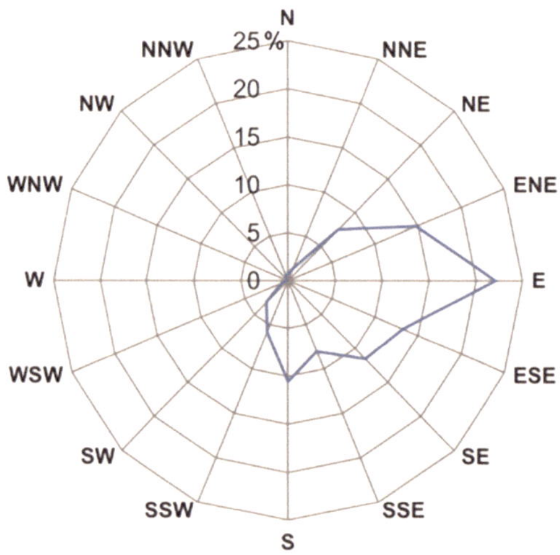

Figure 3.
Probability of occurrence of wave direction offshore of the study area.

direction is E. As for the wave conditions of this coast, an average significant wave height is 1.9 m with a wave period of 8.6 s, and a tidal range is approximately 1.5 m.

3. Field observation at Kimigahama Beach

3.1 Method

First, the shoreline changes on Kimigahama Beach were investigated using the aerial photographs taken in 1963 and 2018, as shown in **Figure 4**. The shoreline changes before and after the construction of the artificial reef can be clarified using these aerial photographs. The longitudinal profiles along transect Nos. 1–5 shown in **Figure 4** were measured using an RTK-GNSS on 9 July 2019. Of the survey lines, transect Nos. 2 and 4 were aligned across the salient formed after the construction of the artificial reef, and transect Nos. 1, 3, and 5 were aligned on both sides of the salient. Then, the beach topography along each transect was compared. The conditions of the coast were also observed on 14 June 2022 and foreshore and backshore materials were sampled along transect No. 4. Furthermore, beach topography around the artificial reefs was investigated in detail using a bathymetric survey map prepared in 2002 by the Chiba Prefectural Government.

3.2 Result

Kimigahama Beach had undergone significant beach changes caused by anthropogenic interventions in the past. In 1963, a natural sandy beach of 100 m width was extended alongshore without any protective measures between the rocky headland and Point Inubo, as shown in **Figure 4**. Then, a seawall was constructed near the coastline without preserving a sufficiently wide sandy beach as a buffer zone and the foreshore was excavated to raise the elevation of the flat land behind the seawall.

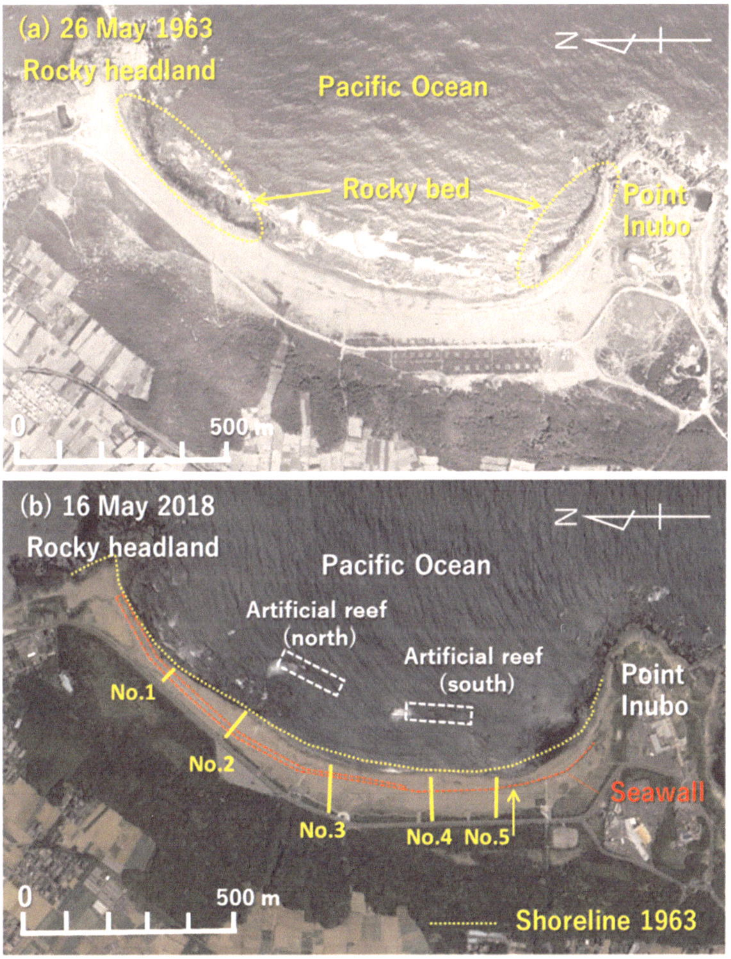

Figure 4.
Aerial photographs taken in 1963 and 2018. (a) 26 May 1963 and (b) 16 May 2018.

These interventions resulted in a reduction in the volume of the sandy beach, causing beach erosion. As a measure against erosion caused by the artificial activity, two artificial reefs (submerged breakwaters) with a length of 50 m and a crown height of 2 m below MSL were constructed offshore of the shoreline until 2012.

After the installation of these artificial reefs, the shoreline locally advanced to form a salient behind these artificial reefs because of their wave-sheltering effect, whereas the shoreline retreated on both sides, and wave overtopping the seawall became significant at the site designated by an arrow in **Figure 4b** immediately south of the south artificial reef. Numerous concrete blocks had to be placed to prevent waves from overtopping the seawall. On the other hand, because of sand deposition behind the south artificial reef, the amount of windblown sand on the widened beaches increased and sand was not only deposited in front of the seawall but also transported up to the hinterland.

Figure 5 shows the longitudinal profiles along transect Nos. 1–5. Along transect No. 1 at the north end, the shoreline is completely covered by a gently sloping seawall with no foreshore left in front of the seawall (**Figure 6**). A walkway with an elevation of 5.5 m above MSL has been constructed immediately landward of the crown of the seawall, and a sand dune with a slope of 1/6.5 is formed landward of this walkway (**Figure 7**). Transect No. 2 crosses the north artificial reef and the beach is wide because of the deposition of sand caused by the wave-sheltering effect of the north artificial reef. In this area, a walkway with an elevation of +5.5 m runs along the shoreline, and a vegetation zone extends a 20 m wide on the seaward side of the walkway and a foreshore with a slope of 1/7 exists near the shoreline. On the other hand, on the landward side of the walkway, a low sand dune with an elevation of +6.0 m is formed (**Figures 8** and **9**).

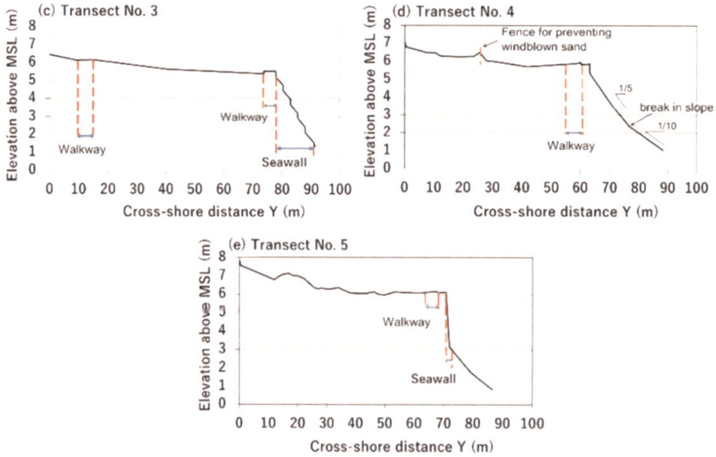

Figure 5.
Longitudinal profiles along transect Nos. (a) Transect No. 1, (b) Transect No. 2, (c) Transect No. 3, (d) Transect No. 4, (e) Transect No. 5.

Figure 6.
Transect No. 1 (near the shoreline).

Figure 7.
Transect No. 1 (hinterland).

Transect No. 3 crosses the south opening of the north artificial reef. At this site, a gently sloping seawall has been constructed near the shoreline and no foreshore exists, as shown in **Figure 10**. Moreover, a walkway with an elevation of +5.5 m has been constructed, and landward of this, a flat land, which was artificially constructed by reclamation using beach sand, extends. Transect No. 4 crosses the north end of the south artificial reef and a walkway with an elevation of +6 m runs in the south–north direction, and a wide foreshore is formed owing to the wave-sheltering effect of the south artificial reef (**Figure 11**). There is a break in the slope at 2 m height in this longitudinal profile, and the foreshore slope of 1/10 and the backshore slope of 1/5 are separated by this break in the slope, which were formed by waves and windblown sand, respectively. Windblown sand was transported from the shoreline up to the walkway along this transect (**Figure 12**). Landward of this walkway, flat land with an elevation of approximately +6 m is formed (**Figure 13**). The longitudinal profile with the combination of a flat land of +6 m height and a steep slope to the shoreline cannot be formed under the natural conditions, implying that this flat land landward of the seawall was artificially formed.

Figure 8.
Transect No. 2 (backshore).

Figure 9.
Transect No. 2 (walkway).

Figure 10.
Transect No. 3.

Figure 11.
Transect No. 4 (foreshore).

Figure 12.
Transect No. 4.

Figure 13.
Transect No. 4.

Since the deposition of windblown sand in front of the seawall along transect No. 4 has occurred since 2012 after the construction of the artificial reefs, the rate of windblown sand deposition can be estimated to be 2.5 m^3/m/yr, because the change in the cross-sectional area in the seaward area of the seawall is 18 m^2. Transect No. 5 crosses the seawall near the south end of the beach. As shown in **Figure 14**, the coastline is protected by numerous concrete blocks from waves overtopping the seawall.

Figure 15 shows the bathymetry around the two artificial reefs measured in 2003. These artificial reefs were constructed at a depth of approximately 4 m. The area between the north artificial reef and the rocky headland is covered by exposed rocks, as designated by complicated contour lines. The area between the south artificial reef and Point Inubo is also covered with exposed rocks, similar to the north artificial reef. In short, the areas between the rocky headland and 450 m south of it, and between Point Inubo and 400 m north of it, as shown by dotted lines in **Figure 15**, are covered by exposed rocks, and sandy beach only extends within the area of the alongshore length of approximately 1.1 km including the artificial reefs.

Figure 14.
Transect No. 5 (6 June 2017).

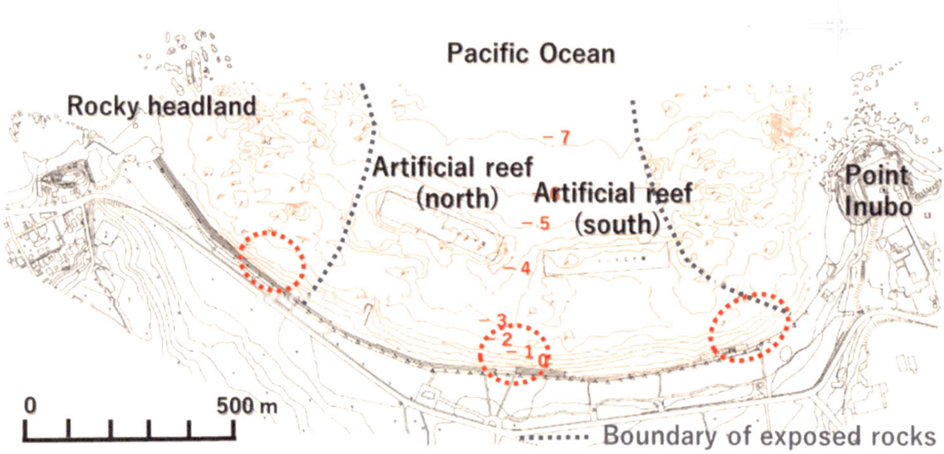

Figure 15.
Bathymetry of Kimigahama Beach measured in 2003 by Chiba Prefectural Government.

Taking these conditions into account, it is clear that this beach has a closed system of littoral drift, and sand transport occurred from outside the wave-shelter zone of the artificial reefs to the inside because of the construction of artificial reefs on such a coast. In the landward area of the south artificial reef, a wide flat seabed of 3 m depth extends and a steep foreshore slope extends in the zone with a depth smaller than approximately 1 m. The alongshore movement of sand in a narrow band in front of the seawall was triggered by the construction of artificial reefs, resulting in the formation of the present beach topography.

4. Model for predicting topographic changes

Topographic changes including the accumulation of sand in the shoreward zones of the artificial reefs caused by waves and the landward transport of windblown sand from the salients after the installation of the artificial reefs were predicted using a model for predicting 3-D beach changes under combined actions of waves and wind [21]. With this model, the beach changes caused by waves occurring in the depth zone between the depth of closure (h_c) and the berm height (h_R) are predicted using the BG model [22], whereas topographic changes caused by windblown sand in the area landward of a berm top are predicted by a cellular automaton method.

The BG model is based on following concepts: (1) the contour line becomes orthogonal to the wave direction at any point at the final stage, similarly, (2) the local beach slope coincides with the equilibrium slope at any point, and (3) a restoring force is generated in response to the deviation from the statically stable condition, and sand transport occurs owing to this restoring force [22].

In the calculation of windblown sand by the cellular automaton method, the two most important processes, saltation and avalanche, are taken into account [20]. Two-dimensional meshes were taken on Cartesian coordinates (x, y), and the elevation at the mesh point was set as $z\ (x, y, t)$. The mesh size is assumed to be sufficiently larger than the size of the sand particles. The saltation distance L_s was defined using Eq. (1) on the basis of the observation results obtained by Andreotti et al. [30] and is the simplest polynomial expression that can be used to evaluate the obtained results of the sand flux on a sand dune including multiphase flow.

$$L_S = L\left[1.0 + b_1\left(\frac{Z}{h}\right) - b_2\left(\frac{Z}{h}\right)^2\right] \tag{1}$$

b_1 and b_2 are the coefficients to control sand transport flux, expressed by the product of L_s defined using Eq. (1) and the mass of moving sand. L is a reference distance (here, we choose 1.0 m), and h is the reference height (here, the berm height).

Eq. (1) shows that the higher the elevation where sand particles are deposited, the longer the distance that the sand particles are transported by wind, but L_s has a limit and the sand flux after the maximum value is reached is regarded as a constant, and the decreasing functional form is not employed. When there is an obstacle in the field, saltation is assumed not to occur, because a vortex is formed behind an obstacle owing to the separation of the flow [31]. Originally, the sand flux is given by the product of the mass of moving sand and the saltation distance, and the sand flux can be expressed by Eq. (1) when the wind velocity is constant, assuming that the mass of moving sand is constant. When the wind velocity changes, the coefficient of Eq. (1) can be changed depending on the wind velocity.

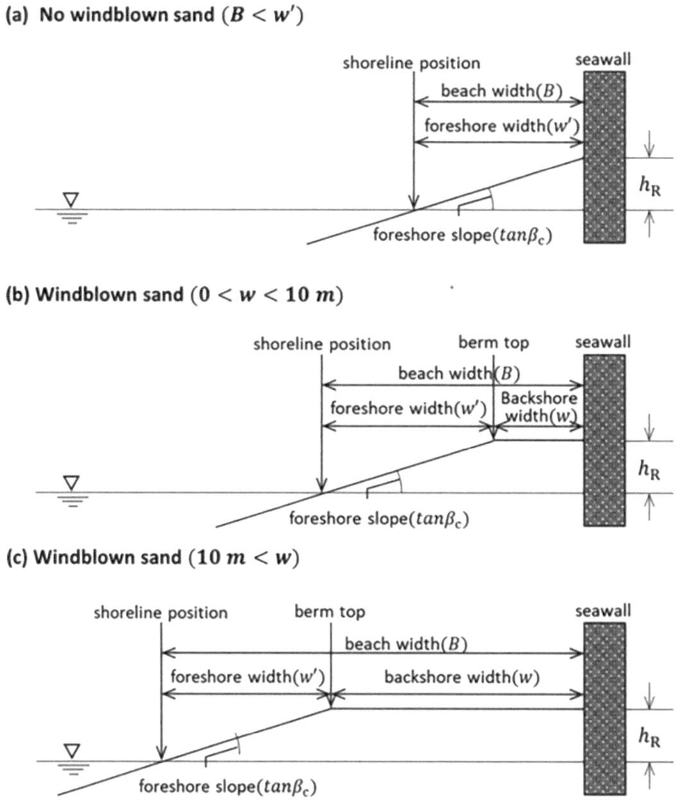

Figure 16.
Schematic view of condition for windblown sand to occur. (a) No windblown sand (B < w’). *(b) Windblown sand* (0 < w < 10 m). *(c) Windblown sand* (10 m < w).

To combine the BG model and the cellular automaton method, the calculation domains were separated at the location of the berm, assigning the landward region of the berm as the domain of windblown sand. The rate of windblown sand is assumed to attain equilibrium at a location distant from the starting point for the approach run in the downwind direction. Here, the condition for the windblown sand to occur was defined, as shown in **Figure 16**, depending on the backshore width, assuming that the minimum approach run is 10 m [32]. No windblown sand is transported when the beach width B is smaller than the foreshore width w’; the mass of moving sand (q) attributable to windblown sand was given by the value multiplied by the coefficient μ shown by Eq. (2) when the backshore width w is smaller than 10 m. When w is greater than 10 m, μ is unity.

$$\mu = \frac{1}{2}\left[\cos\left(\frac{\pi}{10}b\right) + 1\right] \tag{2}$$

5. Numerical simulation

5.1 Calculation conditions

Topographic changes around the artificial reefs built on Kimigahama Beach under the combined actions of waves and wind were calculated using a model in which the

Grain size of sand particle d_{50} (mm)		0.2
Equilibrium slope		1/10
Incident wave conditions	Incident wave height (m)	1.0
	Wave direction α (deg)	15.0
Water level above MSL (m)		0.0
Depth range of beach changes	Depth of closure h_c (m)	5.0
	Berm height h_R (m)	2.0
Coefficient of cross-shore and longshore sand transport $K_y = A/\sqrt{d_{50}} = 0.2/\sqrt{0.2} = 0.45$ [33]		0.45
Depth distribution of longshore sand transport $\varepsilon(Z) = 1/(h_c + h_R)$ (m^{-1})		Uniform
Wind direction α_w (deg) relative to Y-axis		0.0
Moving mass by wind (m^3/m/yr)	2.5×10^{-6}	
Coefficients of Eq. (1)	b_1	10
	b_2	0.5
Critical slope of sand on land and seabed		1/2
Calculation domain	Longshore distance x (m)	1000
	Cross-shore distance y (m)	600
Mesh size	Δx (m)	10
	Δy (m)	10
Time intervals Δt (hr)		0.2
Total calculation steps		2.0×10^5

Table 1.
Calculation conditions.

BG model was combined with a cellular automaton method for predicting the amount of windblown sand. **Table 1** shows the calculation conditions. The grain size of the seabed material was set to be 0.2 mm from the grain size measured on the backshore along transect No. 4. In the bathymetric chart in **Figure 4**, no topographic changes can be seen offshore of the opening of the artificial reefs, so the depth of closure due to waves was set to 5.0 m. The equilibrium slope was assumed to be 1/10 from the foreshore slope along transect No. 4, as shown in **Figure 5**. The berm height was given by the height of the break in the slope of 2 m. Since the predominant wave direction is E, as shown in **Figure 3**, wave direction was set to be E. The coefficients b_1 and b_2 and the mass of moving sand q were assumed to be $b_1 = 10$, $b_2 = 0.5$, and $q = 2.5 \times 10^{-6}$ m^3/m^2/step on the basis of the rate of windblown sand of 2.5 m^3/m/yr measured along transect No. 4. In the numerical simulation, the topographic changes were predicted in Cases 1 and 2 without/with windblown sand, respectively.

5.2 Results of numerical simulation

The initial topography for the calculation was assumed to have a foreshore slope of 1/10 and a 1/50 slope in the offshore zone deeper than 1 m. Then, waves were incident in this assumed initial topography for a sufficiently long time (10 years) to obtain the beach topography before the installation of the artificial reefs. By this calculation, we carried out the matching of beach topography to the given wave conditions. **Figure 17**

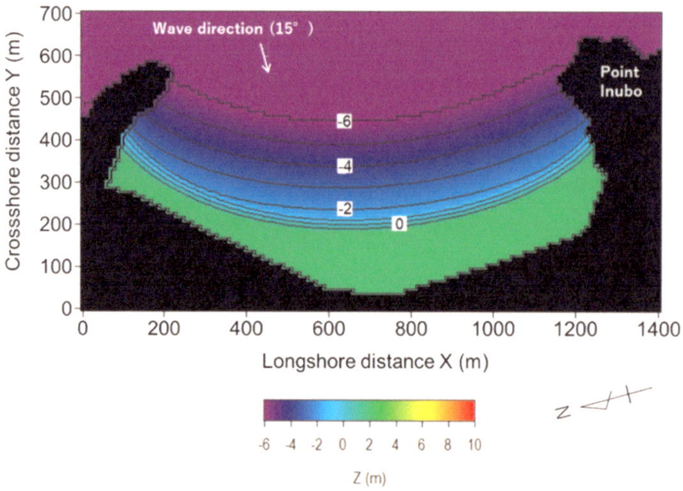

Figure 17.
Beach topography adjusted to the given wave conditions.

shows the result. Then, two artificial reefs and a seawall with a height of 6 m were installed along the shoreline. **Figure 18** is thus the obtained initial topography. Since the seabed with a depth equal to or larger than 2 m is covered with exposed rocks, solid bed was assumed in these areas.

Figure 19 shows the results of the calculation in Case 1, that is, under waves without the wind effect, after the installation of the artificial reefs. The shoreline advanced owing to the alongshore sand transport toward the lee of the two artificial reefs with an increase in the beach width. Moreover, two salients with a berm height of +2 m were formed behind the artificial reefs.

Then, the calculation in Case 2, that is, with the wind effect, was carried out, given the same topography as that in Case 1. The beach topography and topographic changes relative to the initial topography in Case 2 are shown in **Figure 20**. It is seen that the

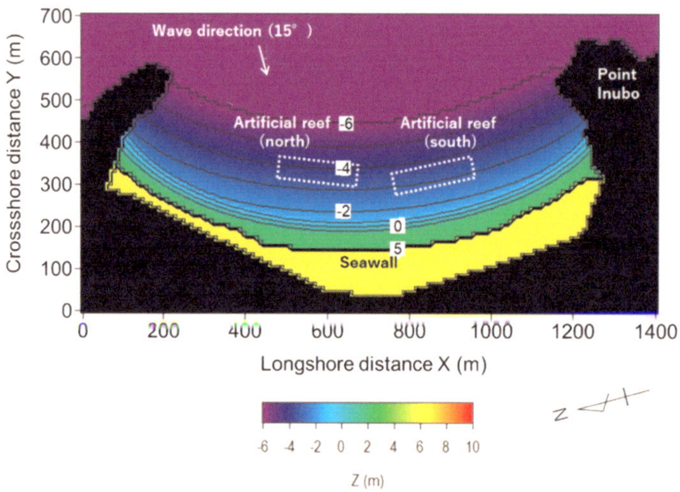

Figure 18.
Initial topography.

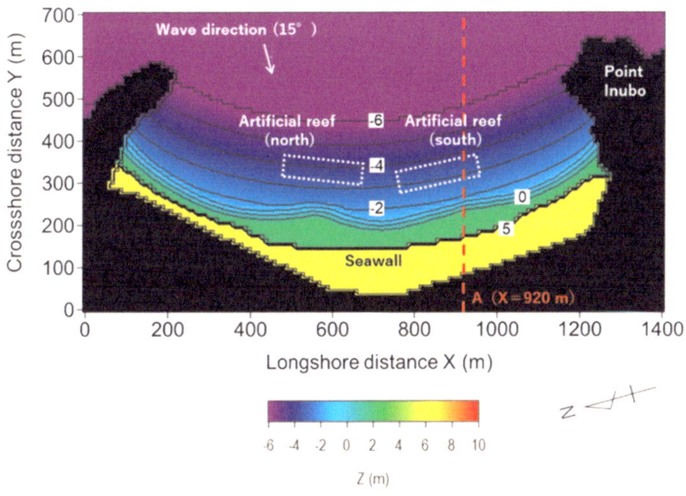

Figure 19.
Predicted topography only under wave action.

beach width increased behind the artificial reefs, and sand was transported by wind from the foreshore to inland causing the deposition of sand in front of the seawall. Furthermore, part of the windblown sand was transported into the hinterland over the seawall.

Figure 21 shows for comparison of the beach profiles in Cases 1 and 2 along the transect across X = 920 m, as shown by dotted lines in **Figures 19** and **20**. In Case 1, only a flat, plane beach with an elevation of +2 m was formed. In Case 2, however, sand was transported landward by wind, forming a steep slope of 1/7.5 in front of the seawall up to the crown height of +6.0 m of the seawall, whereas the shoreline has retreated because part of the beach sand was transported landward by wind. This result well explains the formation of a 1/5 slope (**Figures 5d** and **9**) from the shoreline to the top of the seawall, which was observed in the field. It is concluded that installing a wave-dissipating structure, such as an artificial reef, on a coast composed of fine and medium-size sand induces the concentrated deposition of sand in the lee of the structure and the decrease in sand volume and the shoreline recession in the entire area.

5.3 Prediction of beach changes

In Case 3, topographic changes were predicted under the condition that beach nourishment was carried out using sand of the same grain size as that at the present coast at a location where the beach width decreased after the construction of the artificial reefs. **Figure 22a** and **b** show the initial topography in Case 3 and the change in beach elevation under the condition with/without beach nourishment using sand with a volume of 17,000 m^3 behind the openings of the two artificial reefs.

Figure 23a and **b** show the results of the prediction and the topographic changes relative to the initial topography in Case 3. In this case, alongshore sand transport continues to occur from the nearby beach to the lee of the artificial reefs, and the salients also continued to develop in the lee of the artificial reefs. Therefore, part of beach nourishment sand was transported over the seawall to the hinterland owing to

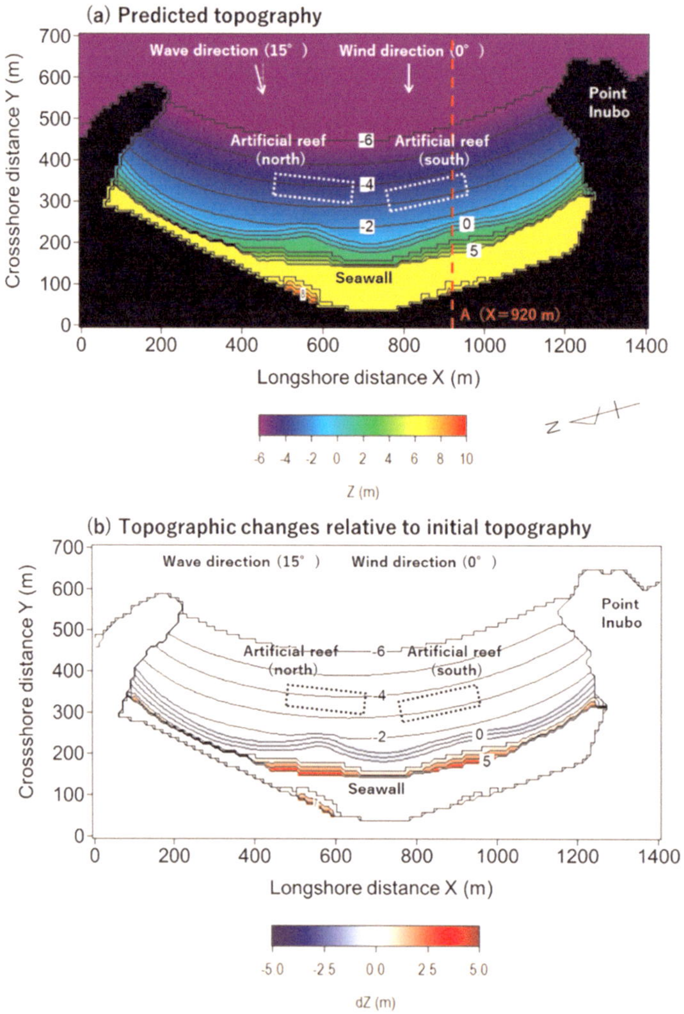

Figure 20.
Beach topography and topographic changes in Case 2 with wind action. (a) Predicted topography. (b) Topographic changes relative to initial topography.

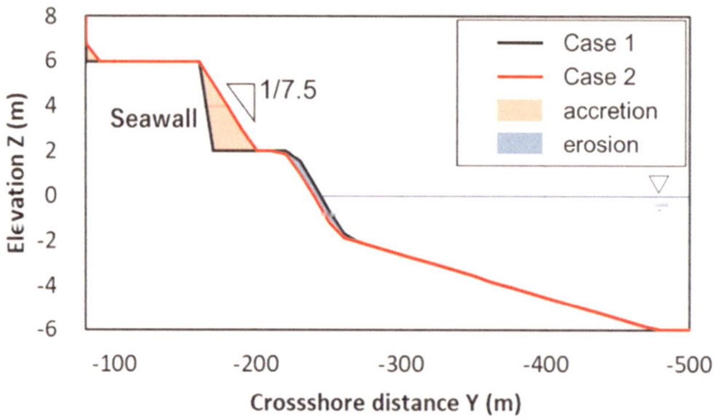

Figure 21.
Longitudinal profiles along transect across X = 920 m in Cases 1 and 2.

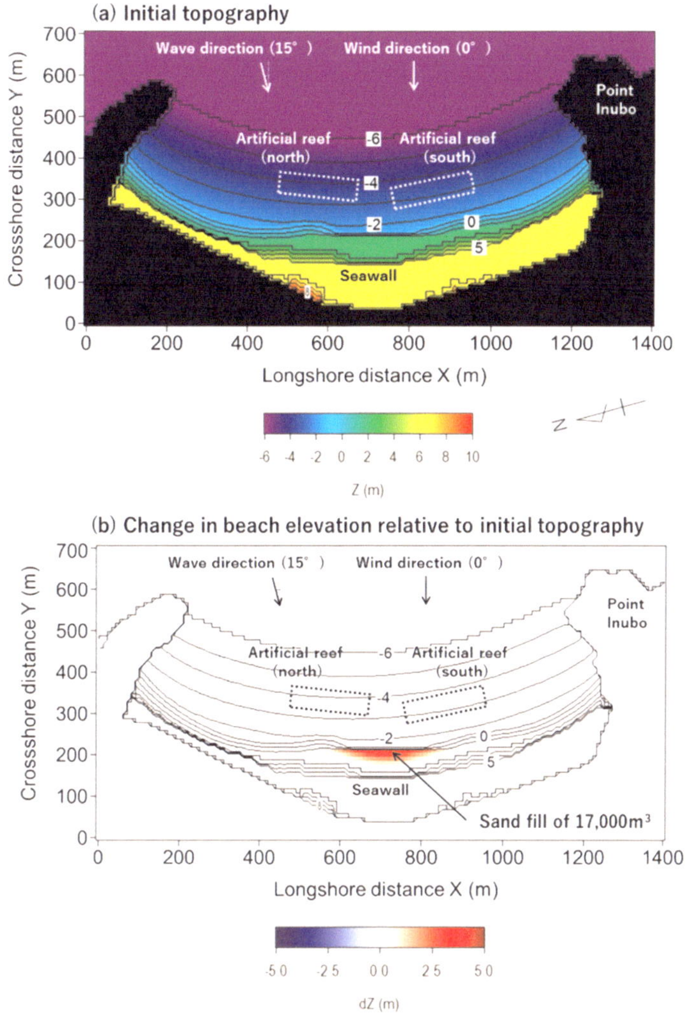

Figure 22.
Beach topography and topographic changes in Case 3 with beach nourishment. (a) Initial topography. (b) Change in beach elevation relative to initial topography.

the wind effect. Because part of beach nourishment sand was transported away to the hinterland, the effect of beach nourishment is minimal in increasing the beach width.

In Case 4, the artificial reefs were removed to investigate their adverse effects using the bathymetry shown in **Figure 22a** as the initial bathymetry. **Figure 24a** and **b** show the results of the calculation in Case 4 when artificial reefs were removed, and the change in topography relative to the initial topography. Owing to the removal of artificial reefs, salients that formed in the lee of the artificial reefs disappeared, and nourishment sand was transported to the entire pocket beach, resulting in the sand deposition in the entire area except behind the artificial reefs. Furthermore, since the sand deposition was no longer concentrated behind the artificial reefs owing to the removal of artificial reefs, the amount of sand blown over the seawall to the hinterland has greatly decreased.

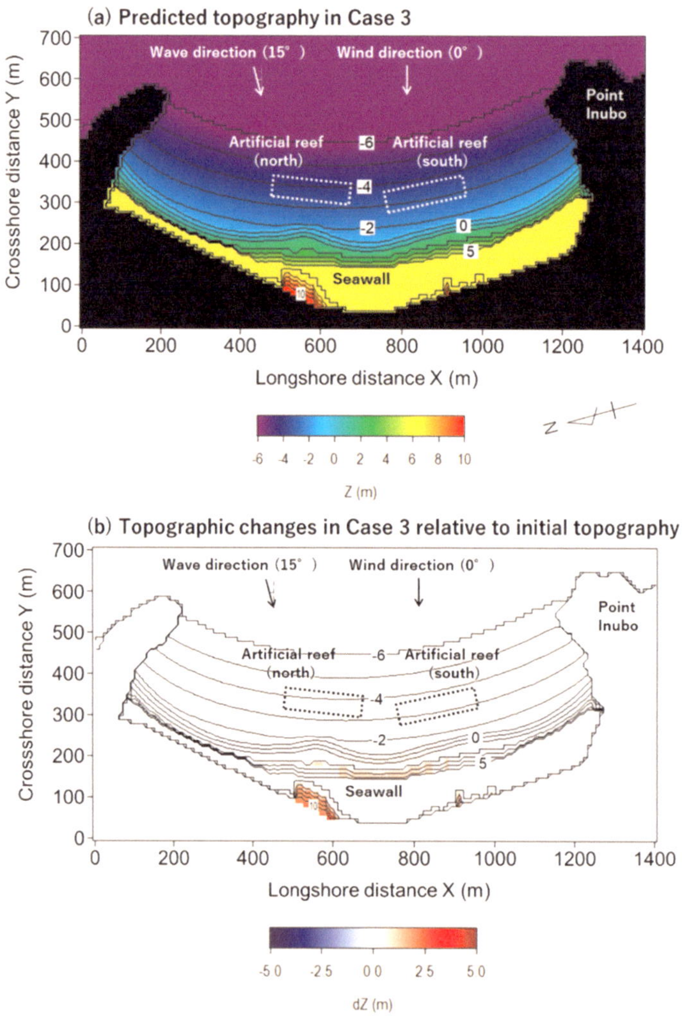

Figure 23.
Predicted topography and topographic changes in Case 3 with beach nourishment. (a) Predicted topography in Case 3. (b) Topographic changes in Case 3 relative to initial topography.

6. Discussion

In general, when a detached breakwater or submerged breakwater is constructed on a coast composed of fine and medium-size sand, sand accumulates in the lee of the structure owing to the wave-sheltering effect, forming a salient behind the structure. Owing to this sand deposition effect, these structures have been widely used in Japan as a measure against beach erosion. However, the beach formed by this accretive effect of waves is also subject to wind action, resulting in a significant amount of windblown sand. On a sandy beach widened in the lee of the structure, therefore, the amount of sand blown to the hinterland is also increased, causing loss of foreshore sand. Accordingly, on such a coast, even though beach nourishment is carried out, nourishment sand is transported to the hinterland, devaluating the effect of beach nourishment, and causing another sand maintenance problem. When considering

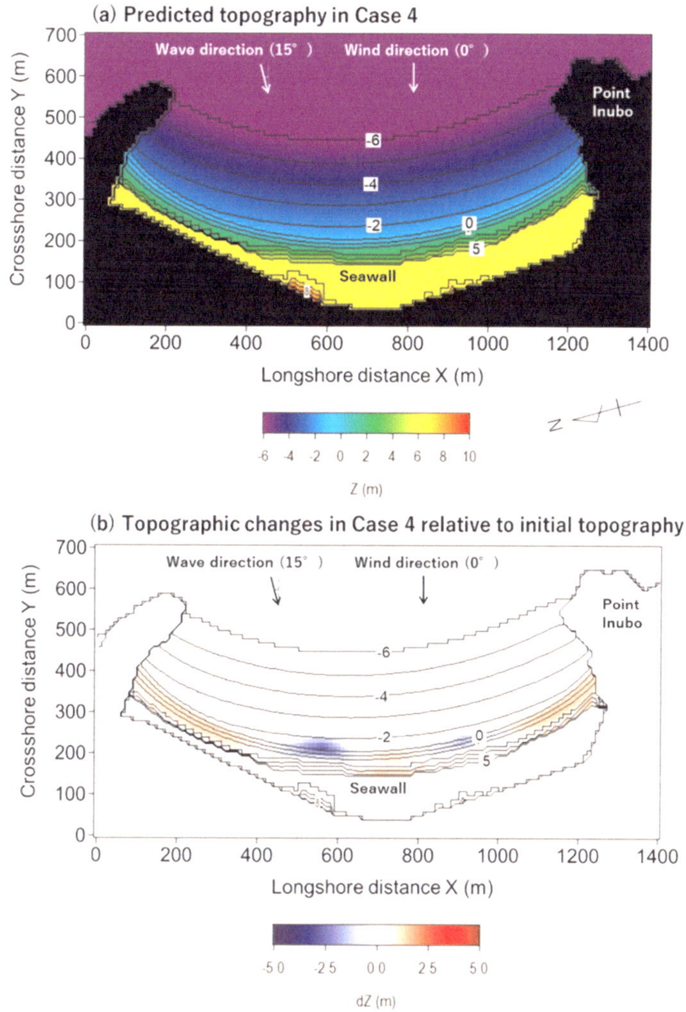

Figure 24.
Calculation results in Case 4 after the removal of the artificial reefs. (a) Predicted topography in Case 4. (b) Topographic changes in Case 4 relative to initial topography.

shore protection measures against beach erosion on a coast composed of fine and medium-size sand, the effect of windblown sand must be taken into account. In such a case, the present model can be useful for predicting future beach changes.

7. Conclusions

Beach changes after the construction of artificial reefs and the increase in the amount of sand blown from the foreshore in the lee of the artificial reefs to inland were investigated, taking Kimigahama Beach in Chiba Prefecture, Japan, as an example. Then, a numerical simulation of these topographic changes was carried out using the combination of the BG model for predicting beach changes caused by waves and a cellular automaton method for predicting the windblown sand. In the field observation, it was found that salients were formed after the construction of two artificial reefs and the amount of sand blown from the widened foreshore in the lee of the artificial reefs increased. Not only was fine sand transported in front of the seawall but also part of sand was further transported inland. The formation of salients in the lee of the artificial reefs and the deposition of windblown sand in the hinterland as observed in the field were numerically predicted well.

As an application of the prediction of beach changes, the effect of beach nourishment at the opening of the artificial reefs was predicted using the same model. When beach nourishment was carried out while maintaining the present condition of the reefs as they are, sand was further transported inland by wind. Instead, if artificial reefs were removed and then beach nourishment was carried out, nourishment sand was distributed on the entire sandy beach, and loss of sand toward the hinterland became minimal. From this, the construction of a facility with a wave dissipating function such as a detached breakwater or an artificial reef in the nearshore zone on a coast composed fine and medium-size sand must be carefully carried out. It is important for a coastal engineer to sufficiently consider not only the effect of the movement of sand due to waves but also the management of windblown sand when shore protection facilities are constructed on a coast composed of fine and medium-size sand.

Author details

Takuya Yokota[1*], Takaaki Uda[2] and Yasuhito Noshi[3]

1 Coastal Engineering Laboratory Co., Ltd., Tokyo, Japan

2 Public Works Research Center, Tokyo, Japan

3 Nihon University, Funabashi, Japan

*Address all correspondence to: ctr.ty.725@gmail.com

References

[1] Hsu JRC, Lee JL, Klein AHF, González M, Medina R. Headland-Bay Beaches. Singapore: World Scientific Publishing Co., Pte. Ltd; 2021. p. 786

[2] Pelnard-Considere R. Essai de Theorie de l'Evolution des Formes de Ravage en Plages de Sables et de Galets. Societe Hydrotechnique de France. IV'eme Journee de L'Hydraulique Question III, Rapport. 1956:289-298

[3] Bailard JA. An energetics total load sediment transport model for a plane sloping beach. Journal of Geophysical Research. 1981;**86**(C11):10938-10954

[4] Perlin M, Dean RG. A Numerical Model to Simulate Sediment Transport in the Vicinity of Coastal Structures. Fort Belvoir, Virginia: U.S. Army, Corps of Engineers, Coastal Engineering Research Center; 1983. p. 119

[5] Watanabe A. 3-Dimensional numerical model of beach evolution. In: Proceedings of Conference on Coastal Sediments '87. ASCE; 1987. pp. 802-817. Available from: http://coastalsediments.cas.usf.edu/documents/cs87.pdf

[6] Horikawa K. Nearshore Dynamics and Coastal Processes. Tokyo: University of Tokyo Press; 1988

[7] Hanson H, Kraus NC. GENESIS: Generalized Model for Simulating Shoreline Change. Vicksburg, USA: Coastal Engineering Research Station; 1989. p. 185

[8] Hanson H, Larson M. Simulating coastal evolution using a new type N-line model. In: Proc. of 27th International Conf. on Coastal Eng. ASCE. Vicksburg; 2000. pp. 2808-2821

[9] Lesser GR, Roelvink JA, van Kester JATM, Stelling GS. Development and validation of a three-dimensional morphological model. Coastal Engineering. 2004;**51**(8–9):883-915

[10] Camenen B, Larson M. A total load formula for the nearshore. In: Proceedings of Coastal Sediments '07 Conf. ASCE. 2007. pp. 56-67

[11] Nam PT, Larson M, Hanson H, Hoan LX. A numerical model of beach morphological evolution due to waves and currents in the vicinity of coastal structures. Coastal Engineering. 2011;**58**(9):863-876

[12] Bagnold RA. The Physics of Blown Sand and Desert Dunes. London: Methuen & Co; 1941

[13] Kok JF, Parteli EJR, Michaels TI, Karam DB. The physics of wind-blown sand and dust. Reports on Progress in Physics. 2012;**75**:10

[14] Sauermann G, Kroy K, Herrmann HJ. Continuum saltation model for sand dunes. Physical Review E. 2001;**64**:031305

[15] Kroy K, Sauermann G, Herrmann HJ. Minimal model for aeolian sand dunes. Physical Review E. 2002;**66**:031302

[16] Andreotti O, Claudin P, Douady S. Selection of dune shapes and velocities Part 2: A two-dimensional modelling. European Physical Journal. 2002;**28**:341-352

[17] Duran O, Parteli EJR, Herrmann HJ. A continuous model for sand dunes: Review, new developments and application to barchans dunes and barchan dune fields. Earth Surface Processes and Landforms. 2010;**35**:1591-1600

[18] Werner BT. Eolian dunes: Computer simulations and attractor interpretation. Geology. 1995;**23**:1107-1110

[19] Nishimori H, Yamasaki M, Anderson KH. A simple model for the various pattern dynamics of dunes. International Journal of Modern Physics B. 1998;**12**:257-272

[20] Katsuki A, Kikuchi M, Nishimori H, Endo N, Taniguchi K. Cellular model for sand dunes with saltation, avalanche and strong erosion: Collisional simulation of barchans. Earth Surface Processes and Landforms. 2011;**36**:372-382

[21] Yokota T, Uda T, Kobayashi A, Hoshigami Y, Katsuki A, Noshi Y. Prediction of topographic changes on Enshu-nada coast considering effect of both waves and windblown sand. Proceedigs of JSCE. 2020;**76**: 469-474

[22] Uda T, Serizawa M, Miyahara S. Morphodynamic model for predicting beach changes based on Bagnold's concept and its applications. London, UK: INTEC; 2018. p. 188

[23] Serizawa M, Uda T, San-nami T, Furuike K. Three-dimensional model for predicting beach changes based on Bagnold's concept. In: Proceedings of the 30th International Conference. San Diego, California, USA: ICCE; 2006. pp. 3155-3167. DOI: 10.1142/9789812709554_0265

[24] Serizawa M, Uda T, San-nami T, Furuike K, Ishikawa T. BG-model predicting three-dimensional beach changes based on Bagnold's concept and applications. In: proceedings 4th International Conference on Asian and Pacific Coasts 2007. Nanjin: China Ocean Press; 2007. pp. 1165-1179. ISBN: 9787502768836 7502768831

[25] Inman DL, Bagnold RA. Littoral processes. In: Hill MN, editor. The Sea. New York: Wiley; 1963. pp. 529-533

[26] Bagnold RA. Mechanics of marine sedimentation. In: Hill MN, editor. The Sea. New York: Wiley; 1963. pp. 529-533

[27] Fujiwara H, Uda T, Onishi T, Miyahara S, Serizawa M. Prediction of beach changes around artificial reef using BG model. In: Proceedings of 33rd Conference on Coastal Engineering. Santander, Spain: ICCE; 2012. pp. 1-12. DOI: 10.9753/icce.v33.sediment.77

[28] NEDO NeoWinds [Internet]. Available from: http://appwdel.infoc.ned.go.jp/Nedo_Webgis/top.html. [Accessed: October 21, 2021]

[29] Japan Meteological Agency [Internet]. Available from: https://www.jma.go.jp/ima/index.html. [Accessed: October 21, 2021]

[30] Andreotti B, Claudin P, Douady S. Selection of dune shapes and velocities Part 1: Dynamics of sand, wind and barchans. European Physical Journal. 2002;**28**:321-339

[31] Pye K, Tsoar H. Aeolian Sand and Sand Dunes. London: Unwin Hyman; 1990. pp. 42-43

[32] Horikawa K, Hotta S, Kubota S, Katori K. Field observation of windblown sand by trench trap. Journal of Japan Coastal Engineering. 1983;**30**: 406-410

[33] Serizawa M, Uda T, San-nami T, Furuike K, Ishikawa T, Kumada T. Model for predicting beach changes on coast with sand of mixed grain size based on Bagnold's concept. Coastal Sediments. 2007;7:314-326

Section 2

Applied Numerical Simulation

Chapter 7

Numerical Simulation of Land and Sea Breeze (LSB) Circulation along the Guinean Coast of West Africa

Amadou Coulibaly, Bayo J. Omotosho, Mouhamadou B. Sylla, Amoro Coulibaly and Abdoulaye Ballo

Abstract

This study uses observed and simulated data to analyze the dynamics LSB rotation along the Guinean Coast of West Africa. A non-hydrostatic fully compressible numerical model is used to simulate LSB circulation. To evaluate the model's ability to capture the LSB kinematics, the study used a modified model code with ERA-Interim and CFS as forcing data. Comparison of observed and simulated LSB patterns shows that the model reliably captures the LSB circulation in the region. The simulated diurnal evolutions of hodographs and onshore/offshore winds also follow the observations. A dynamical analysis performed by extracting individual forcing terms from the horizontal momentum equations at selected regions within the study area showed that the direction of the wind rotation is a result of a complex interaction between surface and synoptic pressure gradients, advection, and horizontal and vertical diffusions forces. However, hourly analysis of the rotation term suggests that surface gradient seems to dominate over oceanic region, while diffusion terms are more important for land area. This may be attributed to the variation of surface roughness due the landscape and urbanization. Therefore, this reveals the link between urbanization and LSB circulation in coastal region of West Africa, where most important cities are located.

Keywords: land and sea breeze, clockwise, anticlockwise, hodograph rotation, numerical simulations, Guinean coast

1. Introduction

The theory of land and sea breezes (LSB) is based on the thermal contrasts between the land and water. During the day, the land is warmer than the sea because of heating from the sun. Hence, the warm air over the land rises and expands forming cumulus clouds, while cooler air from the sea surface therefore flows inland to replace the rising warm air. The resulting circulation can move several kilometers inland as onshore wind circulation, called sea breeze. At night, the land cools faster than the nearby ocean and a shallow mesoscale pressure gradient develops, with a higher surface pressure over the land. The resulting circulation is directed from the land to the sea (offshore circulation) near the surface, called land breeze.

Generally, the land breeze (LB) circulation is weaker than the sea breeze (SB) in both velocity and height of development because the heat source for the land breeze is much weaker than the heat source for the sea breeze circulation [1]. LB fronts tend to only affect a small area of the sea, in comparison with the much larger effect of SB.

In West Africa along the Guinea coast, LSB circulation occurs almost throughout the year although in varying strength [2, 3]. During the day, the circulation is driven by strong heating, while at night it is driven by cooling of the landmasses and resulting pressure anomalies. In tropical West African region, Coulibaly et al. [4] showed the winter frequency and the seasonality behaviors of LSB with both clockwise (anticyclonic) and anticlockwise (cyclonic) hodograph rotations. Kusuda and Alpert [5] and Haurwitz [6] evaluate the diurnal evolution of SB in the Northern Hemisphere with an influence of the Coriolis force in the sense of rotation, while over Sardinia (mid-latitude), the sense of rotation seems to be influenced by the combination of surface and synoptic pressure gradients and Coriolis and advection forces [7].

LSB is more frequently and prominently observed in tropical regions than in higher latitudes due to strong radiative heating, convection, and weak Coriolis force. It is also influenced by the prevailing large-scale wind and topographic friction. When LSB circulation prevails on land, changes in the temperature structure, humidity, and roughness occur in the air adjacent to the coast and lead to formation of a thermal internal boundary layer (TIBL) [8]. This effectively reduces the mixing height in the coastal regions in the daytime. LSB circulation and TIBL are the two important phenomena that influence the pollution plume direction and diffusion in coastal regions. Many factors such as topography, synoptic flow, and latitude are shown to influence the evolution and characteristics of the SB.

With the growing computational power and resulting improved modeling capabilities, numerical simulations of LSB circulation have gained attention since the 1960s. Much of the earlier numerical work was performed using two-dimensional hydrostatic models with coarse grid spacing (≈10 km). While these contributed greatly to our understanding of the mechanics and structure of the LSB circulation, they nevertheless remained highly idealized. Due to the large size of the model horizontal grid (>1 km), it is difficult to differentiate between hydrostatic and non-hydrostatic simulations [9]. While non-hydrostatic effects may weaken the mature nature of LSB, hydrostatic influence tends to overestimate LSB intensity [10]. Therefore, to adequately simulate LSB circulation and its associated features such as planetary boundary-layer (PBL) influence, it is decided to use three-dimensional models, even though two-dimensional models may be adequate for many idealized simulations [7].

Many theoretical and numerical modeling studies have been reported on the overall dynamics of the LSB circulations [7, 11–16]. Also, there have been observational and modeling studies of LSB characteristics over different regions [4, 7, 15–22].

However, the availability of non-hydrostatic numerical models with less than 1 km resolution has highlighted the complex nature of LSB and its associated nonlinear interactions on several scales [23]. Using numerical methods, Estoque [24] showed the development of a zone of low-level convergence at the leading edge of LSB current as it sets in over land, while Pielke [25] showed the effects of topographical friction on LSB evolution using a 3D numerical model. Therefore, numerical simulations can be considered as computational tools in minimizing the identified knowledge gaps by [7, 15] and assessing the governing factors in LSB dynamics over complex terrain and coastlines. Consequently, this study aims to evaluate the dynamics of LSB circulation over the Guinean Coast of West Africa using a fully compressible non-hydrostatic numerical model for simulating.

Based on the study by Moisseeva and Steyn [7] who used WRF-ARW version 3.4, this study will use WRF-ARW version 3.7.1 to examine the dynamics of LSB circulation over the region. The physics and dynamics of WRF-ARW version 3.7.1 allow the extraction of its dynamical factors related to atmospheric radiation, microphysics, planetary boundary layer and surface layer physics, land surface physics, and cumulus options of the model [16].

In winter period, Coulibaly et al. [4] showed both clockwise and anticlockwise rotations of the wind in coastal West Africa identifying some LSB episodes. This has been a good starting point for which the numerical modeling of the LSB circulation and hodograph rotation in the region has been based (see **Figure 1**).

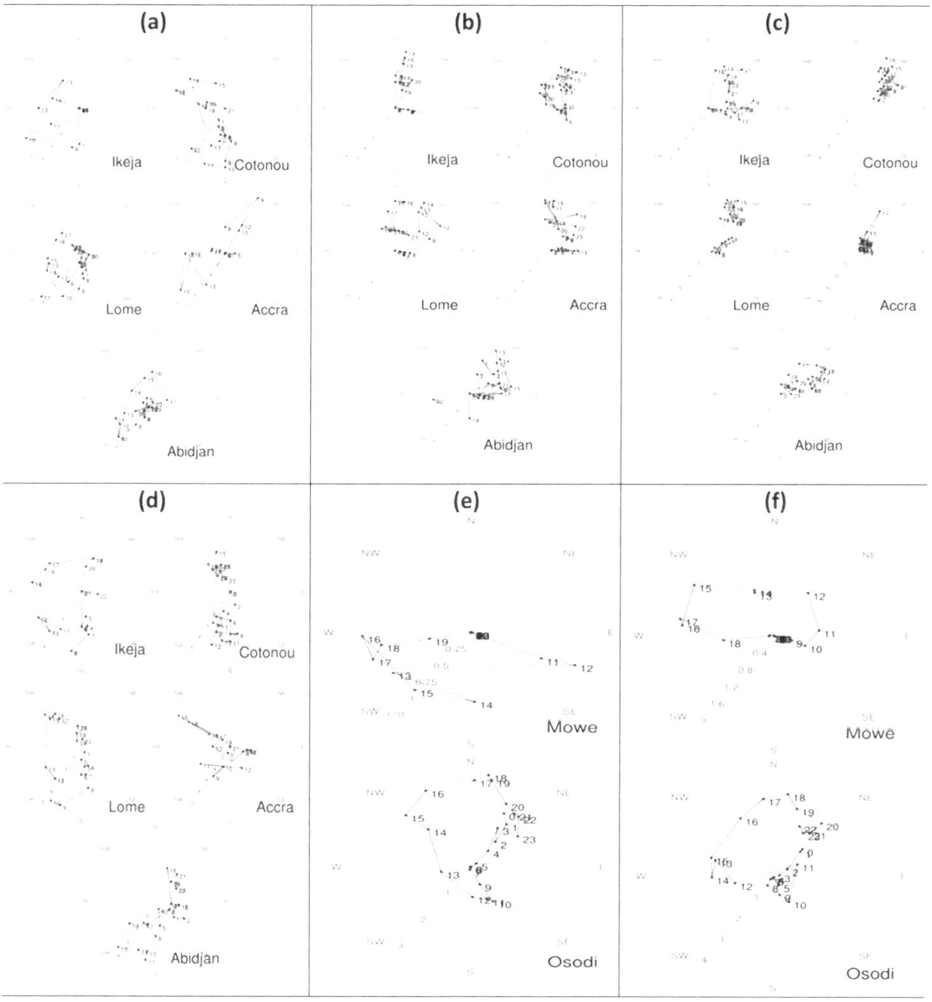

Figure 1.
*Daily hodograph rotations of SB; **a:** January 1, 2011, **b:** May 12, 2011, **c:** August 30, 2011, **d:** December 13, 2011, **e:** December 16, 2014, and **f:** December 17, 2014. The numbers near dots indicate the hour of the day (LST).*

2. Data and methods

2.1 Model description

In this study, numerical simulations of LSB dynamics were performed using Weather Research and Forecasting (WRF) model following the method employed by [7, 16]. While the model offers great operational forecasting capabilities, it is limited in dynamical analysis because the basic dynamical equations are deeply embedded in the solver remaining inaccessible to the model user. To overcome this limitation, this study extracted the individual tendency terms from the momentum equation of the model [16]. Details about the extraction processes are available in my PhD thesis called Coulibaly (2016) published in Lambert Academic Publishing (https://www.lap-publishing.com/extern/listprojects).

Using an observational study, Coulibaly et al. [4] identified some days with favorable atmospheric conditions for the formation of LSB called LSB episodes across the Guinea Coast of West Africa. The monthly occurrence of LSB showed a primary maximum occurrence in December over the study region used to do the numerical simulation.

Based on the model configuration used in the studies by Moisseeva and Steyn [7] and Coulibaly et al. [16], this study configures for all identified LSB episodes the entire study domain. The model was forced using Climate Forecast System (WRF-CFS) and ERA-Interim (WRF-ERA) reanalysis data with 15 km and 3 km as outer and inner domain grid spaces, respectively. In order to avoid the effects of the sub-grid scale on the dynamics of the LSB in WRF model, Steyn et al. [26] used grid sizes of 9 km and 3 km for outer and inner domains, respectively. However, as 9 km is within the grid zone of convection-permitting/non-convection-permitting resolution, grid sizes 15 km and 3 km were used in this study (**Figure 2**).

A 3-year period (2013–2016) reanalysis data of Climate Forecast System Reanalysis version 2 (CFSRv2) from the National Centers for Environmental Prediction (NCEP2) and ERA-Interim were used to initialize the model. According to Wang et al. [27], the reanalysis data are a global, high-resolution, coupled atmosphere–ocean-land surface-sea ice system designed to provide the best estimate of the state of these coupled

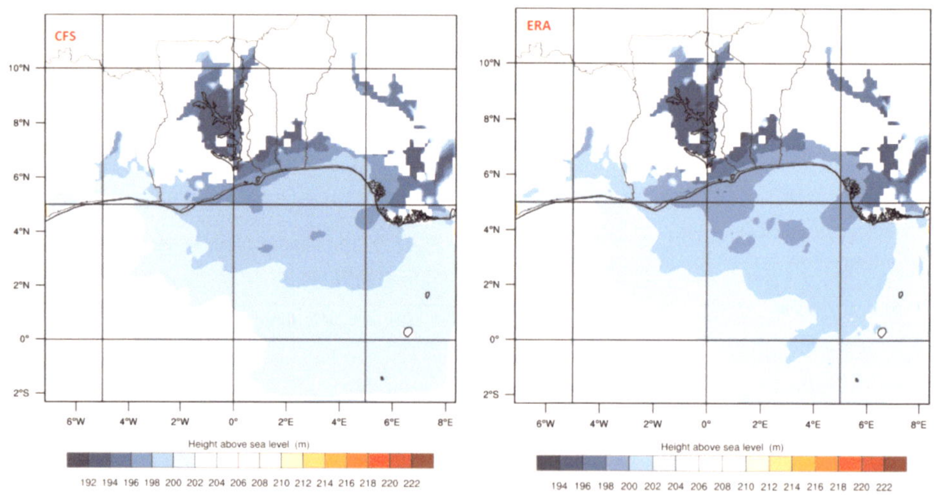

Figure 2.
Outer domain for real case simulations for both WRF-CFS and WRF-ERA.

Stations	Longitude (°)	Latitude (°)	Distance from the closest sea (km)	Height above sea level (m)
Oshodi	3.383E	6.50 N	16	32
Mowe	3.458E	6.81 N	48	67
Cotonou	2.38 E	6.35 N	0.6	7

Table 1.
Coastal stations metadata.

domains. For the identified episodes of LSB, high-resolution pressure-level (0.5 degrees latitude/longitude) and surface and radiative flux (0.3 degree Gaussian grid) 6-hour forecasts were obtained for 0000, 0600, 1200, and 1800 UTC. More details about ERA-Interim and CFSRv2 are available in the studies by Barrisford et al. [28] and Saha [29], respectively.

Most of the LSB episodes identified by Coulibably et al. 4] occurred in the month of highest monthly occurrence (December). Therefore, the simulations were performed, based on the individual LSB episode in December, over 30 hours from 1800LST of the previous day to 0000LST of the following day, with a 6-hour spin-up. As the study is primarily interested in daytime dynamics, the analysis was performed starting 0900 LST, that is, 9 hours after the beginning of each simulation. RK time-step Δt was set to 90 seconds, as recommended $6 \times \Delta x$ (km) [30]. Since WRF allows for output of instantaneous fields only, the history interval was set to 10 minutes. Wind and dynamical tendency fields were hence output six times each hour and subsequently averaged to produce an estimate of hourly averages. Through the analysis of various production runs, it was determined that 50 vertical eta-levels provided sufficient vertical resolution within the boundary layer, and hence, this configuration was adopted for all runs with the pressure top set up to 5000 Pa. In order to reduce the number of figures, this study will only show the results of two LSB episodes (December 16 and 17, 2014). The details of selected stations are shown in **Table 1**.

2.2 Model evaluation

The evaluation of the model was performed using hourly data of two Automated Weather Stations in Oshodi-Lagos and Mowe in Nigeria from 2014 to 2016 [3] and 1-year (2014) 6-hourly data from Cotonou. In this study, the model evaluation processes used by Crosman and Horel [15] are also applied. Based on the findings of Crosman and Horel [15], this study adds the diurnal wind rose analysis to highlight the strength of LSB over the study region [16]. The study is located in the northern part of the Gulf of Guinea; therefore, offshore winds are taken as northerlies (between 330° and 30°), while onshore winds as southerlies (between 150° and 210°). This can facilitate the plots of wind roses and their associated strength at each identified location over a period of time (**Figures 3–5**). To plot the hodographs for both observed and modeled data, the u and v wind components at 10 m are considered in polar coordinates. In numerous studies, daily hodographs have been considered as primary criteria to evaluate the model based on the variability of wind data. In order to reduce the number of figures, this study will only show the results of two LSB days (December 16 and 17, 2014) in **Figures 3**, **4** and **6–9**.

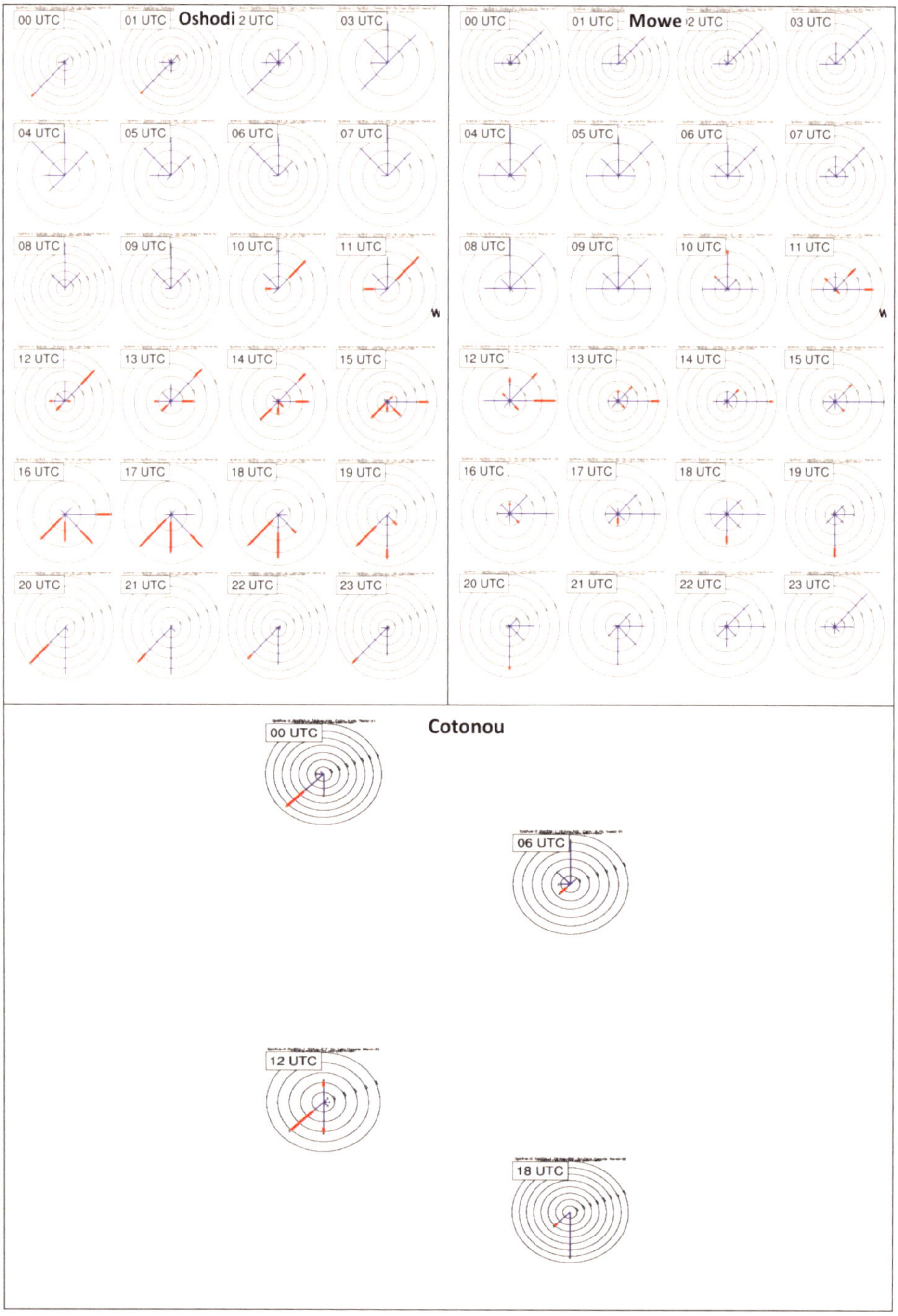

Figure 3.
Three-year period (2014–2016) hourly average wind roses at different locations.

Figures 3 and **4** shows that the West African Monsoon (WAM) prevails on LSB over the region for both observed and simulated wind roses even though there are night/early-morning weak offshore winds (LB) and enhanced daytime onshore winds (SB) in **Figure 3** Cotonou has only 6-hourly observational data.

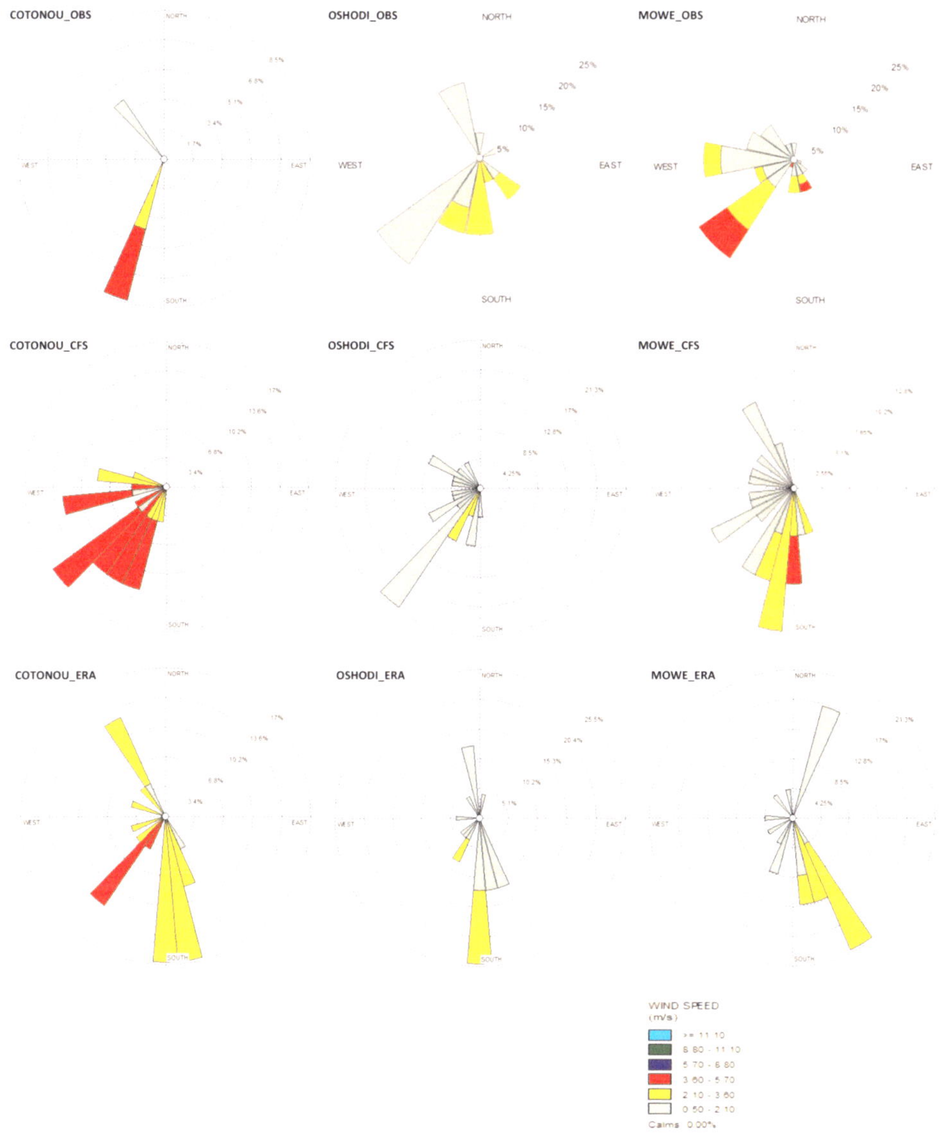

Figure 4.
Wind roses at Cotonou, Oshodi, and Mowe for December 16, 2014, as LSB day. The first row shows observation; the second and third rows display wind roses from WRF-CFS and WRF-ERA, respectively.

Figure 4 shows that CFSRv2 well captured the observed patterns of dominant WAM, night/early-morning weak LB, and enhanced daytime SB, than Era-Interim. The strength of daytime SB depends mainly on the location because it is enhanced for locations close to the sea (Oshodi) than those far away (Mowe).

Figure 4 shows that both simulated and observed wind strength distributions seem to be in good agreement (see Oshodi and Mowe) even though Cotonou has only 6-hourly observational data.

From observational data, the classification of wind strength is followed:

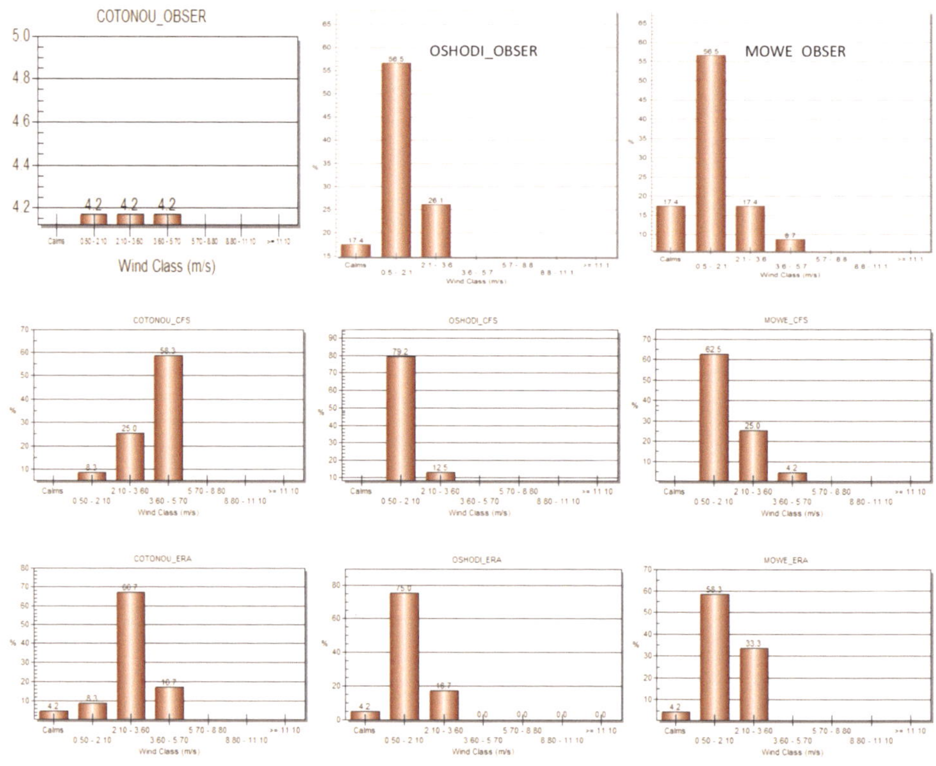

Figure 5.
Wind frequency distribution at Cotonou, Oshodi, and Mowe for December 16, 2014. The first row displays the observation, while the second and third rows are for WRF-CFS and WRF-ERA, respectively.

i. 56.5% were between 0.5 and 2.10 ms^{-1} in both Oshodi and Mowe;

ii. 26.1% were between 2.10 and 3.60 ms^{-1} in Oshodi, while there are only 17.4% in Mowe;

iii. more than 8% between 3.6 and 5.7 ms^{-1} in Mowe; and

iv. 17.4% were calmed in both Oshodi and Mowe.

From the simulated wind data, the following is obtained:

i. more than 58% between 0.5 and 2.10 ms^{-1} in Oshodi and Mowe, while this rate is less than 9% in Cotonou.

ii. more than 25% between 2.10 and 3.60 ms^{-1} in Cotonou and Mowe, which is less than 16% in Oshodi;

iii. Era-Interim showed 4.2% of calm winds, and

iv. There are no calm winds for CFSRv2.

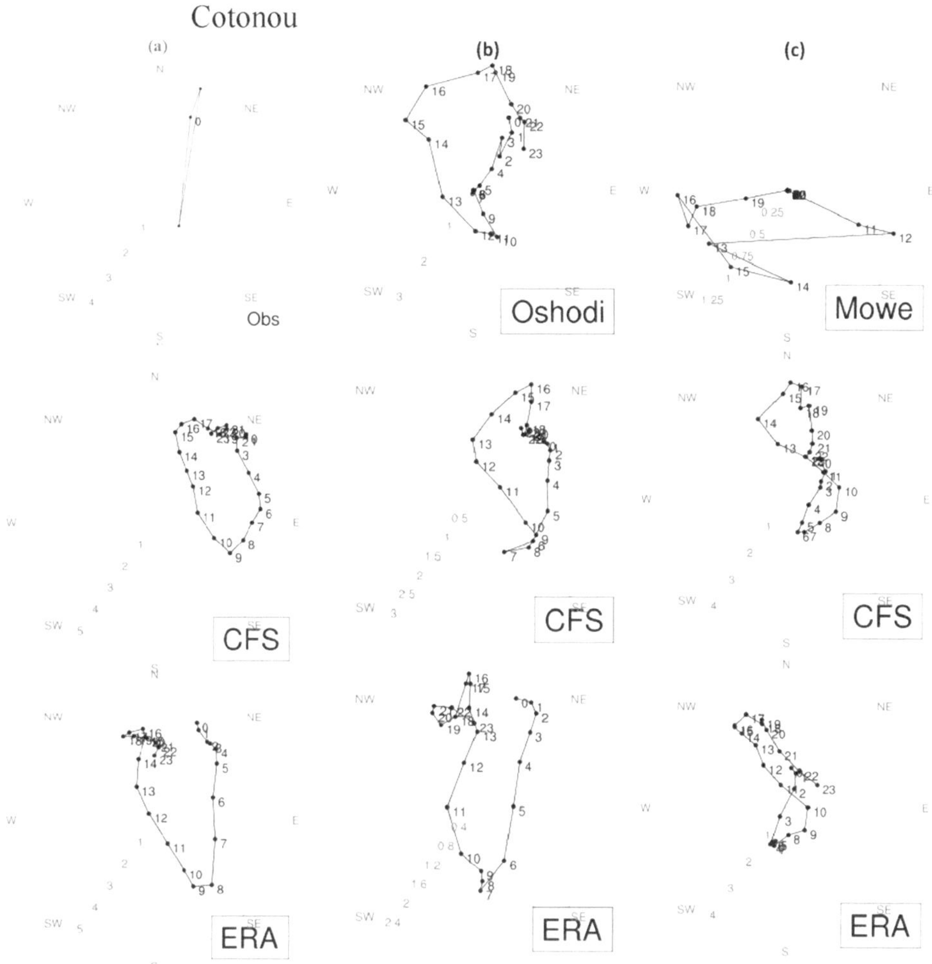

Figure 6.
Wind hodographs on December 16, 2014. (a): Cotonou, (b): Oshodi, (c): Mowe. The first row represents observation, while the second and third rows are, respectively, for WRF-CFS and WRF-ERA.

While both WRF-CFS and WRF-ERA well captured the observed patterns of wind strength distribution, all underestimated calm winds over this study region. Therefore, the WRF model is suitable to evaluate diurnal wind rose behaviors.

Figures 6 and 7 shows that there is agreement between observed and simulated hodographs for selected SB episodes indicating a daily clockwise rotation (see Oshodi and Cotonou locations), even though there is a quite difference in Mowe location where the sense of rotation turns into anticlockwise between 2:00 and 11:00 am on December 16th (**Figure 6**) and between 9:00 am and 7:00 pm on December 17th, 2014, due to the fact that Mowe is far away from the sea (**Figure 7**). A clearly onshore-offshore nature of SB circulation with indeterminate sense of rotation can be also seen.

Meanwhile, the existence of daytime clockwise rotation of the theoretical hodographs over the study region, which is one of the important SB characteristics in the northern hemisphere, due to the variation of Coriolis force [5], has been revealed. While the clear diurnal clockwise rotation can be seen in the nearest locations to the

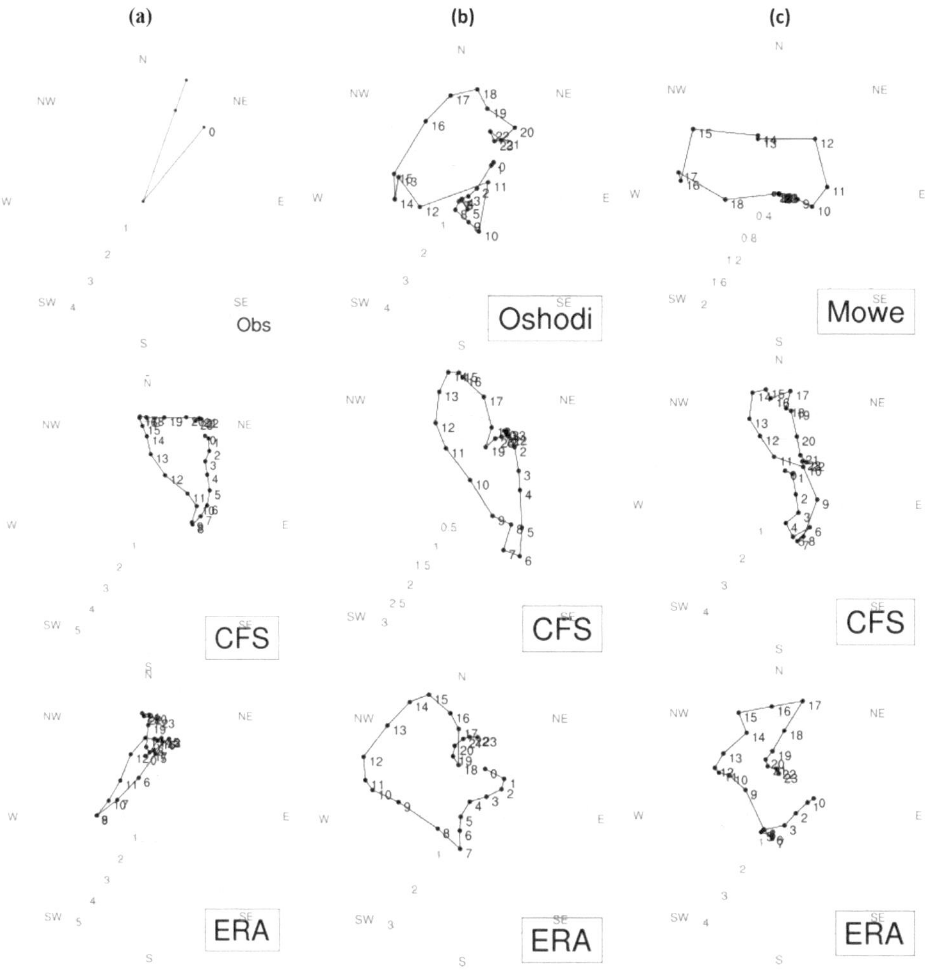

Figure 7.
Wind hodographs on December 17, 2014. (a): Cotonou, (b): Oshodi, (c): Mowe. The first row represents observation, while the second and third rows are, respectively, for WRF-CFS and WRF-ERA.

sea, the slight anticlockwise rotation appears in the farthest locations. This is consistent with the apparition of the enhanced daytime SB at the closest stations (**Figure 3**), stippling that the sense of rotation of diurnal winds may be influenced by the distance separating the location to the sea.

As **Figures 6–9** shows the agreement in the patterns of daily evolution of observed and simulated winds for closest locations. Again the farthest locations show a slight early morning difference.

Hence, **Figures 6–9** show that (for the observed and simulated wind fields):

i. WRF model suitably captured the rotation of LSB in the coastal areas of West Africa.

ii. There is a relationship between the sense of rotation of the LSB and the location distance from the sea.

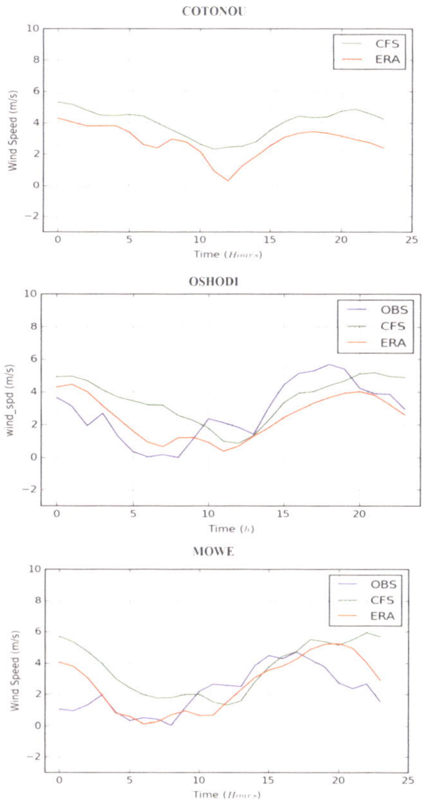

Figure 8.
Diurnal evolution of onshore/offshore wind at 10 m on December 16, 2014. Blue lines represent the observation, green and red, respectively, for WRF-CFS and WRF-ERA.

2.3 Analysis of simulation

2.3.1 Horizontal wind rotation

Here, we follow the method of [15, 16] in representing the horizontal momentum equations of WRF in its simplified vector form as follows:

$$\frac{\partial V_h}{\partial t} = \frac{\partial V_{pg}}{\partial t} + \frac{\partial V_{adv}}{\partial t} + \frac{\partial V_{cor}}{\partial t} + \frac{\partial V_{hdif}}{\partial t} + \frac{\partial V_{vdif}}{\partial t} \tag{1}$$

where V_h is total horizontal velocity vector $V = (u,v)$ and subscripts pg., adv, cor, hdif, and vdif corresponds to forcing due to pressure gradient, advection, Coriolis, horizontal, and vertical diffusion. Taking the 850 mb (hPa) pressure level to be representative of overlying synoptic weather conditions, the pressure gradient term V_{pg} can be further separated into synoptic *(syn)* and surface *(surf)* forcing by assuming that:

$$V_{syn} = V_{pg}(850mb) \tag{2}$$

Therefore, the total wind may be represented as

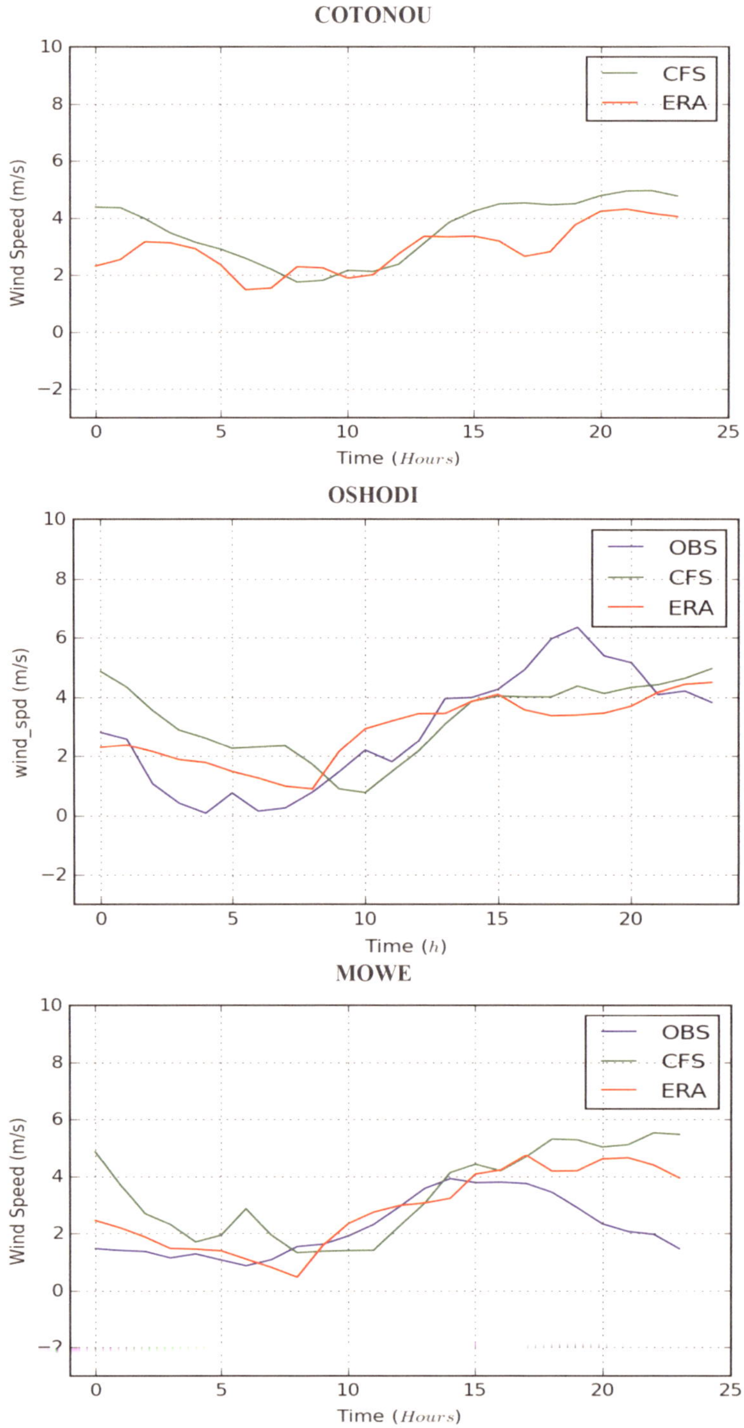

Figure 9.
Diurnal evolution of onshore/offshore wind at 10 m on December 17, 2014. Blue lines represent the observation, green and red, respectively, for WRF-CFS and WRF-ERA.

$$V_{pg} = V_{surf} + V_{syn}$$

So that, the other pressure gradient forcing can be regarded as near ground or surface effects, expressed as follows:

$$V_{surf} = V_{pg} - V_{syn} \tag{3}$$

Now, to express the pressure gradient forcing in terms of all its different components, the full WRF horizontal momentum equations, excluding the effects of curvature and acoustic modes, can now be written as follows:

$$\frac{\partial V_h}{\partial t} = \frac{\partial V_{surf}}{\partial t} + \frac{\partial V_{syn}}{\partial t} + \frac{\partial V_{adv}}{\partial t} + \frac{\partial V_{cor}}{\partial t} + \frac{\partial V_{hdif}}{\partial t} + \frac{\partial V_{vdif}}{\partial t} \tag{4}$$

Neumann [31] showed that any changes in horizontal wind direction can be expressed as:

$$\frac{\partial \alpha}{\partial t} = \frac{1}{V_h^2} k. \left(V_h \times \frac{\partial V_h}{\partial t} \right) \tag{5}$$

Where:

- α represents the angle of wind vector relative to the horizontal axis,
- V_h is the horizontal wind vector,
- k is a vertical unit vector.

Positive values of Eq. (5) represent anticlockwise rotation (ACR), while negative values correspond to clockwise rotation (CR). Using the components of the total wind vector in Eq. (4) to expand the cross product in Eq. (5), it is possible to determine the terms significantly influencing the sense of rotation of LSB.

2.3.2 Hodograph rotation patterns

From Eq. (5), it is possible to create contour maps representing CR and ACR regions for the simulated domain. From bottom to top, the module output has 49 pressure levels.

To take into account surface averages of hodograph rotation, the third model level (~ 989 hPa) was used representing appropriate surface level (10 m above). To identify the possible invariant features of SB circulation, daytime hourly $\frac{\partial \alpha}{\partial t}$ values were averaged between 0900 and 1700 LST to produce $\frac{\partial \alpha_{day}}{\partial t}$. **Figure 10** represents the daytime regional patterns of CR ($\frac{\partial \alpha_{day}}{\partial t} < 0$) and ACR ($\frac{\partial \alpha_{day}}{\partial t} > 0$) for both WRF-CFS and WRF-ERA.

Figure 10 shows two different regions with positive rotation tendency (region 1, over water) and negative rotation tendency (region 2, over land) for both WRF-CFS and WRF-ERA. The choice of these regions was based on fact that they represent two

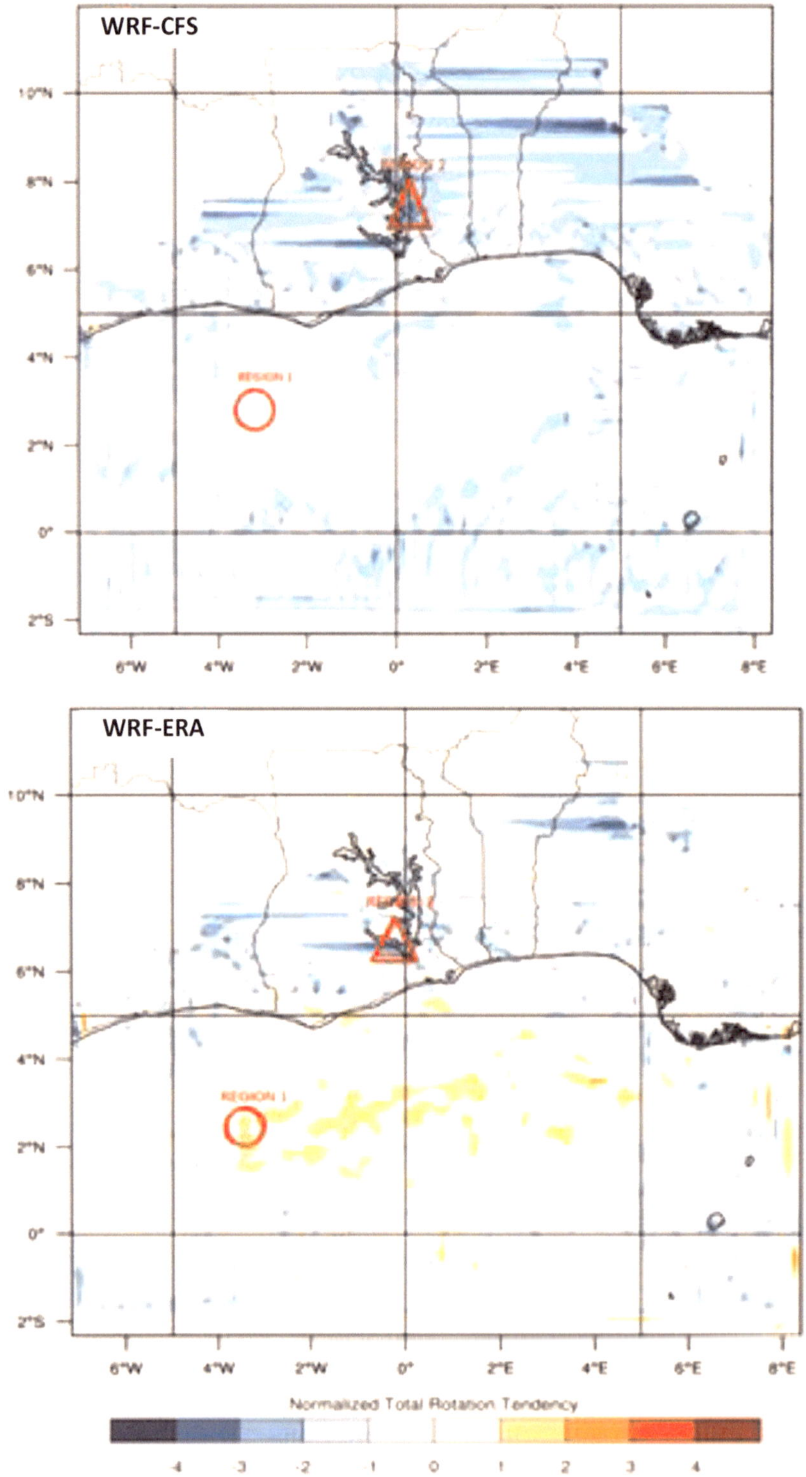

Figure 10.
Patterns of hodograph rotation for the selected SB day, December 16, 2014. Total $\frac{\partial \alpha}{\partial t}$ values are averaged over daytime (0900–1700 LST) showing the subregions 1 and 2 (red circle and triangle) identified.

different topographies, and one on water is dominated by ACR, while the one on land dominated by CR. Therefore, it is obvious that the circulation of SB is influenced by the topographies of the region.

2.3.3 Individual components of surface wind circulation

Using the components of the total wind vector in Eq. (4) to expand the cross product in Eq. (5), Crosman and Horel [15] and Coulibaly et al. [16] showed that the total rate of rotation can be written in terms of individual forcing as:

$$\frac{\partial \alpha_{tot}}{\partial t} = \frac{\partial \alpha_{surf}}{\partial t} + \frac{\partial \alpha_{syn}}{\partial t} + \frac{\partial \alpha_{adv}}{\partial t} + \frac{\partial \alpha_{cor}}{\partial t} + \frac{\partial \alpha_{hdif}}{\partial t} + \frac{\partial \alpha_{vdif}}{\partial t} \qquad (6)$$

2.3.3.1 Influence of the individual tendency terms

In order to investigate the influence of each component of Eq. (6) over each region in **Figure 10**, hourly tendency values for the selected grid points were extracted. These values were normalized by the Coriolis parameter to produce non-dimensional values and also spatially averaged among the selected grid points for each hour. **Table 2** summarizes the daily evolution of the individual forcing terms for WRF-ERA-Region1, (red circle in **Figure 10**).

Table 2 contains the data showing daily occurrence of ACR reaching its maximum around 1100LST before decreasing (**Figure 11**). This daily occurrence of ACR is leading by both surface pressure gradient and advection in contrast to the synoptic pressure gradient. The contrast between surface and synoptic pressure gradients maybe due to the formation of LSB return flow near 850 hPa level [3]. Due to the combined effects of topographies, pressure, and temperature, the nonzero values of the pressure gradient are justified even though the region is over water. The Coriolis, horizontal, and diffusion terms are rather insignificant in this region because the region is, first of all, close to the equator where Coriolis force is generally weak and the

Time (hour)	$\frac{\partial \alpha_{tot}}{\partial t}$	Surface gradient	Synoptic gradient	Advection	Coriolis	Horizontal diffusion	Vertical diffusion
09	0.3832836	−2.482071	1.666516	0.1180504	0.1258831	0.429564	0.2070672
10	2.557241	0.7772687	0.429424	0.3793316	0.566098	0.2907483	0.2197096
11	2.935411	1.85723	−0.6445861	1.042618	0.4937116	0.131484	0.235257
12	1.480419	1.073565	−0.4436242	1.007558	−0.1384531	0.2259261	0.2200678
13	1.112699	0.4998687	0.1314001	1.190483	−0.2559998	0.1615198	0.1972847
14	0.8657498	1.085379	−0.4206522	0.5209342	−0.1935283	0.1714763	0.1016577
15	0.2768821	0.3298867	0.1789315	−0.473791	−0.080399	0.2163928	−0.00345805
16	0.02195143	0.5319856	−0.4363321	−0.2797991	0.06992882	0.09765537	−0.1007054
17	0.1945854	0.5197564	−0.1871125	0.1424769	0.1199066	0.08179352	−0.1505437

Table 2.
Daytime evolution of rotation tendency terms by WRF-ERA Region_1.

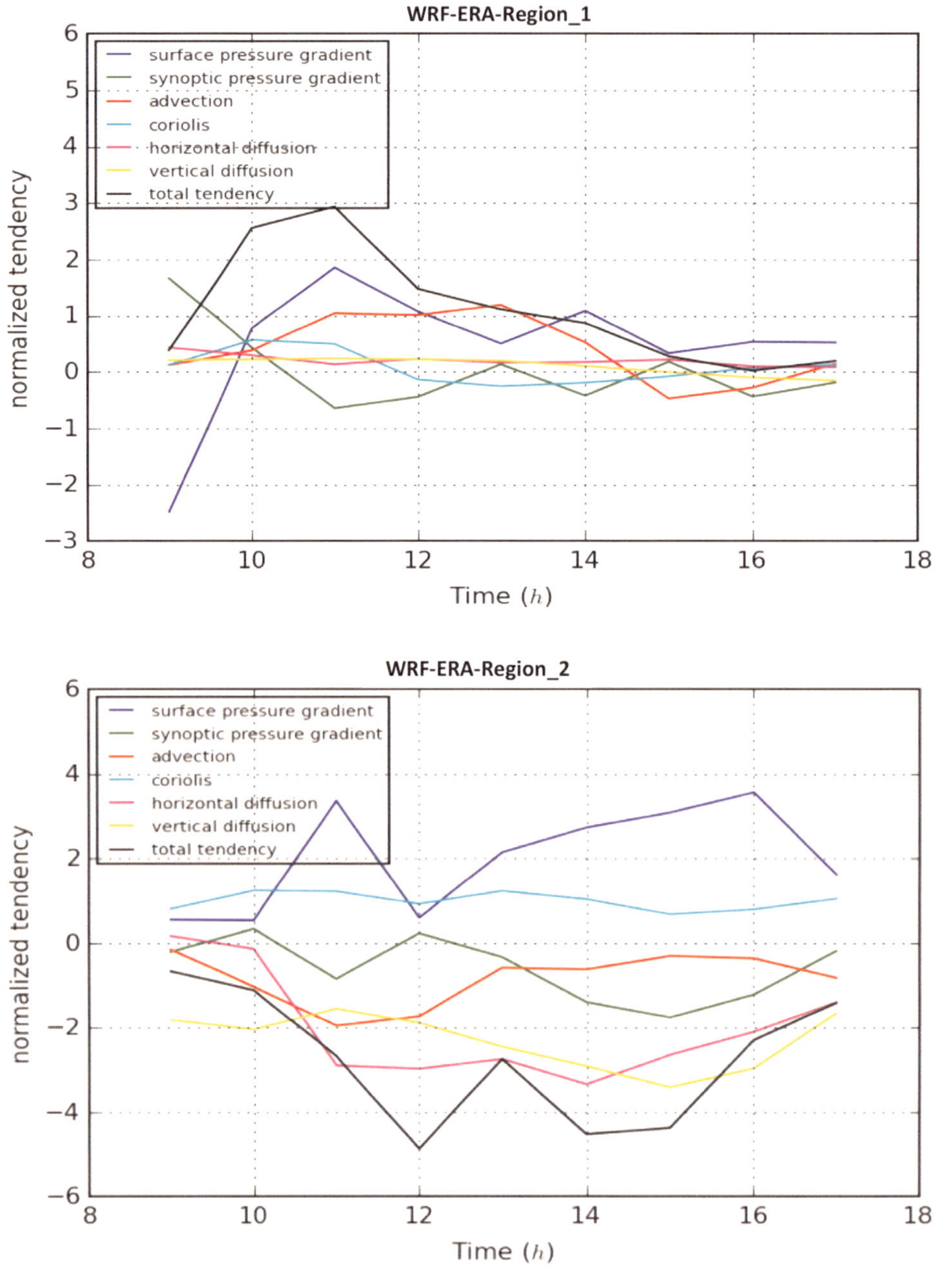

Figure 11.
Diurnal evolution of components of horizontal wind for WRF-ERA regions (1 and 2), on December 16, 2014. Positive and negative values correspond to ACR and CR, respectively.

weakness of friction forces over water, which may minimize the effects of both horizontal and diffusion terms (**Figure 11**).

Table 3 contains the data showing the daily occurrence of CR reaching its maximum at 1200 LST for WRF-ERA simulations. The sense of rotation is the results of combined effects of horizontal and vertical diffusion terms reinforced by both

Time (hour)	$\frac{\partial \alpha_{tot}}{\partial t}$	Surface gradient	Synoptic gradient	Advection	Coriolis	Horizontal diffusion	Vertical diffusion
09	−0.7839572	0.06672001	0.2442949	−0.1528706	0.8046611	0.1641116	−1.910874
10	−1.222245	0.6371607	0.2133593	−1.073048	1.269202	−0.137919	−2.131
11	−2.819007	4.254928	−1.710708	−2.025437	1.250522	−3.025733	−1.562579
12	−5.095439	1.427554	−0.5770028	−1.813763	0.9864119	−3.143581	−1.975057
13	−2.78958	2.692701	−0.7969017	−0.6335011	1.312145	−2.87843	−2.485593
14	−4.572093	2.805323	−1.379125	−0.7210833	1.167803	−3.460328	−2.984682
15	−4.456161	2.866465	−1.481936	−0.3873717	0.7739218	−2.792548	−3.434692
16	−2.493698	3.756342	−1.390541	−0.4019374	0.8376386	−2.309347	−2.985853
17	−1.481868	2.39475	−0.9565901	−0.8265868	1.086807	−1.480353	−1.699895

Table 3.
Daytime evolution of rotation tendency terms by WRF-ERA over land Region_2.

advection and synoptic pressure gradients. Due to the dynamical features of the land, the surface pressure gradient strongly acts in opposition to all above terms. This opposition is reinforced by a relatively important Coriolis force, which is a noteworthy feature of the dynamics of this region (**Figure 12**).

Tables 4 and **5** contain the data of the two different regions (Region 1 with ACR and Region 2 with CR) for WRF-CFS simulations. ACR in Region 1 results from the combined actions of horizontal diffusion and surface pressure gradient tendencies in contrast to the synoptic pressure gradient and vertical diffusion terms. The actions of both Coriolis and advection terms are insignificant in this region (**Figure 12**).

In Region 2, all of the vertical diffusion, advection, and pressure gradient terms are acting to introduce CR in the afternoon in contrast to the synoptic pressure acting to turn in ACR. The effects of both Coriolis and horizontal diffusion terms are relatively insignificant up to mid-day as a result of the land location where it was expected to have a frictional effect due to the spatial distribution of land use, land cover, and topography, which may induce horizontal diffusion effects [16].

The results further show the significantly influence of the tendency terms such as surface and synoptic pressure gradients, horizontal and vertical diffusions, and advection. Over water (Region 1 for all simulations), the total rotation tendency ($\frac{\partial \alpha_{tot}}{\partial t}$) is following the shape of surface pressure tendency ($\frac{\partial \alpha_{surf}}{\partial t}$), while other tendencies are in opposite senses trying to remove themselves. This is consistent with the findings from [16]. This suggests that the evolution of LSB rotation is largely dependent on the topographic and coastal features of the domain, even though these regions are located over water far away from the coast. Since LSB is a mesoscale phenomenon, its scale is not restricted to the immediate coastal region. Hence, the analysis can be performed away from the regions of sharp gradients in topography, roughness, and temperature and still capture the dynamics of the phenomenon [15].

In contrast to the Region 1, Region 2 is showing different scenarios depending on the simulations. While the total tendency term is following the shape of vertical diffusion tendency for WRF-CFS, all tendency terms tend to cancel themselves out

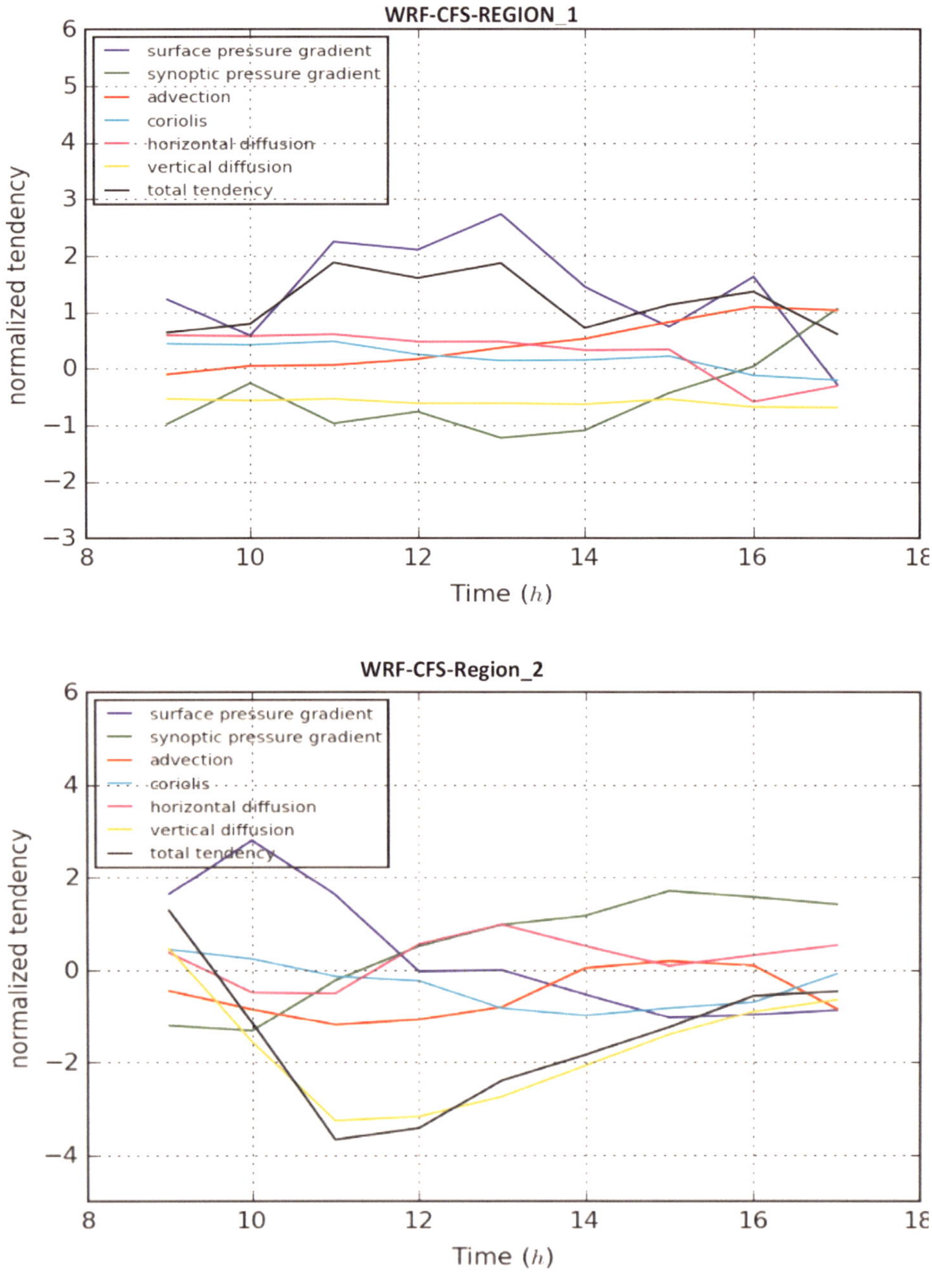

Figure 12.
Diurnal evolution of components of horizontal wind for WRF-CFS regions (1 and 2), on December 16, 2014. Positive and negative values correspond to ACR and CR, respectively.

being generally in opposite senses for WRF-ERA, which demonstrates that WRF-CFS can suitably simulate the dynamics of LSB circulation than WRF-ERA across the coastal West Africa. Nonetheless, the significant influence of the variation of the surface roughness due to the spatial distribution of land use, land cover, and topography on the evolution of LSB circulation over the region [16] can be seen.

Time (hour)	$\frac{\partial \alpha_{tot}}{\partial t}$	Surface gradient	Synoptic gradient	Advection	Coriolis	Horizontal diffusion	Vertical diffusion
09	0.6366872	1.229313	−0.9747893	−0.1039042	0.438543	0.583059	−0.5355338
10	0.7928402	0.5815904	−0.2575232	0.04461657	0.4180464	0.5713866	−0.5652765
11	1.878465	2.245744	−0.9682401	0.05855043	0.4775602	0.6014003	−0.5365493
12	1.606455	2.10231	−0.762126	0.1646468	0.2456327	0.4714136	−0.6154224
13	1.869271	2.732349	−1.224496	0.3645681	0.1375036	0.47354	−0.6141933
14	0.7143283	1.45007	−1.092533	0.5211268	0.1470429	0.3206365	−0.6320145
15	1.133428	0.7353157	−0.4317437	0.821609	0.214676	0.3336835	−0.5401126
16	1.366479	1.628749	0.03543856	1.095926	−0.1230026	−0.5891124	−0.681519
17	0.6067592	−0.2832675	1.060809	1.03419	−0.2068172	−0.3100053	−0.6881502

Table 4.
Daytime evolution of rotation tendency terms for WRF-CFS over ocean (Region_1).

3. Conclusion and discussions

This study examined the dynamics of LSB circulation across the Guinean coast of West Africa. The earlier observational study shows the occurrence of LSB throughout the year with seasonal variability in the region. While LSB circulation formed everywhere along the Guinean coasts, the hodographs exhibited both theoretically expected CR and "anomalous" Counter-Coriolis ACR. Due to the complex, nonlinear nature of LSBs, numerical modeling presented the only possible method by which to understand the underlying dynamics of this mesoscale phenomenon.

A numerical simulation was therefore performed, with ERA-Interim and CFS as forcing data, using the adjusted WRF-ARW model code and subsequently evaluated for accuracy using local observations. The diurnal evolutions of modeled and observed onshore/offshore winds were found to be in good agreement. While WRF model offers great operational forecasting capabilities, effectively no options for dynamical analysis are available. The basic dynamical equations are embedded deeply in the solver and remain inaccessible to the user. This presents a serious limitation to those using WRF to investigate the dynamics driving the LSB circulation, even though the model demonstrates excellent performance and accuracy. In order to overcome this limitation, the model original code was adjusted to allow for the extraction of the individual tendency terms of the horizontal momentum equations [7, 16].

Generally, the terms found to have significant contribution to the total momentum balance of LSB circulation over the domain included pressure gradient (subsequently separated into surface and synoptic components), advection, and horizontal and vertical diffusion. Since the region is close to the equator, Coriolis term generally did not have significant effects on the LSB circulation. The rate of rotation of the total horizontal momentum tendency was plotted for the entire domain. Two regions (for CR and ACR) with one CR and ACR for each of CFS and ERA around the coastline area were selected for term-by-term dynamical analysis. Following [7, 16], the strength of rotation due to each component of the horizontal momentum equations was determined for the selected regions. The direction of rotation was found to be a result of a complex interaction between surface and synoptic pressure gradients,

Time (hour)	$\frac{\partial \alpha_{tot}}{\partial t}$	Surface gradient	Synoptic gradient	Advection	Coriolis	Horizontal diffusion	Vertical diffusion
09	1.290793	1.646445	−1.198673	−0.44654	0.4487459	0.3760073	0.4648078
10	−1.131925	2.804544	−1.30877	−0.8466268	0.245173	−0.482654	−1.543591
11	−3.659746	1.632497	−0.2229338	−1.180996	−0.1330353	−0.5050589	−3.25022
12	−3.409001	−0.02535242	0.516894	−1.068267	−0.229282	0.557843	−3.160837
13	−2.386143	0.001518607	0.9784042	−0.7957155	−0.8215593	0.9847814	−2.733572
14	−1.832128	−0.526906	1.172982	0.04501417	−0.9784316	0.5211309	−2.065917
15	−1.232315	−1.024484	1.710036	0.1978537	−0.8169148	0.09199113	−1.390798
16	−0.5517031	−0.9646802	1.577152	0.09926868	−0.6908129	0.3227653	−0.8953955
17	−0.4564246	−0.864257	1.42138	−0.8347424	−0.07867106	0.5376047	−0.6377391

Table 5.
Daytime evolution of rotation tendency terms for WRF-CFS over land (Region_2).

advection, and horizontal and vertical diffusions. However, higher variability as well as unlikely individual term magnitudes suggests that the simulation requires further improvements to be considered conclusive. For more investigations, an idealized simulation should be carried out using a similar domain configuration as that of a real case.

Consequent upon all the above numerical simulations, it can be concluded that:

i. Both observed and simulated hodograph rotations and wind fields for selected LSB days showed good agreement. Therefore, it can be concluded that WRF model is suitable to evaluate the dynamics of LSB rotations in different regions with different topography features. Over ocean, the model performance was quite good for both reanalysis data (CFS and ERA-Interim). But in land side, for CFS re-analysis data, the rotation tendencies seemed to be affected by the presence of steep topography and spatial distribution of land use, cover, which is not clear for ERA-Interim, demonstrating again CFS is better reproducing the LSB dynamics than ERA-Interim. The overall dynamics of the hodograph rotation tendencies were well captured by WRF model.

ii. The real case simulations showed complex patterns of LSB rotations across the Guinean Coast of West Africa. The regions over land showed more CR, while regions over ocean seem to be significantly influenced by ACR. This can likely be attributed to the variation of surface roughness due to the spatial distribution of land use, land cover change, and topographies.

iii. The sense of rotation seems to be influenced by a complex interaction between pressure gradients, diffusion, and advection terms over regions with different topographies exhibiting different force balance for achieving their sense of rotation. Globally, it is obvious that the balance is not always dominated by the pressure gradient terms in contrast to some earlier studies. The results show different rotation features depending to the locations; thus, when diffusion tendencies are predominantly important over water side, land side seems to be dominated by surface gradient tendencies.

Acknowledgements

This work is supported by the African Institute for Mathematical Sciences, www.nexteinstein.org, with financial support from the Government of Canada, provided through Global Affairs Canada, www.international.gc.ca, and the International Development Research Centre, www.idrc.ca.

This research was carried out through the West African Science Service Centre on Climate Change and Adapted Land Use (WASCAL) initiative supported by the German Federal Ministry of Research and Education.

We acknowledge that the results of this research were achieved using computational resources at the German Climate Computing Center (DKRZ). The ERA-Interim dataset was downloaded from the European Centre for Medium – Range Weather Forecasts (ECMWF) (available online http://apps.ecmwf.int/datasets/data/interim-full-daily/levtype=sfc/). The CFSv2 dataset was downloaded from the Environmental Modeling Center at NCEP (available online ftp://nomads.ncdc.noaa.gov/modeldata/cfsv2_analysis_flxf).

Author details

Amadou Coulibaly[1*], Bayo J. Omotosho[2], Mouhamadou B. Sylla[3], Amoro Coulibaly[1] and Abdoulaye Ballo[4]

1 WASCAL Doctoral Study Programme on Climate Change and Agriculture (DSP-CC and Agric.), Institut Polytechnique Rural de Formation et de Recherche Appliquée (IPR-IFRA), Katibougou, Mali

2 WASCAL Doctoral Study Programme on West African Climate System (DSP-WACS), Federal University of Technology, Akure, Nigeria

3 AIMS-Canada Research Chair in Climate Change Science, African Institute for Mathematical Sciences (AIMS)| AIMS, Kigali, Rwanda

4 WASCAL Competence Center, Ouagadougou, Burkina Faso

*Address all correspondence to: amadou.coulibaly@ipr-ifra.edu.ml

References

[1] Atkinson BW. Meso-scale Atmospheric Circulations. New York: Academic Press; 1989;**495**:125-214

[2] Bajamgnigni GAS, Steyn DG. Sea breezes at Cotonou and their interaction with the west African monsoon. International Journal of Climatology. 2012; **33**:2889-2899. DOI: 10.1002/joc.3637

[3] Abatan AA, Abiodun JB, Omotosho JB. On the characteristics of sea breezes over Nigeria coastal region. Theoretical and Applied Climatology. 2013;**116**:93–102. DOI: 10.1007/ s00704013-0931-z

[4] Coulibaly A, Omotosho BJ, Sylla MB, Coulibaly A, Ballo A. Characteristics of Land and Sea Breezes (LSB) along the Guinea coast of West Africa. Theoretical and Applied Climatology. 2019;**138**: 953-971. DOI: 10.1007/s00704-019-02882-0

[5] Kusuda M, Alpert P. Anti-clockwise rotation of the wind hodograph. Part I: Theoretical Study. Journal of Atmospheric Science. 1983;**40**:487-499

[6] Haurwitz B. Comments on the sea breeze circulation. Journal of Meteorology. 1947;**4**:1-8

[7] Moisseeva N, Steyn DG. Dynamical analysis of sea-breeze hodograph rotation in Sardinia. Atmos. Chemical Physics. 2014;**2014**:13471-13481. DOI: 10.5194/acp-1413471

[8] Stull RB. An Introduction to Boundary Layer Meteorology. Norwell, MA: Kluwer Academic; 1988. p. 670

[9] Avissar R, Pielke RA. Influence of landscape structure on local and regional climate. Landscape Ecology. 1990; **4**(2-3):133-155. DOI: 10.1007/ BF00132857

[10] Martin LC, Pielke AR. The adequacy of the hydrostatic assumption in sea breeze Modeling over flat terrain. American Meteorological Society. 1983;**40**:1472-1481. DOI: 10.1175/ 1520- 0469

[11] Estoque MA. The sea breeze as a function of the prevailing synoptic situation. Journal of the Atmospheric Sciences. 1962;**19**:244-250

[12] Mak KM, Walsh JE. On the relative intensities of sea and land breezes. Journal of the Atmospheric Sciences. 1976;**33**:242-251

[13] Zhong S, Takle ES. An observational study of sea- and land breeze circulation in an area of complex heating. Journal of the Atmospheric Sciences. 1992;**31**: 1426-1438

[14] Simpson JE. Diurnal changes in sea-breeze direction. Journal of Applied Meteorology. 1996;**35**:1166-1169

[15] Crosman ET, Horel JD. Sea and Lake breezes: A review of numerical studies. Boundary Layer Meteorology. 2010;**137**: 1-29. DOI: 10.1007/s10546-010-9517-9

[16] Coulibaly A, Omotosho BJ, Sylla MB, Diallo Y, Ballo A. Numerical simulation of land and sea breeze (LSB) circulation along the Guinean Coast of West Africa. Modelling Earth System Environment. 2020;7:2031–2045. DOI: 10.1007/ s40808-020-00953-0

[17] Abbs DJ, Physick WL. Sea-breeze observations and modelling: A review. Australian Meteorological Magazine. 1992;**41**:7-19

[18] Tijm ABC, Holtslag AAM, Van Delden AJ. Observations and Modeling of the Sea Breeze with the return current. Monthly Weather Review. 1999; **127**:625-640

[19] Fukuda K, Matsunaga N, Sakai S. Behavior of Sea Breeze in Fukuoka. Annual Journal of Hydraulic Engineering. 2001;**44**:85-90

[20] Colby FP Jr. Simulation of the New England Sea breeze: The effect of grid spacing. Wea. Forecast. 2004;**19**:277-285

[21] Drobinski P, Bastin S, Dabas A, Delville P, Reitebuch O. Variability of three dimensional sea breeze structure in southern France: Observations and evaluation of empirical scaling l. Annales de Geophysique. 2006;**24**:1783-1799

[22] Hisada Y, Takayoshi U, Nobuhiro M. A relationship between the characteristics of sea breeze and land-use in fukuoka metropolitan area. In: The Seventh International Conference on Urban Climate. Yokohama, Japan; 2009, 2009

[23] Miller S, Keim B, Talbot R, Mao H. Sea breeze: Structure, forecasting, and impacts. Reviews of Geophysics. 2003; **41**:1011. DOI: 10.1029/2003RG000124

[24] Estoque MA. The sea breeze as a function of the prevailing synoptic situation. Journal of the Atmospheric Sciences. 1962;**19**:244-250

[25] Pielke RA. A three dimensional numerical model of the sea breezes in the South Florida. Monthly Weather Review. 1974;**102**:115-139

[26] Steyn D, Ainslie B, Reuten C, Jackson P. A retrospective analysis of ozone formation in the lower Fraser Valley, British Columbia, Canada. Part I: Dynamical model evaluation. Atmospheric Ocean. 2013;**51**:153-169

[27] Wang W, Xie P, Yoo S, Xue Y, Kumar A, Wu X. An assessment of the surface climate in the NCEP climate forecast system reanalysis. Climate Dynamics. 2011;**37**:1601-1620

[28] Barrisford P, Kallberg P, Kobayashi S, Dee D, Uppala S, Simmons AJ, et al. Atmospheric conservation properties in ERAinterim. Quarterly Journal of the Royal Meteorological Society. 2011;**137**(659): 1381-1399

[29] Saha S. The NCEP climate forecast system reanalysis. Bulletin of the American Meteorological Society. 2010; **91**:1015-1057

[30] Skamarock W, Klemp J, Dudhia J, Gill D, Barker D, Wang W, Powers J. A description of the advanced research WRF version 3, Technical Report. 2008

[31] Neumann J. On the rotation rate of the direction of sea and land breezes, Journal of the Atmospheric Sciences. 1977;**34**(12):1913–1917

Chapter 8

Fluid Dynamics Simulation of an NREL-S Series Wind Turbine Blade

Bharat Ramanathan

Abstract

Wind turbine blades are known for their complex geometry and difficult-to-predict characteristics. So, this chapter aims to look in depth at theory, design, modeling, and simulation of a 1.2 MW wind turbine blade (35 m). Computational fluid dynamics (CFD) will be used to simulate the blade. The design tip speed ratio (TSR), the center point of the design, is optimally chosen as 7. The various parameters like torque vs TSR, Cp, and Ct vs TSR will be found for varying pitch angles. Simulations will be performed on the blade, and the results will be compared with those obtained from blade element & momentum (BEM) theory. Along with this, QBlade and XFoils results are compared with a much more accurate CFD simulation. To conclude, the accuracy of various methods will be compared and evaluated.

Keywords: wind turbine, HAWT, CFD, fluent, BEM, NREL, renewable, mechanical

1. Introduction

Wind turbines have been around for more than a century now. Moreover, the idea of harvesting energy from wind has existed for even longer. Charles F. Brush invented the first automatic wind turbine for power generation in 1887. That does not mean that the technology has stagnated or remained unchanged. Since then, many scientists have improved the early designs, typically made out of wood and later aluminum. These days, the materials used are as exotic as the design, involving manufacturing processes, such as resin-transfer molding to create fiber-reinforced composites [1]. Even the designs themselves have changed radically. We have vertical axis wind turbines (VAWT) and horizontal axis wind turbine (HAWT) [1, 2]. Both have pros and cons, but for this chapter, HAWT will be the main focus owing to their relative popularity, high efficiency, and simplicity. **Figure 1** shows HAWT and VAWT for comparison.

Renewable energy is projected to hit 31% of all energy generation by 2035 across the world. Out of this, a quarter will be from wind power alone [3] with a high projected growth rate. However, the wind turbine blade is one complex piece of engineering that requires its own attention. It is a significant portion of the entire cost of the machine. The blades are indeed immense and have subtle curves. The fundamental aim of this chapter will be to see how these curves affect the wind turbine blade performance. The theory of the blades will be explained first, followed by design

Figure 1.
HAWT and VAWT side by side (image from iStock/purchased for use).

in CAD. Then the blade performance will be estimated roughly using the blade element and momentum (BEM) theory. Then, extensive simulations will be performed in ANSYS Fluent to determine the exact characteristics. The same will be verified and validated by our earlier estimates obtained by BEM theory.

The chapter has been designed concisely and easy to follow, so that it will be comprehensible for newcomers yet contain essential data for the experts. It is divided as follows: Firstly, we will look at BEM and perform the initial estimation of blade characteristics. The exact "curves" of the blades will be determined using simple equations solved in MATLAB. We will use NREL-S Series Airfoils (the "curves"), namely, S815, S825, and S826 for root, primary & tip portions, respectively. The geometry will require specific parameters of the airfoils. Here, QBlade and XFoils will prove particularly handy for determining these parameters and assist in blade design. This step will be followed by CFD logic and ANSYS Fluent working methods. Particular attention will be given to Navier-Stokes Equation in rotational domain and how it differs from the common coordinate form. After this, the CAD modeling will be touched on briefly as it is highly software dependent. After that, the Fluent simulations will be performed, and the various performance parameters will be noted. Finally, the results obtained will be compared with the estimates produced earlier.

2. Theory

The first requirement of any simulation is the initial estimation of quantities of interest. This calculation will give an idea about the expected results and will act as a prerequisite for the last step: verification and validation. Without this, the simulation might produce some other results that need not always be correct.

2.1 One dimensional momentum

One-dimensional momentum theory is one of the oldest theories of wind turbines. Much literature is dedicated to the same. It relates the velocities upstream, at the turbine blades, and downstream with a mathematical induction factor "a." The induction factor is essential and has some implications for the wind turbine as a whole. The exact derivation may be seen in [2, 4]. The formulae alone will be listed here.

Let the velocity of wind upstream be u; at the turbine be y; and downstream be v; then:

DOI: http://dx.doi.org/10.5772/intechopen.107013

$$y = \frac{(u + v)}{2} \tag{1}$$

The maximum power that can be generated from the wind is:

$$P_{wind} = 0.5\rho u^3 A \tag{2}$$

The maximum torque that can be extracted from the wind is:

$$T_{wind} = 0.5\rho u^2 A \tag{3}$$

It is convenient to use a non-dimensional power coefficient (Cp) and torque coefficient (Ct) as it is a ratio from $0 < C < 1$. They are defined as:

$$C_p = \frac{P_{out}}{P_{wind}} \tag{4}$$

$$C_t = \frac{T_{out}}{T_{wind}} \tag{5}$$

"a" is the mathematical induction factor defined as:

$$\mathbf{C_p = 4a(1 - a)^2} \tag{6}$$

$$\mathbf{C_t = 4a(1 - a)} \tag{7}$$

These are the two primary verification and validation formulae. Through Fluent simulation, the velocity distribution and torque/power will be found separately. Notice how the induction factor links them. One can use Eq. (6) to compute the induction factor as a ballpark figure. Eq. (7) can then be used to provide theoretical Ct. But Ct can be directly found through Eq. (5) using Fluent. By comparing these two values, we can check for the correctness of our results.

The maximum power coefficient (percentage) that can be delivered by a wind turbine is 59.3%, and the induction factor must be less than 0.5. Anything above that is impossible or has no practical significance. This limit in power coefficient is known as Betz Limit, and it arises because some energy needs to be present in the wind to move past the turbine blades to prevent local wind accumulation. The same can be figured out by finding the maximum value of Eq. (6). Verification and validation will be performed in the end using these points. Large deviations can be expected as this theory does not account for turbulence and is a significant approximation of the underlying physics.

2.2 Blade element momentum

A wind turbine blade is known for its characteristic twist as one moves along the body. This twist is the characteristic angle defined by β. An airfoil provides max lift only for a particular angle of attack. That is easy in an airplane , exhibiting translational motion where one adjusts the pitch angle for maximum lift. However, a wind turbine exhibits rotational motion. The fundamental problem is that any two points on the circle's radius never move at the same speed. Since the angle of attack is computed keeping the net velocity wind vector as a reference, the angle with this must remain constant. If one looks at this velocity vector, it has a k component of wind velocity (incoming wind) and an i component of the rotational velocity. The net velocity vector is:

$$V_{relativetoblade} = u_{wind}\hat{k} + r\omega\hat{i} \tag{8}$$

One can see that as we move across the blade, the net velocity vector will change and is, in fact, a function of radii. The only way to make a constant angle with this vector at all points is to compensate with a twist angle of our own. Hence, the characteristic curve.

BEM theory will help us with the exact mathematical formulae to compute this twist angle and hence, will be the starting point of our design. For exact derivation, please refer to [1–3, 5–7].

The local-speed ratio for arbitrary radii r and with incoming wind velocity u from the center is defined as:

$$\lambda_r = \frac{r\omega}{u} \tag{9}$$

Note the local speed ratio changes as one moves along the blade. The angle made by the horizontal and the net velocity vector (as defined earlier) is:

$$\phi = \frac{2}{3}\tan^{-1}\frac{1}{\lambda_r} \tag{10}$$

Here, the angles can be written as below for zero pitch. Also, α was assumed to be equal to 5.25 degrees. Please refer to **Figure 2** for visualizing the exact angles.

$$\alpha = \phi - \beta \tag{11}$$

The chord length of the airfoil is given by:

$$c = \frac{8\pi r}{BC_l}(1 - \cos\phi) \tag{12}$$

where B is the number of blades (3 in our case), Cl is the lift coefficient, and r is the radii from the center. However, one can notice that except for Cl, the net blade length and the rotational speed ω, everything else is defined and can easily be computed. Many approaches, such as calculating axial and tangential induction factors have been

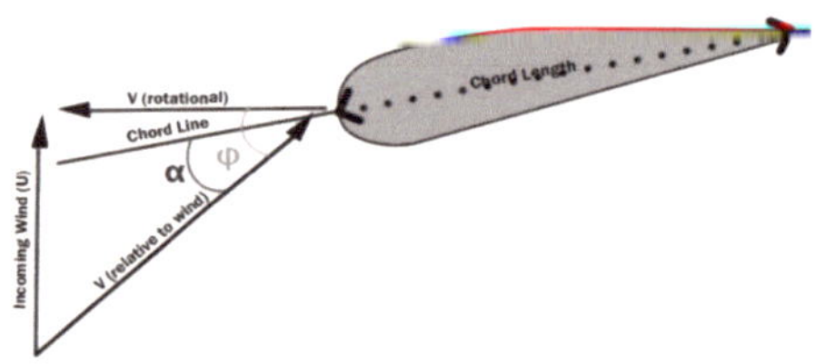

Figure 2.
Airfoil angles (image referred from [1]).

developed [3, 5]. This method suffers from the limitation that the airfoil data must be present, and the airfoil must be uniform. Neither is applicable in our case, as NREL has not released airfoil data separately, and the root, primary, and tip are composed of totally different airfoils. The net blade length can be determined from previous designs for a chosen power level. Furthermore, 35 meters from the hub center to the tip was assumed to be sufficient. NREL has mentioned the maximum obtainable lift coefficient for each of the three airfoils. These values were averaged, and a slightly lower figure of 1.3 was chosen for Cl, and Cd was chosen as 0.1. Finally, the wind turbine was designed for an optimal TSR of 7. The TSR and blade length give an omega of 2.42 rad/sec.

The chord length and twist angle give us everything to design our blade in CAD. As a cross verification, XFoils and QBlade will be used to generate performance curves. **Tables 1** and **2** give the variation of twist angle and chord length as one moves across the blade.

QBlade will take the various airfoil parameters and generate curves, such as Cl vs TSR, Cd vs TSR, and Cl/Cd vs TSR. These are the three curves of interest. Our earlier value of Cl can be cross-checked with this software-generated graph. It is worth noting that QBlade does assume varying airfoils across the blade. The blade itself is designed through a selection tab for different airfoils. After this, the simulation is performed using highly simplified models for turbulence and airfoils. Later, CFD simulations will verify and produce highly accurate results.

3. QBlade and XFoils simulation

QBlade and XFoils are open-source software for the prediction of wind turbine performance. The models generated and performance curves have been shown in the figures below:

We get a maximum power coefficient close to the Betz limit (59.2%, approximately) at the expected TSR of 7. The power coefficient that was obtained verifies our

Distance from center	Twist angle[a]
4	29
8	16
12	9
16	6
20	5
24	3
28	2
30	0.025
35	0

[a] *Approximate values.*

Table 1.
Twist angle vs distance from center.

Distance from center	Chord length[a]
4	4.4
8	3.4
12	2.6
16	2
20	1.8
24	1.4
28	1.2
30	1.1
35	1

[a] *Approximate values.*

Table 2.
Chord length vs distance from center.

Figure 3.
Airfoil shapes superimposed on one another.

design to some extent. **Figure 3** shows the plot of various airfoil curves used for this chapter. **Figure 4** shows the QBlade simulation results.

4. Computational fluid dynamics and fluent

The fluid equations can be solved either in differential form or integral form. The differential form is obtained by applying conservation laws to a fluid particle in an Eulerian frame of reference. The integral form is obtained by applying conservation laws to a control volume.

The following are the differential form of fluid equations:

$$\frac{\partial u}{\partial x} + \frac{\partial v}{\partial y} = 0$$
$$\rho\left(u\frac{\partial u}{\partial x} + v\frac{\partial u}{\partial y}\right) = -\frac{\partial p}{\partial x} + \mu\nabla^2 u$$
$$\rho\left(u\frac{\partial v}{\partial x} + v\frac{\partial v}{\partial y}\right) = -\frac{\partial p}{\partial y} + \mu\nabla^2 v \tag{13}$$

There is also the integral form of fluid equations, as listed below:

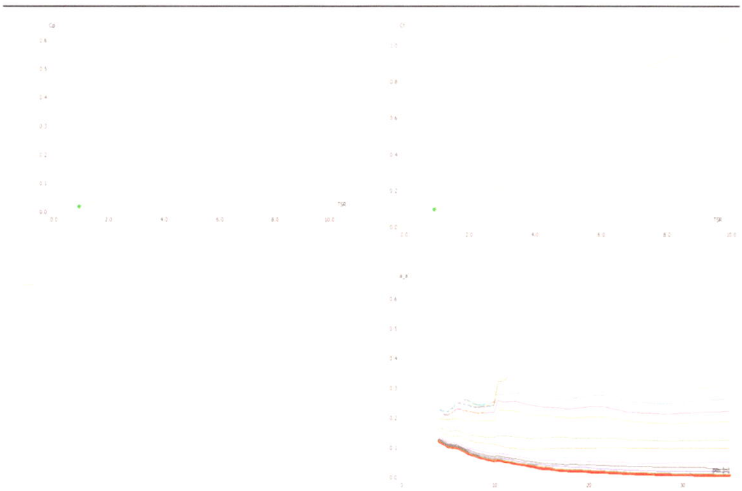

Figure 4.
QBlade simulation graphs.

$$\int_S (v.n)dS = 0$$
$$\int_S \rho(v.\hat{n})\vec{V}dS = -\int_S \rho\hat{n}dS + \vec{F}_{visc} \tag{14}$$

The integral form of fluid equations is mainly preferred since conservation is always valid for any control volume or mesh "chunk." Conservation does not apply to each element in the differential form; hence, in industry, the integral form is preferred. Our software for CFD simulation is ANSYS Fluent, which uses the integral form for solving the problem. Both these equations of integral form are valid for any region or arbitrary shape of a control volume. However, Fluent will produce numeric solutions satisfying these equations for only a particular shape of control volume defined during the meshing step. Meshing breaks the 3D geometry into smaller "chunks" processed as individual control volumes.

The solution methodology introduces errors when solving these equations in Fluent. Two types of errors are introduced, namely, discretization and linearization errors. Discretization error occurs because we assume the value at the interface of adjoining cells is nothing but the average of each of those cell center values. **Figure 5** shows the control volumes and cell centers for a typical uniform meshing. Uniform rectangular chunks will not be the case for complex geometry, as the geometry will be broken into tetrahedrons. The averaging algorithm is also more exotic and will not be covered here.

However, on solving these equations, we end up with a set of nonlinear algebraic equations. These can be solved only by Newton-Raphson (NR) method by assuming a guess value for each cell center. NR will continually iterate until the error falls below a particular threshold. NR method leads to another source of error known as linearization error. The ultimate aim is to reduce both errors as much as possible.

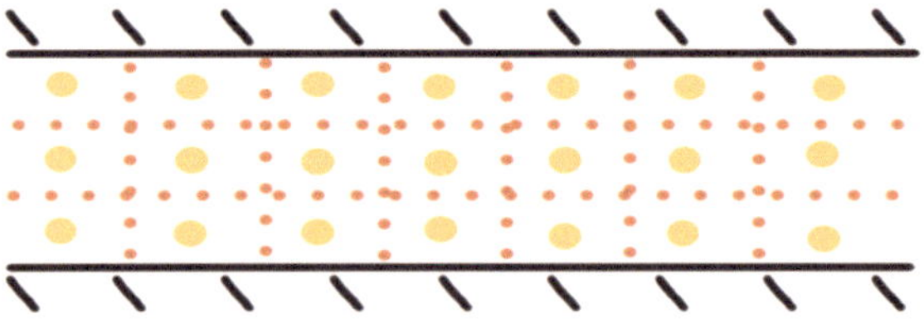

Figure 5.
Discretization I.

The mesh geometry plays an important role here. More "chunks" leads to less discretization error but more linearization error and vice versa. Hence, in Fluent, it is essential to hit the sweet spot for all geometry. Here, we conclude the inner working of Fluent for the translational or inertial frame of reference. In the next section, the Navier-Stokes equation will be modified for the rotational frame of reference.

5. Navier-Stokes in rotational domain

Navier-Stokes/fluid equation(s) are well known by most in the translational domain. However, as described earlier, a wind turbine exhibits rotational motion, and the fluid equations take a different form. At the outset, it is evident that extra forces will act on the fluid particle due to the rotational motion when viewed from the inertial frame of reference. These forces will have to be accounted for in the integral form of fluid equations as used by Fluent.

A vector is assumed to rotate with a radial velocity Ω [8]. From the perspective of the vector, the vector itself is static. However, from the inertial frame of reference, the vector is rotating. For a small time-period "t", the angle subtended by the new vector position with the old vector is:

$$\theta = \Omega \Delta t \tag{15}$$

The magnitude of change in the new vector from the old vector (position-wise) forms a sector of a circle [8]. This gives a net length of:

$$\Delta i_{new} - \Delta i_{old} = r\Omega\Delta t \tag{16}$$

Notice that the triangle formed by these three vectors is a right-angle triangle for a small change in vector in a short period "t." The (change in vector)/(new vector) = sin (ϕ)

Hence, we can write the net vector change with magnitude and direction as the following [8]. Note that the change vector is perpendicular to both the old vector as well as the rotational axis

$$\Delta \hat{i} = |\hat{i}| \sin(\phi)\Omega\Delta t \frac{\vec{\Omega} \times \hat{i}}{|\vec{\Omega} \times \hat{i}|} \tag{17}$$

By definition of cross product:

$$|\vec{\Omega} \times \hat{i}| = |\hat{i}|\Omega \sin(\phi) \tag{18}$$

Therefore, substituting this in the previous equation

$$\frac{d}{dt}\hat{i} = \vec{\Omega} \times \hat{i} \tag{19}$$

This was for a stationary vector. One can extend the analogy to a vector rotating in a rotating frame of reference. As imaginable, the sum of the rate of change of individual vector rotation and the frame rotation will be the rate of change of net rotation [8, 9]. However, the rate of change in frame rotation has been defined earlier. Hence, we obtain Chasle's theorem [9].

$$\frac{d\vec{A}}{dt} = \frac{dA_i}{dt} + \vec{\Omega} \times \vec{A} \tag{20}$$

where $\left(\frac{dA_i}{dt}\right)$ term is the vector rotation as seen by the observer in the rotating frame of reference [9].

For Navier-Stokes, the fluid velocity "u" is rotating and is viewed from a stationary frame of reference. Hence, by applying Chasle's theorem on the fluid velocity "u."

$$\left(\frac{d\vec{u}_{inertial}}{dt}\right)_{inertial} = \left(\frac{d\vec{u}_{inertial}}{dt}\right)_{rotational} + \vec{\Omega} \times \vec{u}_{inertial} \tag{21}$$

Re-substituting Chasle's theorem twice in this equation, we obtain the final equation and assume constant velocity flow:

$$\left(\frac{d\vec{u}_{rotational}}{dt}\right)_{rotational} = 2\vec{\Omega} \times \vec{u} + \vec{\Omega} \times \left[\vec{\Omega} \times \vec{x}\right] \tag{22}$$

It is interesting to note that the following is known as Coriolis acceleration.

$$2\vec{\Omega} \times \vec{u} \tag{23}$$

The following is the centrifugal acceleration. These two equations put together define the Navier–Stokes in the rotational domain.

$$\vec{\Omega} \times \left[\vec{\Omega} \times \vec{x}\right] \tag{24}$$

The exact equation that will be solved in Fluent for our wind turbine is as follows. One can note the similarities between the earlier two equations and this:

$$\frac{\partial \rho}{\partial t} + \nabla.\rho\vec{v}_r = 0$$
$$\nabla.\left(\rho\vec{v}_r\vec{v}_r\right) + \rho\left(2\vec{\Omega} \times \vec{v}_r + \vec{\Omega} \times \vec{\Omega} \times \vec{r}\right) = -\nabla p + \nabla\tau_r \tag{25}$$

The next section will be brief and devoted to CAD modeling of blade.

6. CAD model of blade

The CAD model of the blade can be made using any 3D CAD software. ANSYS Fluent supports many types of 3D file formats and can import a solid-works project directly. ANSYS also has an inbuilt geometry design software named space-claim, and an old one named design-modeller. The actual CAD design steps are out of the scope of this chapter. Fusion360 was used to design the blade model, and the same was imported as a (.stp) STEP file into ANSYS. **Figure 6** shows the blade geometry as a sketch and a body.

It is crucial to note that we are NOT modeling the blade, but rather the fluid surrounding it. Imagine a cylinder (rather a sector of a 3D cylinder spanning 120 degrees) where the blade is left hollow. The actual geometry is the air surrounding the blade, not the blade itself. It is the subtraction of the blade from the cylinder geometry. It is understandable that as air cannot penetrate the blade, that region is left hollow. One may refer to the following figures for the blade design.

One can see the blade's outline in the 3D sector (120 degrees) of a cylinder in **Figure 7**. The subtracted geometry will be used for simulation and imported into

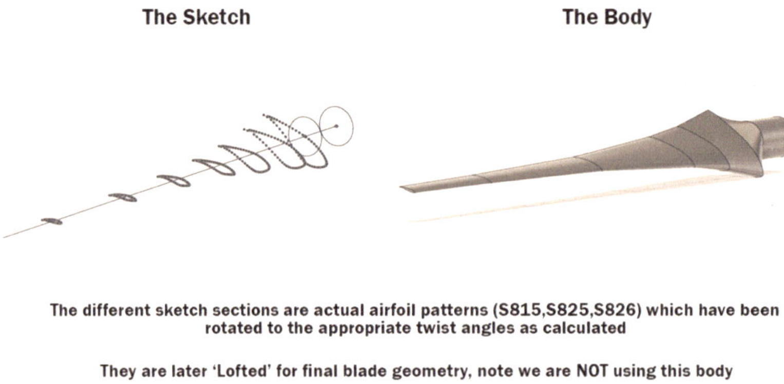

Figure 6.
CAD model of blade.

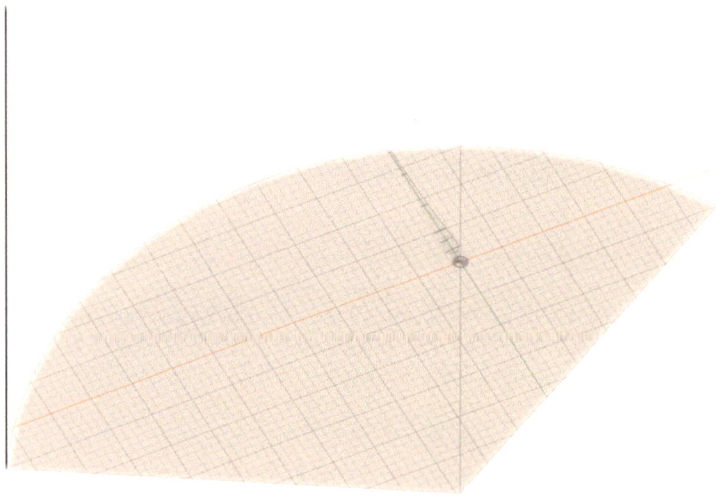

Figure 7.
CAD model for simulation.

ANSYS. The following section will discuss the simulation setup, the mathematical model used, Fluent, boundary conditions, and solution iterations.

7. ANSYS fluent setup: a quick glance

This section will examine how the simulation was run in ANSYS. The various subsections will look at different steps in the execution order. Fluent will need a lot of data and setup before running the simulation and getting results. Before running the simulation, all properties, boundary conditions, and a few other parameters must be defined. One can note that there is symmetry in a wind turbine blade. It repeats every 120 degrees, normal to the rotational axis. We can save much time by simulating only this 120-degree portion and then repeating the graphical "Instances" every 120 degrees. This will be seen in the final results subsection.

Any **ANSYS Fluent** project will go through the same five steps:

- **Geometry Import**: Here you import the geometry (**Figure 8**).
- **Meshing**: Geometry is broken into smaller "chunks" for use as control volume by Fluent (**Figure 9**).
- **Fluent Setup**: Setup all the necessary stuff like mathematical models, boundary conditions, interfaces, and other parameters (**Figure 10**).
- **Solution Iteration**: Model is solved till the solution converges and the error falls below the desired margin.
- **Results**: View your results interactively and graphically (**Figure 11**).

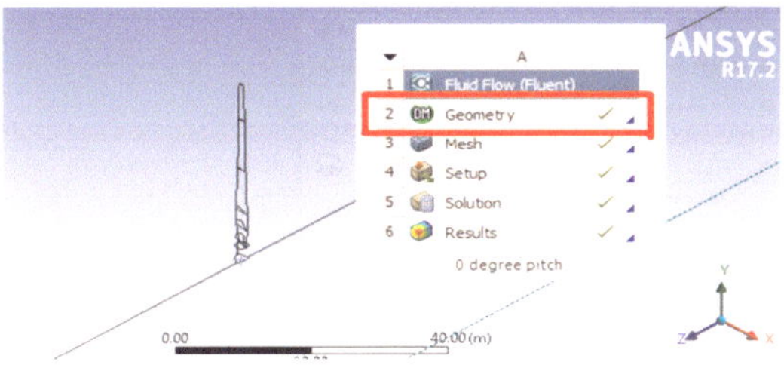

Figure 8.
Geometry import.

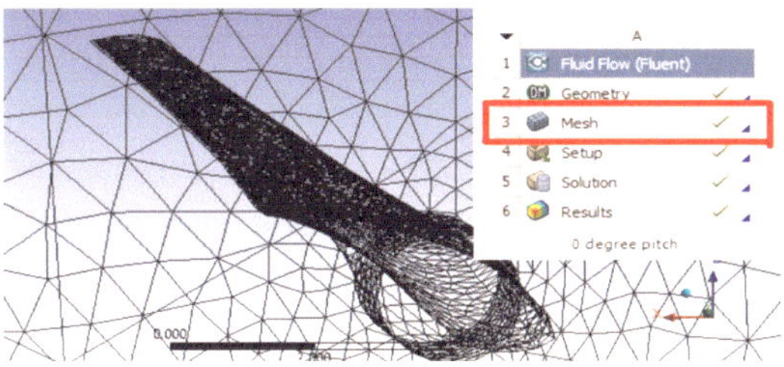

Figure 9.
Meshing of the geometry.

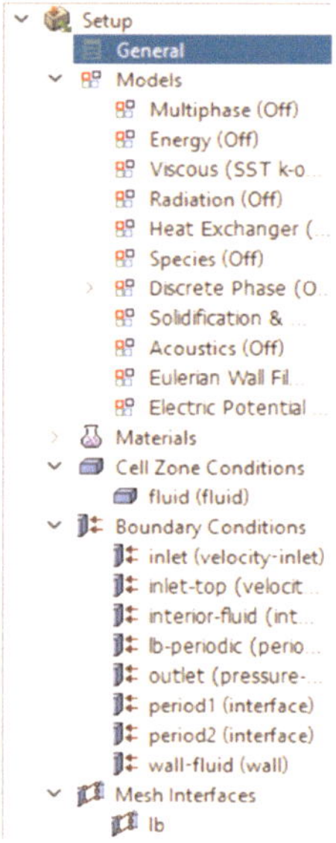

Figure 10.
Fluent options.

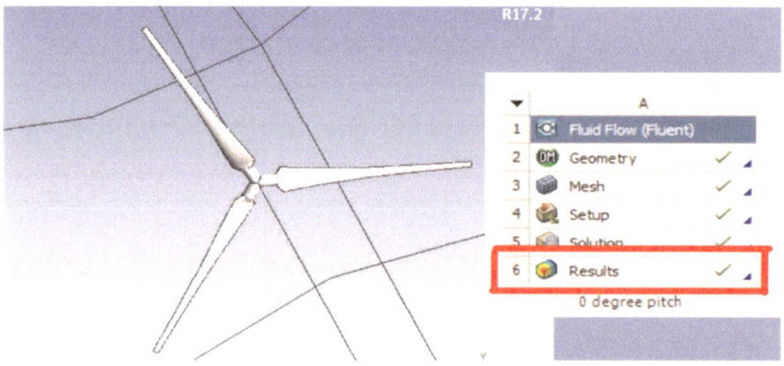

Figure 11.
Results in CFD-Post. Notice the 3-bladed turbine generated from a 1/3rd portion.

7.1 Geometry import and meshing

7.1.1 Fluent setup

The step described in this section is the major one, where one defines the mathematical model to be solved, the accompanying boundary conditions, and the interface setup.

7.1.2 Mathematical model

The primary model to be set up is the turbulence model. **SST k-omega** was used for this chapter. There are multiple turbulence models in **Fluent**, each with pros and cons. The next step is to choose the fluid properties (air was selected). Then one needs to set the cell-zone conditions. Here, the rotational frame of reference must be set. The speed of rotation of the frame is the speed of blade rotation. For multiple simulations, this will take on multiple values. For TSR of 7, it is 2.42 rad/sec.

7.1.3 Boundary conditions

Here, we set the boundary values; the inlet wind velocity is set as 12 m/sec along the z-axis. The fluid was selected to be air. The outlet was set to have a constant atmospheric pressure of 1 atm. **Fluent** uses the gauge pressure system, where the values are computed at the center-point of 1 atm. This system improves the floating point accuracy. Period 1 and period 2 were set as interfaces. The blade itself was set as a wall with no-slip condition.

7.1.4 Need for interfaces

Since we have used a 1/3rd geometry setup, we need to declare the adjoining faces as interfaces. This setup will cause **Fluent** to assume that the geometry repeats. The settings include the periodic angle setup, which will be set at 120 degrees.

7.2 Results

This is the final step post solution iterations. When the errors have fallen below the tolerance value, the simulation is stopped. Finally, the results can be seen in CFD-Post.

8. Results: CFD-Post

8.1 Velocity streamlines

Here, the velocity streamlines for various pitch angles (0,3,5, and 10 degrees) and TSR (5,6,7,8, and 9) are presented. Streamlines track the path of the wind as it flows across the blades. The least obtained velocity was 7.27 m/sec. This velocity (7.27 m/sec) is essential as it directly affects how much energy is extracted from the wind [10] shows some velocity streamlines for non-separated flow, as the turbine is optimized. Our results indicate non-separated flow at optimal TSR. **Figure 12** shows the velocity streamlines for various turbine blade configurations. The five rows of the image are the TSR 5, 6, 7, 8, and 9, and the four columns are pitch angles of 0, 3, 5, and 10 degrees. Every image at location {TSR, Pitch} corresponds to that particular input condition.

8.2 Pressure contours at root of the blade: CFD-Post

The pressure contours are directly responsible for torque generation by the blade. Typically, the bottom portion should have high pressure, followed by the top side, which has low pressure. This net pressure difference creates lift at an angle. A portion of this lift will assist in blade rotation. One can note that as the pitch angle is high and TSR is increased, the low and high-pressure regions shift, and the blade will stop generating torque, or if pitching is increased further, it will generate reverse torque. This unique feature can make the blade stop rotating in stormy or very high TSR conditions by moving the blade to this appropriate zero torque angle. One can also note that the blade produces maximum torque when the angle of attack is close to 5 degrees, which depends on TSR and pitch angle. The two contours shown here are for the root (S815), **Figure 13** and tip (S826), and **Figure 14** of the blade. The five rows of the image are the TSR 5, 6, 7, 8, and 9, and the four columns are pitch angles of 0, 3, 5, and 10 degrees. Every image at location {TSR, Pitch} corresponds to that particular

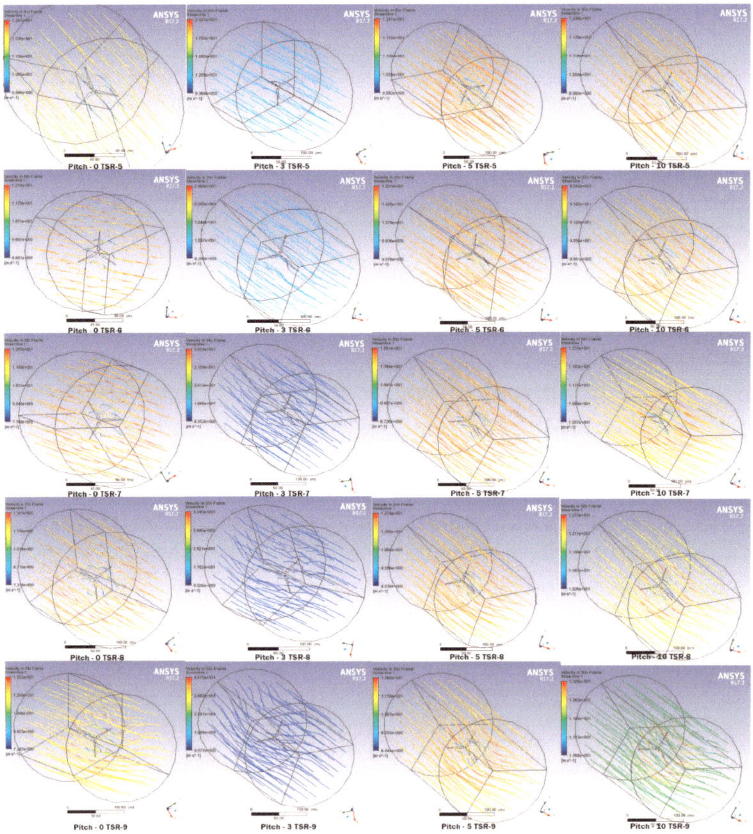

Figure 12.
Velocity streamlines for varying pitch and TSR.

input condition [10] shows pressure contours where the wake (region of low pressure) is visible at the tail end of the airfoil. However, the blade as a whole is not simulated, and this produces slightly differing pressure contours in our case but is still consistent with our results.

9. Results: Torque & Cp

In this section, the numerical results of the generated torque and power coefficient will be presented here. As per our earlier section on one-dimensional momentum theory, the power present in the wind for a wind speed of 12 m/sec and a blade length of 31 m (from root to tip alone, the root to hub center is 4 m) is:

$$P_{wind} = 0.5 * 1.225 * 12^{3} * \pi * 31^{2} = 3193764.336W \tag{26}$$

The results are presented in **Table 3**. The following can be noted from the tabular results:

- The maximum obtained power coefficient is 0.421 for 0-degree pitch and 9 TSR. This value is slightly higher than the one obtained for the pitch of

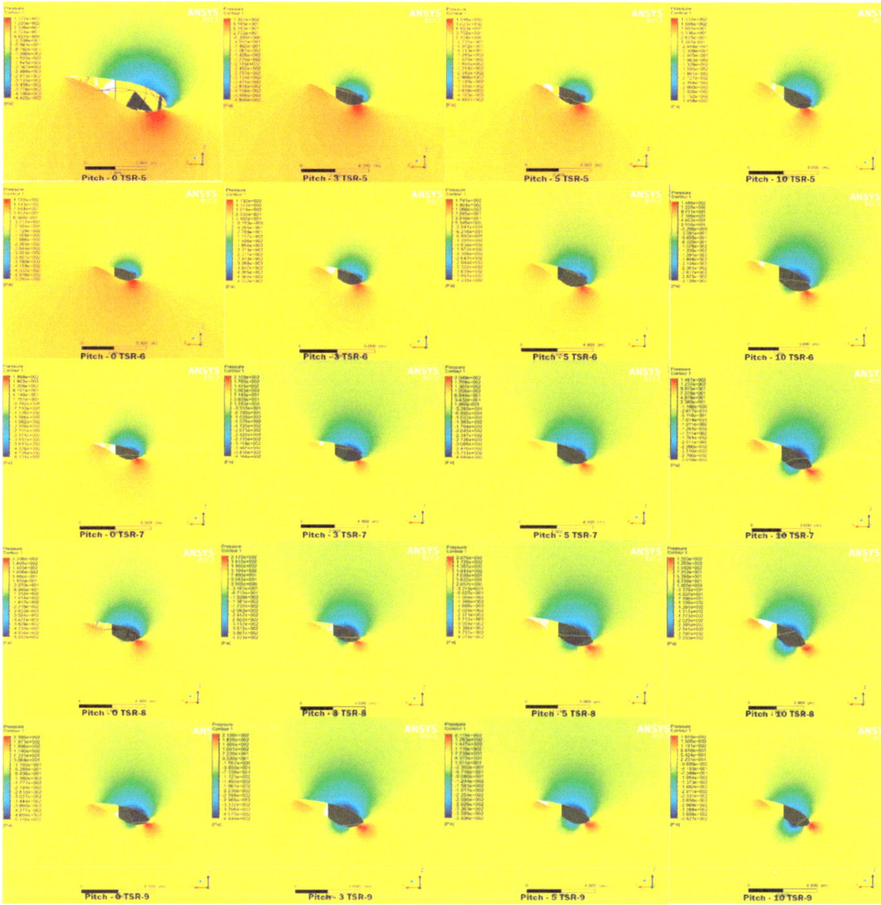

Figure 13.
Pressure contours of blade root (S815) for varying pitch and TSR.

5 degrees, TSR 7, which is our optimal angle of attack (approximately 5.25). The twist angle was computed, and 5.25 degrees was subtracted to give the twist for a 0-degree pitch [11, 12] provides a maximum computed Cp for a 2 MW blade as 0.4436 (not through CFD). Although with minor deviations, our values are still very close.

- The 5-degree pitch gives the most consistent high Cp ($<$ 0.3) for varying TSR. This result is in line with our calculation.

- One hits near zero torque for 10-degree pitch and TSR 8. The correct angle to stop the blade will be around 10 degrees for higher TSR (not simulated).

- The 15-degree pitch gives positive torque only for extremely low TSR of $<$ 5.

- Pitching is a way to control torque generation and plays a significant role in large turbines, unlike small turbines. It is an effective means to control the system as a whole. Graphs in ref. [7] clearly show an increase followed by a decrease in Cp as the pitch angle is linearly increased.

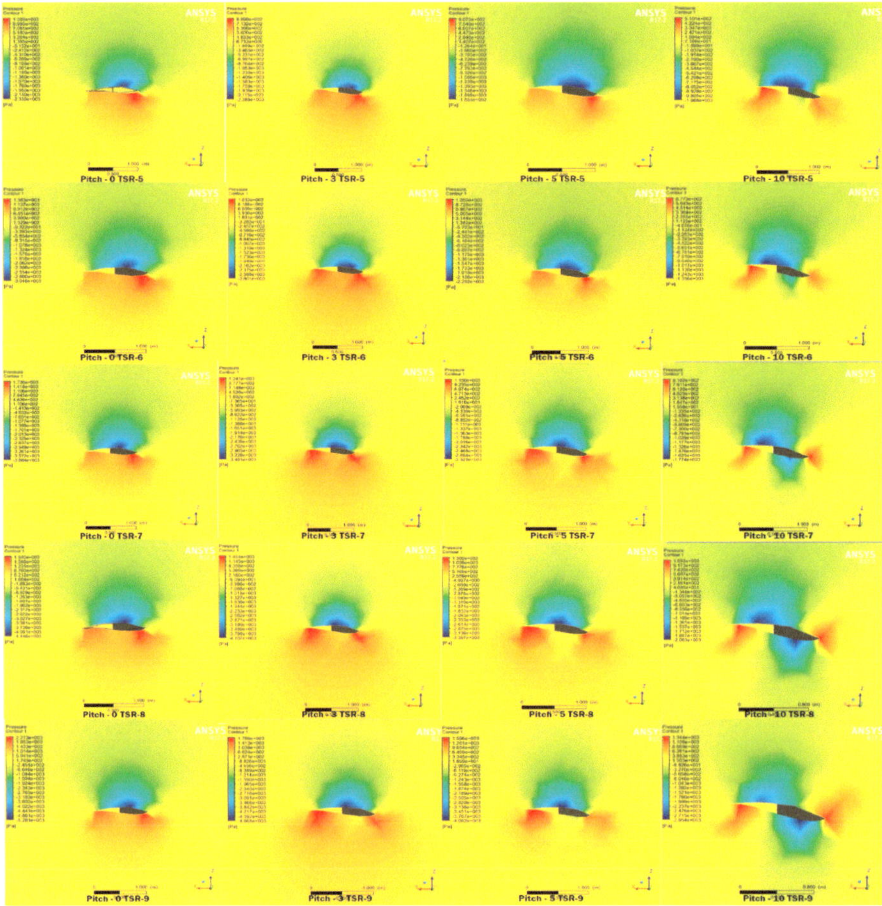

Figure 14.
Pressure contours of blade tip (S826) for varying pitch and TSR.

10. Verification & validation

One can compare the Cp obtained with that predicted by the one-dimensional momentum theory. The induction factor for a Cp (maximum) of **0.421** is **0.121869**

$$C_p = 4a(1-a)^2 = 0.3759 \tag{27}$$

Substituting the induction factor below,

$$C_t = 4a(1-a) = 0.428068 \tag{28}$$

The actual torque coefficient was calculated assuming the net blade length of 35 m to account for tip/hub losses. The computed value below is indeed very close, verifying our solution.

$$C_t = \frac{T_{out}}{0.5\rho u^2 A} = \frac{165373}{339261.3} = 0.487450234 \tag{29}$$

Pitch	TSR	Ω	Torque[a]	Net power	Cp
0	5	1.71	136537	700434.81	0.219313242
0	6	2.06	164431	1016183.58	0.318177384
0	7	2.42	168523	1223476.98	0.383083049
0	8	2.74	157794	1297066.68	0.40612473
0	9	3.09	145060	1344706.2	0.421041147
3	5	1.71	170936	876901.68	0.274566808
3	6	2.06	188062	1162223.16	0.363903857
3	7	2.42	175964	1277498.64	0.399997779
3	8	2.74	160838	1322088.36	0.413959272
3	9	3.09	136504	1265392.08	0.396207092
5	5	1.71	136537	700434.81	0.219313242
5	6	2.06	187062	1156043.16	0.361968836
5	7	2.42	165373	1200607.98	0.375922533
5	8	2.74	143951	1183277.22	0.370496097
5	9	3.09	110525	1024566.75	0.320802239
10	5	1.71	158776	814520.88	0.255034747
10	6	2.06	120066	742007.88	0.232330192
10	7	2.42	52646	382209.96	0.119673814
10	8	2.74	3828	31466.16	0.009852374
15	5	1.71	69318	355601.34	0.111342386

[a]*1/3rd of total torque generated as the simulation was performed for a single blade.*

Table 3.
Pitch, TSR vs Torque (in N-m), Net Power (in W), and Cp.

Now, let us finally compare our solution with QBlade and XFoils. Below is the Cp vs TSR Curve (**Figure 15**). We can see some deviation from the QBlade data and CFD. This deviation is probably due to extra turbulence caused by the rotor hub, which was accounted for in CFD but was left empty in QBlade. The energy dissipation and turbulence cause a lower figure to appear in CFD.

11. Conclusion

The chapter looked in-depth at the theory behind wind turbines, including solving equations and the various theories describing wind turbines in general. QBlade and XFoils were used to get a basic idea about the performance curves. Navier–Stokes equation was derived in rotational form, and the equations to solve for wind turbines by **Fluent** were listed. A brief look at the CAD design of the blade was given. Then, the complete procedure to perform an ANSYS simulation was explained, including **Fluent** Setup. Then, the results were presented graphically and numerically, and inferences were drawn. Finally, verification and validation were performed, verifying the simulation's correctness.

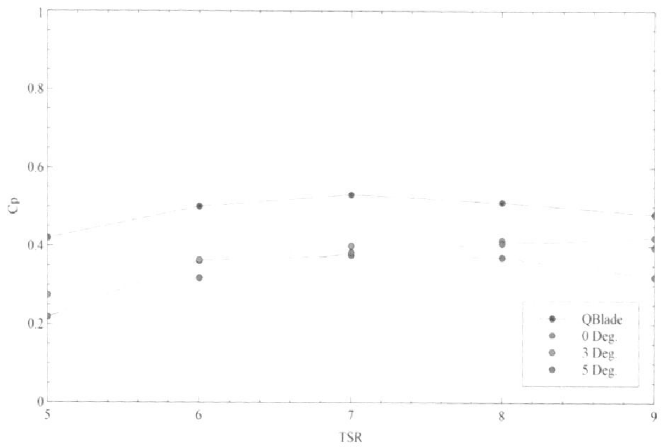

Figure 15.
Cp vs TSR for QBlade and CFD.

The main aim of the chapter was to provide the exact methodology to go from a theory in paper to results in **Fluent**. Every aspect of wind turbine design was covered in this chapter. The correctness of various theories and even QBlade & XFoils was compared to **Fluent**. Multiple simulations were performed in **Fluent**, with each 30 min simulation yielding a single row of tabular results. Finally, pitching inference was drawn for huge blades. Results were reported in a graphical and user-friendly manner without going into raw data. Verification and Validation, the most tricky and essential step, was successful for the simulation, and the errors were within bounds.

Acknowledgements

Major thanks for this chapter, as well as my inspiration, goes to Rajesh Bhaskaran from Cornell University. Please note that I am not affiliated in any way with that university. I just attended an online course on "A Hands-on Introduction to Engineering Simulations" in eDX in 2019. Since then, I was always fascinated by these giant machines and their curved blades.

Furthermore, no small part goes to the Dell R710, Dual Hexa-Core 3.3 GHz Xeon Processor-based server that gave me the raw power to run Fluent. Without it, so many tabular entries would have been impossible.

I am highly thankful to all my teachers, my parents, and others who have helped me along the way, without whom achieving this goal would have been impossible.

Abbreviations

TSR	Tip Speed Ratio
HAWT	Horizontal Axis Wind Turbine
VAWT	Vertical Axis Wind Turbine
CFD	Computational Fluid Dynamics
Cp	Power Coefficient
Ct	Torque Coefficient
a	Axial Induction Factor

BEM	Blade Element & Momentum Theory
NREL	National Renewable Energy Laboratory
a	Axial Induction Factor
u	Upstream Wind Velocity
v	Downstream Wind Velocity
y	Wind Velocity close to turbine
P_{wind}	Maximum Power present in the wind
T_{wind}	Maximum Torque that can be extracted from the wind
λ_r	Local Speed Ratio
ϕ	Angle made by the horizontal and the net velocity vector
α	Angle of Attack
β	Characteristic Twist Angle
c	Chord Length
B	Number of Blades
C_l	Lift Coefficient
C_d	Drag Coefficient
Ω	Radial Coefficient
ρ	Density of Air
t	Time Period
$\vec{v}_r$	Velocity Vector at that point
$\vec{r}$	Position Vector at that point
p	Pressure at that point
τ_r	Torsional Force at that point

Author details

Bharat Ramanathan†
Manipal Institute of Technology Alumnus, Navi-Mumbai, India

*Address all correspondence to: bharatcircdes@gmail.com

† These authors contributed equally.

References

[1] Clausen PD, Reynal F, Wood DH. Advances in Wind Turbine Blade Design and Materials. 2nd ed. Cambridge: Woodhead Publishing; 2013. pp. 413-430. DOI: 10.1533/9780857097286.3.413

[2] Schaffarczyk AP. Introduction to Wind Turbine Aerodynamics. 2nd ed. Switzerland: Springer; 2020. p. 522. DOI: 10.1007/978-3-030-41028-5

[3] Bouhelal A, Smaili A, Guerri O, Masson C. Comparision of BEM and Full-Navier Stokes CFD methods for prediction of aerodynamics performance of HAWT rotors. In: Proceedings of the International Renewable and Sustainable Energy Conference (IRSEC '17). Tangier, Morocco: IEEE; 2018. pp. 1-6

[4] Ignacio C, Quereda L. Design of a Two Bladed Wind Turbine. Luis Manuel MochÃ3n Castro: ICAI - Universidad Pontificia Comillas & NTNU – Norwegian University of Science and Technology. 2018

[5] Hamlaoui MN, Smaili A, Fellouah H. Improved BEM method for HAWT performance predictions. In: Proceedings of the Wind Energy and Applications in Algeria (ICWEAA '18). Algiers, Algeria: IEEE; 2018. pp. 1-6

[6] Koc E, Gunel O, Yavuz T. Comparison of QBlade and CFD results for small scaled horizontal axis wind turbine analysis. In: Proceedings of the Renewable Energy Research and Applications (ICRERA '17). Tangier, Morocco: IEEE; 2018. pp. 1-6

[7] El-Okda Y, Emeara MS, Abdelkarim N, Adref K, Hajjar HA. Performance of a small horizontal axis wind turbine with blade pitching. In: Proceedings of the Advances in Science and Engineering Technology International Conferences (ASET '20). Dubai, United Arab Emirates: IEEE; 2020. pp. 1-5

[8] Whoi. Rotating Coordinate Systems & Equations of Motion [Internet]. 2020. Available from: https://www.whoi.edu/cms/files/12.800 _Chapter_4_'06_25333.pdf. [Accessed: 07 July, 2022]

[9] cfdisrael. Navier-Stokes Equation in Moving Reference Frame (MRF) [Internet]. 2020. Available from: https://cfdisrael.blog/2021/09/22/navier-stokes-equation-in-moving-reference-frame-mrf/. [Accessed: Accessed: 14 July, 2022]

[10] Nouioua A, Dizene R. Modeling of flow around a wind rotor HAWT Application to the dynamic stall. In: Proceedings of the International Renewable and Sustainable Energy Conference (IRSEC '14). Ouarzazate, Morocco: IEEE; 2014. pp. 827-830

[11] Yang C, Lv X, Tong G, Song X. Aerodynamic optimization design and calculation of a 2MW horizontal axial wind turbine rotor based on blade theory and particle swarm optimization. In: Proceedings of the Asia-Pacific Power and Energy Engineering Conference (APPEEC '11). Wuhan, China: IEEE; 2011. pp. 1-6

[12] Zhang J, Zhou Z, Lei Y. Design and research of high-performance low-speed wind turbine blades. In: Proceedings of the World Non-Grid-Connected Wind Power and Energy Conference (WNWEC '09). Nanjing, China: IEEE; 2009. pp. 1-6

Chapter 9

Methods of the Perturbation Theory for Fundamental Solutions to the Generalization of the Fractional Laplaciane

Mykola Ivanovich Yaremenko

Abstract

We study the regularity properties of the solutions to the fractional Laplacian equation with perturbations. The Harnack inequality of a weak solution $u \in W_s^p(R^l)$ to the fractional Laplacian problem is established, and the oscillation of the solution to the fractional Laplacian is estimated. We show that let $1 \leq p < \infty$ and $s \in (0, 1)$, and let $u \in W_s^p$ be a weak solution to $Lu = 0 \quad \text{in} \ \Omega$, with the condition $u = f \quad \text{in} \ R^l \backslash \Omega$, where function f belongs to the Sobolev space $W_s^p(R^l)$. Then, the function $u \in W_s^p$ is locally Holder continuous and oscillation of the function satisfies the estimation $\underset{B(x_0, r)}{osc} \ u \leq C\delta^{\frac{sp}{p-1}} \sqrt[p]{(|u|^p)_{B(x_0,\rho)}} + +C\delta^{\frac{sp}{p-1}} \left\langle |\max\{u, 0\}|^{p-1} \frac{1}{|\cdot - x_0|^{l+sp}} \right\rangle_{R^l \backslash B(x_0,\rho)}^{\frac{1}{p-1}}$ holds for $\delta \in (0, \ 1)$ and for all r, ρ such that $0 < r < \rho$. Also, let $u \in W_s^p(R^l)$ be a weak solution to the boundary problem for the fractional Laplacian $\left((-\Delta)_p^s - b \cdot \nabla\right) u = 0 \quad \text{in} \ \Omega$, and on the boundary $u = f \quad \text{in} \ R^l \backslash \Omega$, where $1 \leq p < \infty$ and $s \in (0, \ 1)$, and let u be non-negative in a ball function with the center x_0 with the radius ρ, then the following estimation $\sup_{B(x_0, r)} \ u \leq C \inf_{B(x_0, r)} \ u + C\left(\frac{r}{\rho}\right)^{\frac{sp}{p-1}} \left\langle |\max\{-u, 0\}|^{p-1} \frac{1}{|\cdot - x_0|^{l+sp}} \right\rangle_{R^l \backslash B(x_0,\rho)}^{\frac{1}{p-1}}$ holds for r, ρ such that $0 < r < \rho$.

Keywords: Holder continuous, partial differential equation, perturbation method, analysis, calculus, mathematics, function, functional analysis, Harnack inequality, Sobolev space, singular integrals, differentiability class, semigroup, interpolation theorem, fractional Laplacian, partial differential equation, Holder inequality, Laplace operator, general solution, regularity, nonlocal model

1. Introduction

The fractional Laplacian is an integrodifferential operator, which can be defined by a formula

$$Lu(x) = c_{l,s}\left\langle \frac{u(x) - u(\cdot)}{|x - \cdot|^{l+2s}} \right\rangle, \tag{1}$$

where $c_{l,s} = \frac{4^s}{\pi^{\frac{l}{2}}} \frac{\Gamma\left(s+\frac{l}{2}\right)}{|\Gamma(-s)|}$ is a constant dependent on the dimension of the space $l > 2$. [1–3].

Let Ω be a bounded domain, we consider the integrodifferential problem

$$\begin{aligned} Lu &= 0 \quad \text{in } \Omega, \\ u &= f \quad \text{in } R^l \backslash \Omega, \end{aligned} \tag{2}$$

where function f belongs to a certain functional class, for instance, to the Sobolev space $W_s^p(R^l)$.

Similarly, we can consider the problem

$$\begin{aligned} (-\Delta)_s^p u &= 0 \quad \text{in } \Omega, \\ u &= f \quad \text{in } R^l \backslash \Omega, \end{aligned} \tag{3}$$

where the $(-\Delta)_s^p$ sits for the fractional p-Laplace operator. The weak solutions to these problems coincide with the class of the minimizers to the functionals

$$Fu = \tilde{c}_{l,s}\left\langle \left\langle \frac{|u(x) - u(y)|^p}{|x - y|^{l+sp}} \right\rangle_y \right\rangle_x, \tag{4}$$

which is defined over suitable Sobolev space.

Let us assume $u \in W_s^p(R^l)$ is a positive weak solution to (3) then inequality

$$\sup_K u(t - \tau, \cdot) \le C \ \inf{}_K u(t, \cdot) \tag{5}$$

holds for any compact set $K \subset \Omega$ and a positive constant C depends only on K, τ, t, s, p.

The fractional Laplace operator is intrinsically connected with the fractional Sobolev spaces $W_s^p(R^l)$ or, more specifically, $W_s^p(R^l)$ can be defined by using the fundamental solution of the fractional Laplace operator. So, for any $s \in (0, 1)$ and any $p \in [1, \ \infty)$, the set W_s^p of all functions u such that

$$W_s^p = \left\{ u \in L^p(R^l) \ : \ \frac{|u(x) - u(y)|}{|x - y|^{\frac{l}{p}+s}} \in L^p(R^l \times R^l) \right\} \tag{6}$$

is called the fractional Sobolev space, which can be equipped with its natural norm

$$\|u\|_{W_s^p(R^l)} = \left(\langle |u|^p \rangle + \left\langle \left\langle \frac{|u(x) - u(y)|^p}{|x - y|^{l+sp}} \right\rangle \right\rangle \right)^{\frac{1}{p}}. \tag{7}$$

Employing the perturbation theory, we can consider the fractional Laplace operator $(-\Delta)_s^p$ with the perturbation b, in the form

$$\Lambda \equiv (-\Delta)_s^p + b \cdot \nabla, \tag{8}$$

where $b(x) = c|x|^{-2s}x$. Since the vector b at $x = 0$ and at $x = \infty$ indicates a stronger singular attitude than permitted by the definition of the Kato classes, this vector does not belong to any Kato class, and the standard upper estimation of its heat kernel $e^{-t\Lambda}(x, y)$ via the heat kernel of the correspondent Laplacian operator is not valid in this case. However, the weighted estimations are still holding.

The integral inequality

$$\left\langle \frac{|\nabla u|^p}{|x|^p} \right\rangle \leq const(l, p)\langle |\Delta u|^p \rangle \tag{9}$$

holds for any $u \in C_0^\infty(R^l), \quad l > 2, p \geq 2$, and the constant in the right part depends only on the dimension of the Euclidian space and on the convexity of the functional space. This estimation can be generalized to abstract spaces with convex norms. If in (9) we take $p = 2$, then (9) becomes a well-established Hardy-Relich inequality with a sharp constant in the form

$$\left\langle \frac{|\nabla u|^2}{|x|^2} \right\rangle \leq c(l, p)\left\langle |\Delta u|^2 \right\rangle, \tag{10}$$

which can be proven by the methods developed in [4, 5].

Applying arguments of the perturbation theory, the operator $\Lambda \equiv (-\Delta)_s^p + b \cdot \nabla$ can be considered a perturbation of the fractional Laplace operator $(-\Delta)_s^p$ and utilizing the Duhamel formula the upper and lower bounds can be easily proven. There is a limiting constant $\tilde{k} > 0$ such that the contraction semigroup $\exp(-t\Lambda)(x, y)$ exists.

Let us consider the following example:

$$\frac{\partial}{\partial t}u = \sum_{k,j=1,\ldots,l} \nabla_k a_{kj}(x)\nabla_j u - \sum_{k=1,\ldots,l} b_k(t, x)\nabla_k u$$

under the integral condition of perturbation

$$\int_{R_+} \left\langle b(t,\cdot) \cdot a^{-1}(t,\cdot) \cdot b(t,\cdot)|\varphi(t,\cdot)|^2 \right\rangle dt \leq$$

$$\leq C\int_{R_+} \langle a(t,\cdot) \cdot \nabla\varphi(t,\cdot), \nabla\varphi(t,\cdot)\rangle dt + M\int_{R_+} \langle \varphi(t,\cdot), \varphi(t,\cdot)\rangle dt,$$

where $C < 4$ and $M < \infty$. So, b can be a function such that

$$\sum_{k=1,\ldots,l} b_k^2(t, x) \leq \iota^2 C\left(\frac{l-2}{2}\right)^2 \frac{1}{|x|^2} + M\frac{1}{|t|}\left(\ln\left(e + \frac{1}{|t|}\right)\right)^{-\frac{3}{2}}.$$

If the matrix a_{ij} is diagonal, then the differential operator is $-\Delta + b \cdot \nabla$, and $b = \frac{l-2}{2}\sqrt{\beta}\frac{x}{|x|^2}$ and for some $0 < \beta < 4$.

Let us consider the elliptic equation

$$a\circ d^2u\equiv\sum_{i,j=1}^{l}a_{ij}\frac{\partial}{\partial x_i}\frac{\partial}{\partial x_j}u=0,$$

where the matrix a is $a_{ij}=\delta_{ij}+b\frac{x_ix_j}{|x|^2}$, $b=-1+\frac{l-1}{1-\chi}$, $\chi<1$, $l\geq 3$. We calculate matrices

$$\nabla a=b(l-1)\frac{x}{|x|^2},\ (a_{ij})^{-1}=\delta_{ij}-\frac{b}{b+1}\frac{x_ix_j}{|x|^2},$$

and we have for the multiplication of the gradients of the matrices

$$\nabla a\circ a^{-1}\circ\nabla a=(1+b)^{-1}\left(\frac{l-1}{|x|}\right)^2$$

then we have

$$\langle\nabla\varphi\circ a\circ\nabla\varphi\rangle\geq(1+b)\frac{l-2}{2}\left\|\frac{\varphi}{|x|}\right\|_2^2\quad\forall\varphi\in W_1^2(R^l),\ l\geq 3,$$

so, if $\beta=4\left(1+\frac{\chi}{l-2}\right)^2$ then $\nabla a\circ a^{-1}\circ\nabla a\in PK_\beta(A)$ with the constant $c(\beta)=0$, for $\beta<4$ it is necessary $\chi\in(-2(l-2),\ 0)$..

Let us assume that $u(|x|=1)=1$. As the solutions, we can consider two functions: the first is $u\equiv 1$ - tautological constant and the second is $u=|x|^\chi$. If parameter $\chi=-\frac{l-2}{s}$ then $\beta=4\left(1+\frac{\chi}{l-2}\right)^2$ and $\beta\leq 4$ for $p>s$ in the ball $K_1(0)$ function $u=|x|^\chi\in L^p(K_1(0))$ on another hand must hold the following estimation

$$\|\exp(-t\Lambda_p)\|_{p\to s}\leq C\exp\left(\frac{c(\beta)t}{\sqrt{\beta}}\right)t^{\frac{-(s-p)l}{2ps}},\quad\frac{2}{2-\sqrt{\beta}}<p<s\leq\infty,$$

where semigroup $\exp(-t\Lambda_p)$ is generated by a linear operator $\Lambda_p=A+b(l-1)\frac{x}{|x|^2}\bullet\nabla$. That means $|x|^\chi\in L^{\frac{pl}{l-2}}(K_1(0))$ but it is impossible because $|x|^\chi\notin L_{loc}^{\frac{pl}{l-2}}K_1(0)$ so the function $|x|^\chi$ cannot be a solution and there is only one trivial solution. If $\beta>4$, then the equation $a\circ d^2u=0$ always has two bounded solutions. Parallel with this equation, we can consider a Cauchy problem for a parabolic equation with the same differential operator. Let us assume that the linear operator $-\Lambda_p\supset\nabla a\nabla-b\nabla$ defines over $D(A_p)$ generates holomorph semigroup in $L^p\left(R^l,d^lx\right)$-space. Let $b\circ a^{-1}\circ b\in PK_\beta(A)$, we denote $b_n=\chi_nb$, where χ_n is an indicator of $\{x\in R^l:\ (b\circ a^{-1}\circ b)(x)\leq n\}$ and $\lim_{n\to\infty}\exp(-t\Lambda_p(b_n))=\exp(-t\Lambda_p(b))$ uniformly at $t\in[0,\ 1]$. If $\beta<1$, $p\in\left[\frac{2}{2-\sqrt{\beta}},\ \infty\right[$ then there is C_0 - contraction semigroup, which is generated by the operator $A+b\nabla$ and the estimates

$$\|\exp(-t\Lambda_p)\|_{p\to p}\leq\exp\left(\frac{c(\beta)t}{p-1}\right),$$

then we estimate the operator norm by the exponential function

$$\| \exp(-t\Lambda_p) \|_{p\to s} \le C \exp\left(\frac{c(\beta)t}{\sqrt{\beta}}\right) t^{\frac{-(s-p)l}{2ps}}, \quad \frac{2}{2-\sqrt{\beta}} < p < s \le \infty$$

hold for $1 \le \beta < 4$, $p < s \in \left[\frac{2}{2-\sqrt{\beta}}, \ \infty\right[$, operator sum $A + b\nabla$ cannot be defined correctly, however, semigroup exists and can be defined as a limit $\exp(-t\Lambda_p(b)) \equiv \lim_{n\to\infty} \exp(-t\Lambda_p(b_n))$, $t \ge 0$ in this case it is a definition of the semigroup [6].

Let us remark that for any smooth enough function $f(x)$, $x \in R^l$, we can write the equations

$$\begin{aligned}
(-\Delta)_{\frac{1-b}{2}} f(x) &= C\left\langle \frac{f(x)-f(\cdot)}{|x-\cdot|^{l+1-b}} \right\rangle = \\
&= C \lim_{t\to 0} \left\langle \frac{f(x)-f(\cdot)}{\left(|x-\cdot|^2 + (1-b)^2 |t|^{\frac{2}{1-b}}\right)^{\frac{l+1-b}{2}}} \right\rangle = \\
&= \lim_{t\to 0} \frac{1}{t} \langle \hat{K}(t, \cdot - x)(f(x) - f(\cdot)) \rangle \\
&= \lim_{t\to 0} \frac{u(t,x) - u(0,x)}{t} \equiv \frac{\partial}{\partial t} u(0,x) \equiv u_t(0,x),
\end{aligned} \tag{11}$$

where the function $\hat{K}(t,x) = C(l,b) \frac{t^{1-b}}{\left(|t|^2 + |x|^2\right)^{\frac{l+1-b}{2}}}$ is the fundamental solution to the associated extension problem. More precisely, the results, which concern the fractional Laplace problems, can be applied to the extension problem in the following form. A function $u : [0,\infty) \times R^l \to R$ is a solution to the initial problem

$$\begin{aligned}
&Div\left(t^b \nabla u\right) = 0 \quad \text{in } R^l, \\
&u(t,0) = f(x), \quad x \in R^l,
\end{aligned} \tag{12}$$

or in expanded form

$$\begin{aligned}
&(-\Delta)_s u + \frac{b}{t} u_t + u_{tt} = 0 \quad \text{in } R^l, \\
&u(t,0) - f(x), \quad x \in R^l.
\end{aligned} \tag{13}$$

Below, we are going to construct the semigroup of contraction, which generator coincides with the realization Λ of the operator $(-\Delta)^s - b \cdot \nabla$ in $L^p(R^l)$, $1 \le p < \infty$, $l > 2$, where the vector $b(x) = c|x|^{-2s} x$ is singular; and prove the Harnack inequality for a weak solution to the boundary problem $\left((-\Delta)_p^s - b \cdot \nabla\right) u = 0$, outside boundary $u = f$, and presuming $u \in W_s^p(R^l)$ is nonnegative in a ball with the center in point x_0 with a radius ρ for all r, ρ such that $0 < r < \rho$.

Let us denote

$$(|u|^p)_{B(x_0,\rho)} = \frac{1}{mes(B(x_0,\rho))} \int_{B(x_0,\rho)} |u(y)|^p dy$$

then we can formulate the next theorem.

Theorem 1. Let $1 \le p < \infty$ and $s \in (0,1)$, and let $u \in W_s^p$ be a weak solution to (2), where its fundamental solution satisfies condition (15).

Then, the function $u \in W_s^p$ is locally Holder continuous and oscillation of the function satisfies the estimation

$$\underset{B(x_0,r)}{osc} \quad u \le C\delta^{\frac{sp}{p-1}} \sqrt[p]{(|u|^p)_{B(x_0,\rho)}} +$$

$$+C\delta^{\frac{sp}{p-1}} \left\langle |\max\{u,0\}|^{p-1} \frac{1}{|\cdot - x_0|^{l+sp}} \right\rangle_{R^l \backslash B(x_0,\rho)}^{\frac{1}{p-1}}$$

holds for $\delta \in (0, \ 1)$ and for all r, ρ such that $0 < r < \rho$.

There exist extensive literature dedicated to the partial differential fractional Laplacian operator, general questions can be found in [4, 7–9], a wide review is presented in [4], an interesting approach to nonlinear heat equations in modulation spaces and Navier-Stokes equations can be found in [10]; some aspects of weights inequalities are described in [2, 11]; fractional Laplacian is considered in [2, 12], the list of selected works consists of 29 works [1–29]. In the recent works [1, 2], authors proved sharp two-sided estimations on the heat kernel of the fractional Laplacian with the perturbation of drift having critical-order singularity, also authors show that the operator with the heat kernel of the fractional Laplacian can be expressed as a Feller generator so that the probability measures uniquely determined by the Feller semigroup admits description as weak solutions to the corresponding SDE.

2. The semigroup generated by $\boldsymbol{\Lambda \equiv (-\Delta)^s - b \cdot \nabla}$, $\boldsymbol{b(x) = c|x|^{-2s}x}$ in $\boldsymbol{L^p, \ 1 \le p < \infty}$

Let us introduce the following mollifiers:

$$b_\varepsilon(x) = c(|x| + \varepsilon)^{-2s}x$$

and

$$\Lambda_\varepsilon \equiv (-\Delta)^s - b_\varepsilon \cdot \nabla,$$

with the domain defined as

$$D(\Lambda_\varepsilon) \equiv (1 + (-\Delta)^s)^{-1} L^p$$

for small positive numbers $\varepsilon > 0$.

Since the following inequality

$$|\nabla(\varsigma + (-\Delta)^s)| \le M(l,s)(\varsigma + (-\Delta)^s)^{-\frac{2s-1}{2s}}$$

holds for all complex numbers such that

$$Re\varsigma > 0,$$

the operator norm $\left\| b_\varepsilon \cdot \nabla (\varsigma + (-\Delta)^s)^{-1} \right\|_{L^p \to L^p}$ is bounded by

$$M(\|b_\varepsilon\|_\infty)(Re\varsigma)^{-\frac{2s-1}{2s}}.$$

So, the Liouville-Neumann series for resolvents converge in L^p and the inequality

$$\left\| (\varsigma + (-\Delta)^s - b_\varepsilon \cdot \nabla)^{-1} \right\|_{L^p \to L^p} \le M(\varepsilon)|\varsigma|^{-1}$$

holds for $Re\varsigma > c(\varepsilon)$. Thus, according to the Hille-Kato perturbation theorem, the operator $-\Lambda_\varepsilon$, $\varepsilon > 0$ generates a holomorphic semigroup.

The next step is to show that there is the strong L^p limit

$$\lim_{\varepsilon \downarrow 0} e^{-t\Lambda_\varepsilon} \overset{def}{=} e^{-t\Lambda}, L^p, \quad 1 \le p < \infty$$

which defines a contraction continuous semigroup in L^p.

First, let us establish that a discretization of the set $\{e^{-t\Lambda_\varepsilon}\}_\varepsilon$ satisfies the Cauchy condition for at least one Lebesgue space. Let us compose the integral identity

$$\left\langle \left(e^{-T\Lambda_{\varepsilon(n)}} - e^{-T\Lambda_{\varepsilon(m)}}\right)^2 \right\rangle + \int_0^T \left\langle \left((-\Delta)^{\frac{s}{2}}\left(e^{-T\Lambda_{\varepsilon(n)}} - e^{-T\Lambda_{\varepsilon(m)}}\right)\right)^2 \right\rangle dt -$$

$$Re \int_0^T \left\langle b_{\varepsilon(n)} \nabla \left(e^{-T\Lambda_{\varepsilon(n)}} - e^{-T\Lambda_{\varepsilon(m)}}\right), e^{-T\Lambda_{\varepsilon(n)}} - e^{-T\Lambda_{\varepsilon(m)}} \right\rangle dt -$$

$$Re \int_0^T \left\langle \left(b_{\varepsilon(n)} - b_{\varepsilon(m)}\right) \nabla \left(e^{-T\Lambda_{\varepsilon(m)}}\right), e^{-T\Lambda_{\varepsilon(n)}} - e^{-T\Lambda_{\varepsilon(m)}} \right\rangle dt = 0,$$

so that we obtain

$$\left\langle \left(e^{-T\Lambda_{\varepsilon(n)}} - e^{-T\Lambda_{\varepsilon(m)}}\right)^2 \right\rangle + \int_0^T \left\langle \left((-\Delta)^{\frac{s}{2}}\left(e^{-T\Lambda_{\varepsilon(n)}} - e^{-T\Lambda_{\varepsilon(m)}}\right)\right)^2 \right\rangle dt +$$

$$c\frac{l - 2s}{2} \int_0^T \left\langle (|x| + \varepsilon)^{-2s}, \left(e^{-T\Lambda_{\varepsilon(n)}} - e^{-T\Lambda_{\varepsilon(m)}}\right)^2 \right\rangle dt \le$$

$$\int_0^T \left| \left\langle \left(b_{\varepsilon(n)} - b_{\varepsilon(m)}\right) \nabla \left(e^{-T\Lambda_{\varepsilon(m)}}\right), e^{-T\Lambda_{\varepsilon(n)}} - e^{-T\Lambda_{\varepsilon(m)}} \right\rangle \right| dt.$$

Since

$$\int_0^T \left| \left\langle \left(b_{\varepsilon(n)} - b_{\varepsilon(m)}\right) \nabla \left(e^{-T\Lambda_{\varepsilon(m)}}\right), e^{-T\Lambda_{\varepsilon(n)}} - e^{-T\Lambda_{\varepsilon(m)}} \right\rangle \right| dt \overset{m,\ n \to \infty}{\to} 0$$

we have the following limit

$$\left\| e^{-T\Lambda_{\varepsilon(n)}} - e^{-T\Lambda_{\varepsilon(m)}} \right\|_2^2 \overset{m,\ n \to \infty}{\to} 0$$

uniformly at $T \in [0,\ 1]$.

Thus, the discretization of set $\{e^{-t\Lambda_\varepsilon}\}_\varepsilon$ is a Cauchy sequence in $L^\infty([0,\ 1],\ L^2)$, and applying the contraction of $e^{-t\Lambda_\varepsilon}$, we estimate

$$\|e^{-t\Lambda}u\|_2 \le \|u\|_2,$$

besides, we have the equality

$$\lim_{\varepsilon\downarrow 0}\|e^{-t\Lambda_\varepsilon}u - e^{-t\Lambda}u\|_2 = 0, \quad u\in L^2,\ t\in[0,\ 1]$$

and applying group property, we have a continuous semigroup of contraction for $t\ge 0$. So, the continuous semigroup of the contraction is constructed in L^2.

Lemma 1. *Let the set of functions $\{f_a(x)\}$ converges in measure to function f then following estimation*

$$\langle f\rangle_E \le \sup_a\langle f_a\rangle_E$$

holds.

Now, let us take $p\in[1,\ \infty)$, then, from the lemma follows estimation

$$\|e^{-t\Lambda}u\|_p \le \|u\|_p, \quad t\ge 0$$

for any $u\in C_0^\infty$. Next, by continuity we extend the semigroup from C_0^∞ over L^p so the contraction semigroup can be defined as L^p- closure of $e^{-t\Lambda}$, thus, there is a L^p-strong limit

$$e^{-t\Lambda} = strong - L^p - \lim_{\varepsilon\downarrow 0}e^{-t\Lambda_\varepsilon}, \qquad t\ge 0,$$

which defined continuous semigroup of the contraction in L^p.

Thus, the contraction semigroup $e^{-t\Lambda}$, $t\ge 0$ in L^p can be defined as a strong limit of holomorphic semigroups $e^{-t\Lambda_\varepsilon}$. Under accepted assumptions, for all $1\le p\le q$, the semigroup $e^{-t\Lambda}$ satisfies natural conditions on its growth

$$\|e^{-t\Lambda}\|_{p\to q} \le C(l)t^{-\frac{l}{2}\left(\frac{1}{p}-\frac{1}{q}\right)}, \quad t\ge 0.$$

This estimation can be deduced from the next inequality

$$\begin{aligned}
&\frac{1}{p}\left\langle\left(\nabla e^{-T\Lambda_\varepsilon}u\right)^p\right\rangle+\\
&+\frac{4(p-1)}{p^2}\int_0^T\sum_i\left\langle\left((-\Delta)^{\frac{s}{2}}\left(\nabla_i e^{-T\Lambda_\varepsilon}u|\nabla e^{-T\Lambda_\varepsilon}u|^{\frac{p-2}{2}}\right)\right)^2\right\rangle dt+\\
&+c\frac{l-2s-p}{p}\int_0^T\left\langle(|x|+\varepsilon)^{-2s}|\nabla e^{-T\Lambda_\varepsilon}u|^p\right\rangle dt+\\
&+2sc\int_0^T\left\langle(|x|+\varepsilon)^{2s-2}|x\nabla e^{-T\Lambda_\varepsilon}u|^2|\nabla e^{-T\Lambda_\varepsilon}u|^{p-2}\right\rangle dt\le\\
&\le\frac{1}{p}\langle|\nabla u|^p\rangle
\end{aligned}$$

that holds for all $T>0$.

3. The Harnack inequality

For any measurable set $E \subset R^l$ and any integrable function f, we denote the mean value

$$(f)_E = \frac{1}{mes(E)} \int_E f(y)dy$$

where $mes(E)$ is the Lebesgue measure of the set $E \subset R^l$.

Let us assume that $\Omega \subset R^l$ is a bounded open set. We are going to consider the Harnack inequality for a weak solution to the following differential problem with the bounded condition

$$\begin{aligned} \Lambda u &= 0 \quad \text{in } \Omega, \\ u &= f \quad \text{in } R^l \backslash \Omega, \end{aligned} \tag{14}$$

where $\Lambda \equiv (-\Delta)^s - b \cdot \nabla$ acts on L^p, $1 \leq p < \infty$ under the condition that its kernel satisfies the following inequality

$$\frac{\mu}{|x-y|^{l+sp}} \leq K(x, \ y) \leq \frac{\lambda}{|x-y|^{l+sp}} \tag{15}$$

for almost all $x, y \in R^l$, $|x-y| \leq 1$ and some μ, λ such that $0 < \mu \leq \lambda < \infty$.

The general information can be found in [3, 18]. By the standard method, the following statements (1–3) can be proven.

Statement 1. *Assuming that* $u \in W_s^p$ *is a weak solution to* (14) *and nonnegative in a ball with the center in point* x_0 *with radius* ρ. *Then, the following inequality*

$$r^{\frac{sp}{p-1}} \left\langle |\max\{u, 0\}|^{p-1} \frac{1}{|\cdot - x_0|^{l+sp}} \right\rangle_{R^l \backslash B(x_0, r)}^{\frac{1}{p-1}}$$

$$\leq C \sup_{B(x_0, r)} u + C\left(\frac{r}{\rho}\right)^{\frac{sp}{p-1}} \left\langle |\max\{-u, 0\}|^{p-1} \frac{1}{|\cdot - x_0|^{l+sp}} \right\rangle_{R^l \backslash B(x_0, \rho)}^{\frac{1}{p-1}}$$

holds for $0 < r < \rho$.

Statement 2. *Assuming that* $u \in W_s^p$ *is a weak solution to* (14) *and nonnegative in a ball with the center in point* x_0 *with radius* ρ. *Then, for all* r, ρ *such that* $0 < r < \rho$, *the following inequality*

$$\sqrt[\delta]{(f^\delta)_{B(x_0, r)}} \leq C \inf{}_{B(x_0, r)}\ u +$$

$$+ C\left(\frac{r}{\rho}\right)^{\frac{sp}{p-1}} \left\langle |\max\{-u, 0\}|^{p-1} \frac{1}{|\cdot - x_0|^{l+sp}} \right\rangle_{R^l \backslash B(x_0, \rho)}^{\frac{1}{p-1}}$$

holds for any $\delta \in (0, \ 1)$.

Statement 3. *Assuming that $u \in W_s^p$ is a weak solution to* (14) *and nonnegative in a ball with the center in point x_0 with radius ρ. Then, for all r, ρ such that $0 < r < \rho$, the following inequality*

$$\sup_{B(x_0,r)} u \le C\varepsilon^{-\frac{p-1}{sp^2}l} \sqrt[p]{((\max\{u,0\})^p)_{B(x_0,\rho)}} +$$
$$+C\varepsilon r^{\frac{sp}{p-1}} \left\langle |\max\{u,0\}|^{p-1} \frac{1}{|\cdot - x_0|^{l+sp}} \right\rangle_{R^l \backslash B(x_0,r)}^{\frac{1}{p-1}}$$

holds for $\delta \in (0, \ 1)$.

Now, we can prove the next theorem.

Theorem 2. Let $u \in W_s^p(R^l)$ be a weak solution to the problem

$$\begin{aligned} \left((-\Delta)_p^s - b \cdot \nabla\right)u &= 0 \quad \text{in } \Omega, \\ u &= f \quad \text{in } R^l \backslash \Omega, \end{aligned} \tag{16}$$

where $1 \le p < \infty$ and $s \in (0, \ 1)$. Assuming $u \in W_s^p(R^l)$ is nonnegative in a ball with the center in point x_0 with the radius ρ, then

$$\sup_{B(x_0,r)} u \le C \inf_{B(x_0,r)} u + C\left(\frac{r}{\rho}\right)^{\frac{sp}{p-1}} \left\langle |\max\{-u,0\}|^{p-1} \frac{1}{|\cdot - x_0|^{l+sp}} \right\rangle_{R^l \backslash B(x_0,\rho)}^{\frac{1}{p-1}}$$

for r, ρ such that $0 < r < \rho$.

Proof. From statement 3, we can write the estimation

$$\sup_{B\left(x_0,\frac{r}{2}\right)} u \le C\varepsilon^{-\frac{p-1}{sp^2}l} \left(((\max\{u,0\})^p)_{B(x_0,\rho)}\right)^{\frac{1}{p}} +$$
$$+\tilde{C}\varepsilon r^{\frac{sp}{p-1}} \left\langle |\max\{u,0\}|^{p-1} \frac{1}{|\cdot - x_0|^{l+sp}} \right\rangle_{R^l \backslash B\left(x_0,\frac{r}{2}\right)}^{\frac{1}{p-1}},$$

and using statement 1, we have the following inequality

$$r^{\frac{sp}{p-1}} \left\langle |\max\{u,0\}|^{p-1} \frac{1}{|\cdot - x_0|^{l+sp}} \right\rangle_{R^l \backslash B\left(x_0,\frac{r}{2}\right)}^{\frac{1}{p-1}}$$
$$\le C \sup_{B(x_0,r)} u + \tilde{\tilde{C}} r^{\frac{sp}{p-1}} \rho^{-\frac{sp}{p-1}} \left\langle |\max\{-u,0\}|^{p-1} \frac{1}{|\cdot - x_0|^{l+sp}} \right\rangle_{R^l \backslash B(x_0,\rho)}^{\frac{1}{p-1}}.$$

So, we obtain the estimation

$$\sup_{B\left(x_0,\frac{r}{2}\right)} u \le C\varepsilon \sup_{B(x_0,r)} u + C\varepsilon^{-\frac{p-1}{sp^2}l} \left(((\max\{u,0\})^p)_{B(x_0,\rho)}\right)^{\frac{1}{p}}$$
$$+C\varepsilon r^{\frac{sp}{p-1}} \left\langle |\max\{u,0\}|^{p-1} \frac{1}{|\cdot - x_0|^{l+sp}} \right\rangle_{R^l \backslash B(x_0,r)}^{\frac{1}{p-1}}.$$

Choosing $\frac{1}{2} < \eta < \tilde{\eta} < 1$ applying the standard argument and the Young inequality, we obtain

$$\sup_{B(x_0, r\tilde{\eta})} u \leq \frac{1}{2} \sup_{B(x_0, r\eta)} u + \tilde{c} \sqrt[\delta]{\left(f^{\delta}\right)_{B(x_0,r)}} +$$

$$+C\left(\frac{r}{\rho}\right)^{\frac{sp}{p-1}} \left\langle |\max\{-u, 0\}|^{p-1} \frac{1}{|\cdot - x_0|^{l+sp}} \right\rangle_{R^l \backslash B(x_0,\rho)}^{\frac{1}{p-1}},$$

now, iterating this argument and applying statement 2, we have proven Theorem 2.

Using Theorem 2, we write

$$\underset{B\left(x_0, \varepsilon^i \frac{r}{2}\right)}{osc} u \equiv \sup_{B\left(x_0, \varepsilon^i \frac{r}{2}\right)} u - \inf_{B\left(x_0, \varepsilon^i \frac{r}{2}\right)} u \leq$$

$$\leq C\left(\varepsilon^i \frac{r}{\rho}\right)^{\frac{sp}{p-1}} \sqrt[p]{\left(|u|^p\right)_{B(x_0,r)}} +$$

$$+C\left(\varepsilon^i \frac{r}{\rho}\right)^{\frac{sp}{p-1}} \left\langle |\max\{u, 0\}|^{p-1} \frac{1}{|\cdot - x_0|^{l+sp}} \right\rangle_{R^l \backslash B\left(x_0,\frac{r}{2}\right)}^{\frac{1}{p-1}}$$

for all $i \in N$, thus Theorem 2 is a consequence of Theorem 1.

4. Conclusion

This chapter is dedicated to studying of the fundamental solutions to the generalization of the fractional Laplaciane by the methods of the theory of perturbation. We establish the regularity properties of the solutions to the fractional Laplacian equation with perturbations of different types. The Harnack inequality of a weak solution in the Sobolev space to the fractional Laplacian problem is studied, and the oscillation of the solution to the fractional Laplacian is estimated.

Mathematics subject classification

46B70, 43A15, 43A22, 44A05, 44A10, 44A45

Author details

Mykola Ivanovich Yaremenko
Department of Partial Differential Equations, The National Technical University of Ukraine, Igor Sikorsky Kyiv Polytechnic Institute, Kyiv, Ukraine

*Address all correspondence to: math.kiev@gmail.com

References

[1] Kinzebulatov D, Madou KR. On admissible singular drifts of symmetric alpha-stable process. Preprint. 2020

[2] Kinzebulatov D, Semenov Yu, Szczypkowski K. Heat kernel of fractional Laplacian with Hardy drift via desingularizing weights. Journal of the London Mathematical Society. 2019;**104**(4):1861-1900

[3] Kinnunen J, Shanmugalingam N. Regularity of quasiminimizers on metric spaces. Manuscripta Mathematica. 2001; **105**:401-423

[4] Evans LC. Partial differential equations. In: Graduate Studies in Mathematics. Vol. 19. Providence, RI: American Mathematical Society; 1998

[5] Ghoussoub N, Moradifam A. Bessel pairs and optimal Hardy and Hardy-Rellich inequalities. Mathematische Annalen. 2011;**349**(1):1-57

[6] Moser J. A Sharp Form of an Inequality by N. Trudinger Indiana University Mathematics Journal. 1971; **20**(11):1077-1092

[7] Acerbi E, Mingione G. Gradient estimates for a class of parabolic systems. Duke Mathematical Journal. 2007;**136**:285-320

[8] Caffarelli LA. Interior a priori estimates for solutions of fully nonlinear equations. Annals of Mathematics. 1989; **130**(1):189-213

[9] Kato T. Perturbation Theory for Linear Operators. Berlin Heidelberg: Springer-Verlag; 1995

[10] Iwabuchi T. Navier-Stokes equations, and nonlinear heat equations in modulation spaces with negative derivative indices. Journal of Differential Equations. 2010;**248**:1972-2002

[11] Cruz-Uribe D, Fiorenza A. Weighted endpoint estimates for commutators of fractional integrals. Czechoslovak Mathematical Journal. 2007;**57**:153-160

[12] Xiong C. Comparison of Steklov eigenvalues on a domain and Laplacian eigenvalues on its boundary in Riemann manifolds. Journal of Functional Analysis. 2018;**275**:3245-3258

[13] Adimurthi K, Phuc NC. Global Lorentz, and Lorentz-Morrey estimates below the natural exponent for quasilinear equations. Calculus of Variations and Partial Differential Equations. 2015;**54**(3):3107-3139

[14] Beni A, Grochenig K, Okoudjou KA, Rogers LG. Unimodular Fourier multiplier for modulation spaces. Journal of Functional Analysis. 2007;**246**:366-384

[15] Cafarelli LA, Peral I. On W1,p estimates for elliptic equations in divergence form. Communications on Pure and Applied Mathematics. 1998; **51**(1):1-21

[16] Hamamoto N, Takahashi F. Sharp Hardy-Leray and Rellich-Leray inequalities for curl-free vector fields. arXiv: 1808.09614

[17] Ortiz-Caraballo C, Perez C, Rela E. Exponential decay estimates for singular integral operators. Mathematische Annalen. 2018;**357**:1217-1243

[18] Palatucci G, Castro AD, Kuusi T. Local behavior of fractional p-minimizers. Annales de Institut Henri Poincare, Analyse nonlineaire. 2016; **33**(5):1279-1299

[19] Polimeridis AG, Vipiana F, Mosig JR, Wilton DR. DIRECTFN: Fully numerical algorithms for high precision computation of singular integrals in Galerkin SIE methods. IEEE Transactions on Antennas and Propagation. 2013;**61**(6):3112-3122

[20] Wang Q, Xia C. Sharp bounds for the first non-zero Steklof eigenvalues. Journal of Functional Analysis. 2009;**257**: 2635-2644

[21] Stein EM. Singular integrals and differentiability properties of functions. In: Princeton Mathematical Series. Vol. 30. Princeton, N.J.: Princeton University Press; 1970

[22] Trudinger N. On Imbeddings into Orlicz Spaces and Some Applications. Indiana University Mathematics Journal. 1968;**17**(5):473-483

[23] Wang B, Huo Z, Hao C, Guo Z. Harmonic Analysis Method for Nonlinear Evolution Equations I, Hackensack. NJ: World Scientific; 2011

[24] Wang FY. Distribution dependent SDEs for Landau type equations. Journal of Stochastic Processes and their Applications. 2018;**128**:595-621

[25] Xia P, Xie L, Zhang X, Zhao G. Lq (Lp)-theory of stochastic differential equations. Stochastic Processes and their Applications. Vol. 130. No. 8. Elsevier; 2020. pp. 5188-5211

[26] Zhang QS. Gaussian bounds for the fundamental solutions of $\nabla(A\nabla u) + B\nabla u - ut = 0$. Manuscript Mathematica. 1997;**93**:381-390

[27] Zhang X. Stochastic homeomorphism flows of SDEs with singular drifts and Sobolev diffusion coefficients. Electronic Journal of Probability. 2011;**16**:1096-1116

[28] Zhang X, Zhao G. Singular brownian diffusion processes. Communications in Mathematics and Statistics. 2018;**6**: 533-581

[29] Zhang X, Zhao G. Communications in Mathematical Physics. 2021;**381**: 491-525

Chapter 10

The Analysis on the Effects of COMT, DRD2, PER3, eNOS, NR3C1 Functional Gene Variants and Methylation Differences on Behavoiral Inclinations in Addicts through the Decision Tree Algorythm

Inci Zaim Gokbay, Yasemin Oyaci and Sacide Pehlivan

Abstract

The aim of this study was to analyze the effects of Catechol-O-methyltransferase (COMT), Dopamine Receptor D2 (DRD2), Period Circadian Regulator 3 (PER3), Endothelial Nitric Oxide Synthetase (eNOS), Nuclear Receptor Subfamily 3 Group C Member 1 (NR3C1) functional gene variants on possible inclinations of the individuals with Substance Use Disorder (SUD) by using decision trees algorithm and to evaluate the similarities with former studies. The decision trees classification was structured by confirming the effects of genetic and epigenetic sequences of gene variants through 10-fold cross-validation under subtitles of the criminal history, continuum of substance use, former polysubstance abuse, attempted suicide, and inpatient treatment. Performance criteria were evaluated with the similarities of former studies' accuracy, sensitivity, and precision values. The branching structure of gene variants obtained by tree classification is consistent with the studies in the literature. Our study serves to be the first to show that there is a need for further comprehensive studies with data from different ethnic groups to increase the predictive accuracy rates and to state that machine learning may guide in predicting the effect of gene variants on behavior in the future.

Keywords: substance use disorder, COMT, DRD2, PER3, eNOS, NR3C1 gene variants, decision tree analysis, 10-fold cross validation

1. Introduction

The *adolescent* and *younger adult* periods defined by the World Health Organization (WHO) as the transition from childhood to adulthood play a vital role in the lives of individuals. These periods are mental and physical developmental processes in which tendency to substance abuse, nutritional disorders, mental problems, and risky behaviors are common. Gorker et al. [1] stated in their study that patients who referred to the child and adolescent psychiatry clinic were diagnosed with anxiety disorder, mood disorder, mental retardation, expulsion disorder, disruptive behavior disorder, borderline intellectual functioning, communication disorder, somatoform disorder, and tic disorders, respectively. It is stated in the aforesaid study that these diagnoses are accompanied by mental retardation, adjustment disorder, attention deficit, and hyperactivity, as well as substance use disorder (SUD). There are studies demonstrating that substances with addiction potential are preferred by adolescents and young people due to their euphoric effect, which is generally seen as a positive effect and to relieve negative effects such as pain, pain-reducing, stress-relieving, and relaxing [2]. However, chronic use of a limited number of classes of substances that begin with such justifications causes physical, psychological, and behavioral changes in humans. Substance Use Disorder (SUD) appears under the influence of multiple factors and the persistence is accompanied by these factors. Environmental factors including family, peer relations, neighborhood relations, and physical conditions of neighborhood and educational environment play various roles in addiction-related situations such as molecular pathways, cellular mechanisms, tolerance in addiction, recurrence of addiction, and substance seeking.

The family is closely involved with individual's developmental behavior disorders, with respect to its cultural roots and family attitudes. Biological effects can be classified under many factors, such as genetical, physiological, temperamental, and impulsive behavior tendencies. Nonetheless, family is the first social unit that the individual belongs to. The baby is born in a family and learns the first social rules from the family. Due to this reason, it's said that family has an impact on individuals both in biological and sociological personality development. Biological disposition—namely temperament—is transferred by genetical heritage while sociological disposition—namely character—is related with upbringing attitudes, attachment, cultural heritage. Studies have brought out that in SUD, having a substance user family member plays a crucial role in individual's life [3].

The SUD refers to tolerance and withdrawal that occurs after chronic use of the substance, which is a tool that an individual uses to cope with many factors that he calls negative effects especially between 11 and 16 years of age or to feel himself belonging to a group; however, behavioral addiction refers to uncontrollable, permanent use despite negative physical, psychological, social, or legal consequences. Bozkurt [4] shared the findings in their study that child-raising attitudes of the parents are effective. Among substance addicted individuals, 20.9% of the participants stated that they were exposed to physical violence and 40.9% stated that there was physical violence in the family. Furthermore, it was determined that 69.8% of the participants had a substance use disorder in their families. Ünal [5] stated in his study that participants had family members with SUD including fathers by 25%, siblings by 50%. The family is also the society in which the temperament structure of the individual, namely genetic tendencies, is also effective. An individual is born with genetic predispositions in the family. There are studies showing that individuals with COMT, DRD2, PER3, eNOS, NR3C1 functional gene variants are prone to develop MID.

The gene variant is a term used to describe the variation in the DNA sequence in the genome. The term variant may be used to describe a change that may be benign, pathogenic, or with an unknown significance.

DNA methylation is the reaction of covalent attachment of a methyl group from the 5-carbon of cytosine in a CpG dinucleotide to the structure, altering gene expression and altering the cell functions.

The aim of this study was to analyze the effects of COMT, DRD2, PER3, eNOS, NR3C1 functional gene variants and COMT, DRD2, and NR3C1 methylation status on the tendencies that have the potential detected in individuals with MID through decision trees algorithm. The criminal record history, continuum of substance use, former polysubstance abuse, attempted suicide, and inpatient treatment will be analyzed by using decision trees.

2. Materials and method

There are studies commenting on the association between statistical results and variables including sociological, psychological, neurological, and genetic fields. Statistical analyses are mathematical operations that process and summarize data based on probability and guide researchers who review the association between variables in this direction. Machine learning techniques are also based on probabilistic models, yet the outcome of these models provides prediction on mutually exclusive events in a widen event set. Regarding these features, machine learning studies have the ability to analyze and obtain a prediction in different perspectives about data. Nevertheless, despite technological developments and improvements in enhancements of accuracy rates of prediction results of these systems, there are limited studies that have been done about substance abuse.

The aim of this study designed in this context was to analyze the criminal record history, continuum of substance use, former polysubstance abuse, attempted suicide, and inpatient treatment by using decision trees and to discuss the association of the findings obtained with the literature.

2.1 Machine learning methods

Machine learning is the structure that includes learning in artificial intelligence applications. On the one hand, it can be defined as the whole of algorithms that imitate human intelligence, and on the other hand, do not need rules that people can interpret and enter manually.

Machine learning applications learn the desired task by assimilating the presented datasets, just as people learn the concepts they see and hear on their own. They can make predictions about the outcome of the new data entry that is out of the data they have learned over time. The training set used in machine learning is used in the machine learning process, and the test set is used in the prediction process. For example, in the design of a system that will ensure that an orchid is selected from a vase with different flowers and taken into a separate vase, as much data about the orchid genus as possible are included in the training set for learning, and other flowers such as height, color, leaf shape, color, curl, flower shape, color distribution, folds are selected. It is ensured that the distinguishable features can be created by the machine. After making these distinctive classes in the training set, the predictive ability of the machine is tested. In this dataset, called the test set, there are flowers in a vase with a new arrangement that the machine has not seen before. It is expected from the machine to find out if there are orchids in this newly encountered cluster and to take it (differentiate) if it is. Machine learning processes are very similar in principle with learning processes of human. In the developmental processes of people, learning is divided into behavioral and cognitive approaches and many sub-branches under it. Machine learning methods are also divided into branches within themselves such as supervised and unsupervised learning [6].

In the most general form, supervised learning is in which the relationship between input and output is learned by matching under the supervision of a supervisor. Unsupervised learning is learning by finding the regularities between the inputs entering the system without a supervisor and producing output. Problems solved using supervised learning are generally divided into classification and regression problems. The important thing in supervised learning methods is to include a target attribute in the dataset. Depending on the type of problem to be addressed, the type of target attribute can be of different type. For target attribute classification problems, there may be class labels, while for regression problems it may be a numerical value.

2.2 The classification method and the decision tree algorithm

The classification method, which is among the machine learning methods, is one of the commonly preferred methods, especially in the field of medicine. Learning in classification algorithms is based on learning and classifying the distribution form from the given training set. Support Vector Machine (SVM), Nonlinear Supporter Vector Machine, Naive Bayes Classifier, Decision Tree Classifier, and Nearest K-neighbor classification algorithms are examples. For instance, Zaim Gokbay et al. [7] presented a decision support system design in their study in order to support the diagnosis of endocrine disease. The classification rules used in the model were created depending on the investigation of visible changes, which has started along with complaints on the physical appearance as well as the laboratory results. The patient complaints were stated by the patient by filling in a questionnaire. Physical changes were investigated by the endocrinologist during the exam, and findings on the mandible evagination, skin cracks were clinically evaluated and entered into the system. Every three data entered into the system were used to classify the prediction model of three different endocrinological diseases. Each class represents a disease. The individual falls into a class in the sum of his answers to the questions, and it is concluded that he has the potential to have the disease indicated by that class. Your symptoms that cause you to come to the doctor serve as an indicator. The formation of such rules allows to make predictions in the next steps. There is the logic of separating data belonging to common features into certain classes in a dataset in classification methods. Numerous algorithms have been developed for this purpose. Examples including the entropy-based classifications, regression and decision trees, memory-based algorithms, Bayesian classifiers may be given.

In the decision tree classification method, the data are classified by separating from the root to the leaf. The if-then rule is implemented in this separation. If the condition is 1, then a chain-like condition 2 and then 3 are formed in order to establish a branching structure from root to leaf. The decision tree method was chosen for the classification performed in this study, because it is based on rules that may be understood by people, both visually and because of the convenience that would provide to multidisciplinary work in comparing the results with the literature.

The model was established with 10-fold cross-validation in the study. The success rates of the decision tree classes were interpreted through the accuracy, class recall, and class precision values, and the association of the sequences obtained with the literature was discussed.

2.3 Classification model performance criteria

The simplest and most common method used to measure the performance of classification models is the accuracy rate, precision, and recall rates.

$$(Accuracy - Rate) = \frac{TP + TN}{TP + FP + TN + FN} x100 \tag{1}$$

$$Class - Precision = \frac{TP}{TP + FP} x100 \tag{2}$$

$$\text{Class} - Recall = \frac{TP}{TP + FN} x100 \tag{3}$$

Parameters in formulas (1)-(3) are defined as true positive (TP), false negative (FN), true negative (TN), and false positive (FP) as follows;

- TP: The number of values actually in the positive class and predicted in the positive class.
- FN: The number of values actually in the positive class but in the negative class in prediction.
- FP: The number of values actually in the negative class but in the estimation positive class.
- TN: Number of values actually in the negative class and predicted in the negative class.

In other words, the accuracy refers to the percentage of samples classified as correct. The measure of how many of the positively predicted outputs are positively predicted is expressed as Recall, and the measure of how many of the positively predicted outputs are positive is expressed as Precision.

2.4 The dataset characteristics

This study was conducted with retrospective data of 211 male participants known to be addicted to at least one substance, obtained from studies completed and published with the approval of the ethics committee (2019/87) of the Ethics Committee for Clinical research within Istanbul Faculty of Medicine [8–10]. The average age of the individuals is 28.67, and the age varies between 18 and 51 years of age. The educational level of the individuals include 2 college graduates, 61 secondary school graduates, 49 primary school graduates, and 99 literate or illiterate participants. The marital status of the participants was as follows: 147 were single, 43 were married, 12 were divorced, and 9 were married but living separately. There were three students among the participants; 56 individuals are employed, whereas 152 individuals are unemployed. Individuals who use at least one of cigarettes and alcohol and use at least one of the addictive substances including cannabinoids, synthetic cannabinoids, cannabis, cocaine, ecstasy, and heroin were included in the study. The first age of start of the individuals for one of these substances varies between 10 and 30 years of age.

There is no missing information in the dataset used within the scope of the study. Therefore, the data were not exposed to a preliminary procedure. Descriptions of attributes, values, and variable names are presented in **Table 1**.

Description	Entry variable name	Type	Values
eNOS is an important mediator of cardiovascular homeostasis due to its role in nitric oxide (NO) production. Nitric oxide has well-known vascular effects, and significantly affects autonomic nervous system activity. A significant gene-environment interaction between eNOS and behavioral risk factors such as chewing tobacco and consuming alcohol was detected in a previous study. In particular, the "GG" genotype was associated with an increased risk of hypertension among individuals who use tobacco or consume alcohol [11]. It was concluded in a previous study that lower nitric oxide levels may increase the dopamine turnover or decrease the dopamine release [12]. The suggestion shared was that these two effects are not mutually exclusive and may appear together, and the reward mechanism based on the release of less usable dopamine in metabolism strengthens the addictive behavior by directing the individual to consume more cannabis.	eNOS- Intron 4a/b VNTR	Nominal	Values: AA (Data count: 128), BA (Data count: 66), BB (Data count: 17)
	eNOS-rs1799983	Nominal	Values: GG (Data count:129), GT (Data count:73) ,TT (Data count: 9)
PER3, which is located on chromosome 1p36.23, contains a polymorphic domain that expresses 4 or 5 copies of the 54-bp tandem repeat sequence (variable number tandem repeat, VNTR). This variation results in the addition/deletion of 18 amino acids and it is linked to sleep and mood disorders as well as circadian preference in humans [13]. Studies have shown a clear association between poor sleep patterns and a range of negative health behaviors such as substance use, suicide attempts, and unintentional injury [14].	PER3-rs57875989	Nominal	Values: 4R/4R (Data count:78), 4R/5R (Data count:97), 5R/5R (Data count:36)
Dopamine receptors, which are divided into two classes including D1-like receptors and D2-like receptors, regulate the effects of dopamine and dopamine components. The cyclic AMP (cAMP) is stimulated or suppressed in the regulation of adenylate cyclase activity which is the most important of these regulation mechanisms. The first study demonstrating that a dopaminergic gene encoding DRD2, the dopamine receptor, may show population variants that may predispose to alcoholism was performed by Blum ve ark. [15, 16]. Many studies were conducted especially on DRD2 after then, and it was stated that a certain form of the DRD2 gene may be associated with a seven-fold increase in susceptibility to alcohol abuse, and when environmental effects are taken into account, individuals carrying this gene are 60% prone to substance use [17] when compared to others.	DRD2-rs1799732	Nominal	Values: Ins/Ins (Data count:168), Ins/Del (Data count:41), Del/Del (Data count:2)
	DRD2-METHYLATION	Nominal	Values: PARTIAL (Data count:126), UNMETILE (Data count:85)

Description	Entry variable name	Type	Values
The COMT enzyme is responsible for degradation and elimination of dopamine neurotransmitters in the prefrontal cortex of the brain. The COMT enzyme and COMT gene functional variants which play a role in the metabolism of catecholamine and catecholamine-containing substances such as dopamine and are thereby important elements of the dopaminergic system have been the focus of many studies. Findings of the study conducted by Delisi et al. [18] suggest that COMT plays a role in cerebral areas that modulate self-regulation and expression of negative emotions, influencing antisocial personality disorder (ASPD) and delinquency.	COMT-METHYLATION	Nominal	Values: PARTIAL (Data count:143), UNMETHYLATED (Data count:68)
	COMT-rs4680	Nominal	Values: Val/Met (Data count:97), Val/Val (Data count:63), Met/Met (Data count: 51)
NR3C1 plays a critical role in HPA axis regulation and it is thereby considered as a possible cause of stress-related disorders. NR3C1 consists of 8 introns and 9 exons on chromosome 5q31-32, encoding the glucocorticoid receptor. Previous studies have reported that altered NR3C1 methylation may cause various psychopathologies including major depressive disorder, bipolar disorder, suicidal behavior, and substance use disorder [19–21].	NR3C1-rs41423247	Nominal	Values: CC (Data count: 139), GC (Data count:60), GG (Data count: 12)
	NR3C1-METHYLATION	Nominal	Values: PARTIAL (Data count:192), UNMETHYLATED (Data count:19)

Table 1.
Attribute descriptions, variable names, types, and values.

3. Findings

In the study, the presence of gene variants of eNOS- Intron 4a/b VNTR, eNOS-rs1799983, PER3-rs57875989, DRD2-rs1799732, COMT-rs4680, NR3C1- rs41423247 and DRD2, COMT, and NR3C1 gene methylation status, the criminal record history, continuum of substance use, former polysubstance abuse, suicidal behavior, and inpatient treatment were analyzed with decision trees, which are the classification algorithms. The accuracy, sensitivity, and precision performance rates of each model are presented in **Table 2**.

3.1 The criminal record history

The tree structure established in order to review the effect of eNOS-Intron 4a/b VNTR, eNOS-rs1799983, PER3-rs57875989, DRD2-rs1799732, COMT-rs4680, NR3C1-rs41423247 gene variants, and DRD2, COMT, and NR3C1 gene methylation states on tendency of decriminalization was presented in **Figure 1**; the effect of input variables (gene variants) on weight distribution is presented in **Table 3**. The tendency of delinquency was evaluated by questioning the forensic history of the participant. The existence of criminal history, the tendency delinquency, and absence of the tendency to delinquency were interpreted that this tendency has been more suppressed and there is not any tendency to action. Therefore, the decision tree structure was

Behavior tendency	Performance scales				
The Criminal Record History	Accuracy Rate	Recall and Precision Rates			
			True: Exists	True: Not Exists	Precision
	52.68 %	Prediction: Exists	71	55	47.06 %
		Prediction Not Exists	45	40	56.35 %
		Recall	42.11 %	61.21 %	
Continuum Of Substance Use	Accuracy Rate	Recall and Precision Rates			
			True: Continous	True: Not Continous	Precision
	49.76 %	Prediction: Continous	71	76	51.70%
		Prediction Not Continous	29	35	45.31%
		Recall	29%	68.47%	
Former Polysubstance Abuse	Accuracy Rate	Recall and Precision Rates			
			True: Exists	True: Not Exists	Precision
	51.21 %	Prediction: Exists	59	67	%46.83
		Prediction Not Exists	36	49	%57.65
		Hassasiyet (recall)	% 62.11	%42.24	
Suicidal Behavior	Accuracy Rate	Recall and Precision Rates			
			True: Exists	True: Not Exists	Precision
	65.00 %	Prediction: Exists	16	28	% 36.36
		Prediction Not Exists	46	141	% 72.46
		Recall	% 25.81	% 81.21	
Inpatient Treatment	Accuracy Rate	Recall and Precision Rates			
			True: Exists	True: Not Exists	Precision
	70.56 %	Prediction: Exists	142	46	% 75.53
		Prediction Not Exists	16	7	% 30.43
		Recall	% 89.87	% 13.21	

Table 2.
Behavioral tendencies and performance rates in classification of gene variants by decision tree method.

established as "Criminal Record History -Yes" or " Criminal Record History -No." There are 116 individuals in the Criminal Record History class of the dataset; however, the absence of criminal record history included 95 data. The tree root is established over the NR3C1 gene, as may be seen in **Figure 1**. The presence of CC genotype of NR3C1-rs41423247 following partial methylation of the NR3C1 gene is an effective sequence in predicting a tendency of an individual to delinquency.

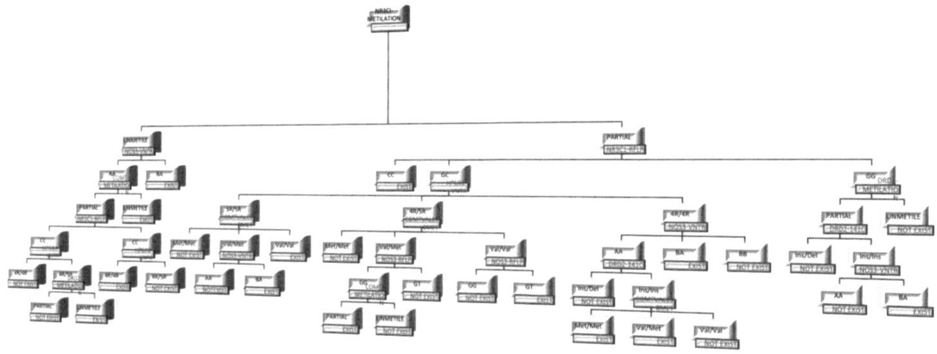

Figure 1.
The decision tree structure constructed with 10-fold cross-validation for the evaluation of gene variant effect of Criminal History.

Attribute	Average weight
COMT-rs4680	0,013
NR3C1- rs41423247	0,010
DRD2- METHYLATION	0,008
eNOS-rs1799983	0,007
NR3C1- METHYLATION	0,005
PER3-rs57875989	0,002
eNOS – Intron 4a/b VNTR	0,002
COMT- METHYLATION	0,001
DRD2-rs1799732	0,000

Table 3.
Average weight values of gene variants in the criminal record history within the context of knowledge acquisition.

The review of **Table 3** reveals that the attributes with the highest information gain are COMT-rs4680 with an average weight value of 0.013, and the lowest variable is COMT-METHYLATION with an average weight value of 0.001. This is concluded that the COMT-rs4680 variant contains the highest information gain on delinquency.

3.2 The continuum of substance use

The tree structure established to investigate the effect of the tendency toward continuous use of the substance without interruption in individuals with SUD. The weight distributions of the input variables (gene variants) on the output are presented in **Table 4**. The tendency for continuous use of the substance without interruption was evaluated by answers of the participants to the question "Do you use the substance intermittently?". The decision tree structure was created as "Intermittent" or "Continuous." The number of individuals who have declared intermittent substance use in the dataset was 100, whereas 111 individuals stated continuous substance use.

Attribute	Average weight
COMT- METHYLATION	0,018
COMT-rs4680	0,011
DRD2-rs1799732	0,010
NR3C1- rs41423247	0,002
DRD2- METHYLATION	0,001
NR3C1- METHYLATION	0,001
eNOS-rs1799983	0,001
eNOS-Intron 4a/b VNTR	0,001
PER3-rs57875989	0,000

Table 4.
Average weight values of gene variants in the trend to continuous substance use within the context of knowledge acquisition.

The tree root starts as part of COMT-METHYLATION in the form of methylated or non-methylated and forms a wide branching structure. In case of the partial methylation of COMT-METHYLATION, the branching continues through the DRD2-rs1799732 gene variant; however, if unmethylated, it continues with the NR3C1-rs41423247 gene variant.

The review of reveals that the variable with the highest information gain is COMT-METHYLATION with an average weight value of 0.018, and the lowest variables with an average weight value of 0.001 are DRD2-METHYLATION, NR3C1 METHYLATION, eNOS-rs1799983, and eNOS-Intron 4a/b VNTR. This is concluded that the COMT-METHYLATION variant contains the highest information gain on delinquency.

3.3 The former polysubstance abuse

Individuals with SUD may use more than one substance such as alcohol, cigarettes, cannabis, heroin, cocaine, toluene, ecstasy, etc. Combined use of at least two of cannabinoid, synthetic cannabinoid, ecstasy, heroin, cocaine, and toluene is investigated within the scope of this sub-assessment. In addition to any of the aforementioned substances, tobacco and/or alcohol use, the use was not included in the multiple substance use, because all of the participants already use these two substances in combination with the other substances mentioned. The reason for exclusion of this situation from the analysis is that it is clear that there will be no meaningful results.

The weight distributions of the input variables (gene variants) on the output are presented in **Table 5**. The decision tree structure class was created as "Polysubstance Use - Yes" or "Polysubstance Use - No." The number of people who declared combined use of at least two of the substances mentioned was 95 in the dataset, and the number of people who declared use of one substance was 116.

The tree root starts by COMT-rs4680 and forms a wide branching structure. The Met/Met and Val/Val genotype branches to NR3C1-METHYLATION and to Val/Met NR3C1- rs41423247 and continues. It is concluded that there is not any tendency to polysubstance use when the Met/Met genotype is unmethylated in the NR3C1-

Attributes	Average weight
COMT-rs4680	0,032
NR3C1- rs41423247	0,021
eNOS-rs1799983	0,018
DRD2-rs1799732	0,008
NR3C1-METHYLATION	0,005
PER3-rs57875989	0,003
COMT-METHYLATION	0,002
DRD2-METHYLATION	0,001
eNOS- İntron 4a/b VNTR	0,000

Table 5.
Average weight values of gene variants in the polysubstance use within the context of information gain.

METHYLATION branch; however, there is a tendency in partial methylation follow-up of the Val/Val genotype.

The review of **Table 5** reveals that the variable with the highest information gain is COMT-rs4680 with an average weight value of 0.032, and the lowest variable is DRD2-METHYLATION with an average weight value of 0.001. This is concluded that the COMT-rs4680 variant contains the highest information gain in multiple substance use.

3.4 The suicidal behavior

The individuals were asked whether they had attempted suicide at least once in order to evaluate the suicidal behavior in individuals with SUD. One-hundred and forty-nine individuals who had never attempted suicide were classified under "No Suicide Attempt," and 62 individuals who had at least one or more suicide attempts were classified under "Suicide Attempts."

The tree root starts by NR3C1-rs41423247 and forms a wide branching structure. It is concluded that suicidality is not seen in the GG genotype, and it branches to NR3C-METHYLATION in the GC and CC genotypes and continues. There is not any tendency when NR3C1-METHYLATION is unmethylated in the CC genotype; however, the same pathway shows the tendency in the GC genotype.

The review of **Table 6** reveals that the variable with the highest information gain is PER3-rs57875989 with an average weight value of 0.013, and the lowest variable is DRD2-rs1799732 with an average weight value of 0.005. This is concluded that the PER3-rs57875989 variant contains the highest information gain on suicidality in individuals with SUD.

3.5 The inpatient treatment

Uzbay (Uzbay 2015) defined substance addiction in general as "a brain disease characterized by some behavioral disorders and the desire to take a substance continuously or periodically in order to feel the pleasurable effects of the substance, or to avoid the discomfort caused by its absence." Based on this definition, hospitalization

Attribute	Average weight
PER3-rs57875989	0,013
NR3C1- rs41423247	0,012
DRD2- METHYLATION	0,008
eNOS-rs1799983	0,008
eNOS-İntron 4a/b VNTR	0,007
COMT- rs4680	0,005
DRD2-rs1799732	0,005
NR3C1- METHYLATION	0,000
COMT- METHYLATION	0,000

Table 6.
Average weight values of gene variants in the suicidal behavior within the context of information gain.

of an individual with SUD to be treated voluntarily may be evaluated as a desire to get rid of substance use or to avoid substance use. The participants were classified under two groups depending on the history of hospitalization. There were 158 individuals under the class of "History of Hospitalization-yes" and 53 individuals under the class of "History of Hospitalization-No." These numbers suggest that the majority of the participants tend to avoid the substance addiction. The tree root starts by NR3C1-METHYLATION and forms a wide branching structure. The branching continues in partial methylation and unmethylated pathways through NR3C1-rs41423247. The individuals with the NR3C1-rs41423247 CC genotype bound in the unmethylated pathway have a tendency to avoid the substance, binding to the eNOS-rs1799983 gene variant occurs when the same pathway is followed in the partial methylation pathway.

The review of **Table 7** reveals that the variable with the highest information gain is NR3C1-METHYLATION with an average weight value of 0.017, and the lowest variable is COMT-METHYLATION with an average weight value of 0.001. This causes to conclude that NR3C1-METHYLATION status provided the highest information about transforming the tendency to avoid or to get rid of the substance into the behavior.

Attributes	Average weight
NR3C1- METHYLATION	0,017
eNOS-rs1799983	0,015
DRD2-rs1799732	0,011
NR3C1-rs41423247	0,007
eNOS-Intron 4a/b VNTR	0,003
DRD2- METHYLATION	0,002
COMT-rs4680	0,001
COMT- METHYLATION	0,000
PER3-rs57875989	0,000

Table 7.
Average weight values of gene variants in the inpatient treatment within the context of information gain.

3.6 Discussion and conclusion

According to the data of the Ministry of Justice, the number of people in prison for crimes related to substance addiction was 57,674 in 2018 corresponding to 21.78% of all convicts [22]. It is detected in some studies on substance addiction that substance users have higher rates of prison history [23]. There is a similar pattern in the database used within the scope of the study, and 53.7% of individuals with SUD have a forensic history. Thirteen of the 28 decision leaves obtained in the established tree structure ended with the existence of a criminal story; however, 15 ended with the absence of a criminal story. The tree root is established over the NR3C1 gene, as seen in **Figure 1**. In a recent study on convicted male individuals, results obtained indicated that the NR3C1 gene is associated with violent behavior in adult males [24]. For instance, the presence of CC genotype of NR3C1-rs41423247 following partial methylation of the NR3C1 gene is an effective sequence in predicting a tendency of an individual to delinquency. It is detected that the dominant variable in the tree is the COMT-rs4680 functional gene variant, which is in line with the studies of the literature [9, 25]. When the tree success rates in **Table 2** are examined, it is seen that the accuracy rate of the model is 52.68%. The lack of information about more individuals in the dataset, lower diversity of the dataset such as the absence of individuals without SUD and with same genetic information as input information have prevented the higher learning.

Substance addiction is a process that causes many systems to change physiologically and the desire to use the substance continuously by withdrawal [26]. Therefore, the tendency to continuous substance use is the natural expected result of the substance addiction. However, the desire to get rid of the substance is effective in getting away from this situation for a while. From this point of view, the root of the COMT-METHYLATION part, which starts in the form of methylated (Partial) or unmethylated, establishes a wide branching structure in the tree structure established. In case of the partial methylation of COMT-METHYLATION, the branching continues through the DRD2-rs1799732 gene; however, if unmethylated, it continues with the NR3C1-rs41423247 gene variant. Forty-four decision leaves were formed on the tree, 20 of which resulted in intermittent use and 24 of them in continuous use. It was detected that the variable with the highest information gain was the COMT-METHYLATION. In the review of the performance rates in **Table 2**, the accuracy rate was detected as 49.76%.

When the COMT-rs4680 genotypes and allele distributions were compared with clinical parameters in the statistical results of the dataset used in the study with X et al., it was observed that multiple substance use was significantly lower in individuals with Met/Met genotype than in individuals with Val/Met and Val/Val genotypes, and multiple substance use was found statistically significantly higher in carriers of the Val allele. Vandenbergh et al. also found in their study that the high-activity Val allele was significantly higher in individuals with multiple substance use [27]. It was observed in the classification of decision trees that there was not any tendency for multiple substance use in the separation of the tree root with COMT-rs4680 and the sequencing of the Met/Met separation with NR3C- METHYLATION to unmethylated. Sequencing continued with DRD2-rs1799732 in partial methyl cleavage of the same pathway. It was concluded that a tendency to multiple substance use appeared in the partial methylation of the sequence with NR3C1 METHYLATION in the Val/Val separation of the main root. Findings of the tree provide similar results when compared with previous studies. Thirty-five result leaves appeared on the tree. Fifteen of these leaves ended that there was a tendency, and 20 ended without any tendency.

It was detected that the variable with the highest information gain was the COMT-rs4680. In the review of the performance rates in **Table 2**, the accuracy rate was detected as 51.21%.

It is stated in studies conducted on individuals with substance use that individuals are suicidal due to their inability to cope with the economic difficulties due to the substance, inadequacy, family problems, exclusion from society, mood disorders, depression experienced during substance withdrawal. It was observed in consideration of the statistical results of the dataset in previous studies (reference) that suicide attempts were at higher levels in individuals who have started to use substances before the age of 15 years. It was commented that individuals at and below 15 years of age have not yet completed their physical and mental development, they may be easily affected by their friends, and the emotional and hormonal changes due to adolescence may have caused these differences. The tree root starts by NR3C1 rs41423247 and forms a wide branching structure. It is concluded that suicidality is not seen in the GG genotype, and it branches to NR3C-METHYLATION in the GC and CC genotypes and continues. There is not any tendency when NR3C1-METHYLATION is unmethylated in the CC genotype; however, the same pathway shows the tendency in the GC genotype. Findings of the tree provide similar results when compared with previous studies. Forty-three result leaves appeared on the tree. Fifteen of these leaves ended that there was a tendency, and 22 ended without any tendency. It was detected that the variable with the highest information gain was the PER3-rs57875989. In the review of the performance rates in **Table 2**, the accuracy rate was detected as 65%.

It was seen that the root formed a wide branching structure from NR3C1-METHYLATION when considering the tree structure established for the trend analysis for the desire to get rid of the substance examined under the presence or absence of inpatient treatment history. The branching continues in partial methylation and unmethylation pathways through NR3C1-rs41423247. The individuals with the NR3C1-rs41423247 CC genotype bound in the unmethylation pathway have a tendency to avoid the substance, binding to the eNOS-rs1799983 gene variant occurs when the same pathway is followed in the partial methylation pathway. Twenty-eight result leaves appeared on the tree. Fifteen of these leaves ended with the decision that there was a tendency, and eight ended without any tendency. It was detected that the variable with the highest information gain was the NR3C1-METHYLATION. The review of the performance rates in **Table 2** reveals that the accuracy rate was detected as 70.56%.

In this study, the effects of genetic and methylation differences of COMT, DRD2, PER3, eNOS, NR3C1 functional gene variants on the potential trends in individuals with SUD were analyzed by decision trees algorithm, and their similarities with previous studies in the literature were evaluated. The tendencies are grouped in a structure suitable for dual classification under the subgroups of tendency to delinquency, tendency to use of the substance, tendency to use of multiple substances, tendency to suicide, and tendency to abandon the substance, respectively. There is not any deficient data in the dataset. The 10-fold cross-validation was used in the model. This method creates k discrete pieces in a dataset with m samples, each containing m/k samples. This method allocates a different dataset for testing each time and uses the remaining k-1 dataset for training purposes. It is trained k times by changing the classifier in this way. In the last step, it estimates the classifier performance by the average of the k errors obtained.

The decision trees are a model that may provide effective results in binary classes among machine learning classification methods. Our study is the first in the literature to examine the effects of gene variants on behavioral tendencies through machine learning methods. However, the lower accuracy rates obtained in this study indicate that the dataset needs to be more diverse and comprehensive.

Conflict of interest

The authors declare no conflict of interest.

Author details

Inci Zaim Gokbay*, Yasemin Oyaci and Sacide Pehlivan
Istanbul University Informatics Department, Istanbul University Faculty of Medicine, Istanbul, Turkey

*Address all correspondence to: inci.gokbay@istanbul.edu.tr

References

[1] Işık G, Korkmazlar Ü, Durukan M, Aydoğdu A. Symptoms and diagnoses of first-time adolescent applications to a child and adolescent psychiatry out-patient clinic. The Journal of Clinical Psychiatry. 2004;**7**(2):103-110

[2] Kılıç FS. Addiction and stimulant drugs. Osmangazi Journal of Medicine. 2016;**1**:55-60

[3] Derin G, Okudan M, Aşıcıoğlu F. Alkol ve madde kullanım bozukluklarında ailevi risk faktörleri. In: Öztürk E, editor. Aile Psikopatolojisi. Ankara: Türkiye Klinikleri; 2021. pp. 118-126

[4] Bozkurt O. Madde Bağımlısı Bireylerin Bağımlılık Süreçlerinde Ailenin Etkisi. Sosyal Bilimler Enstitüsü; 2015

[5] Ünal M. Madde Bağımlılığı ve Alkolizmde Aile. Sosyal Politika Çalışmaları Dergisi. 2004;**2**(2):80-86

[6] Gökbay İZ. Artificial Intelligence Applications In Medicine – An Overview Of The Evolution Of Clinical Decision Support Systems In The Development Process Of Diagnostic And Treatment Methods From Antiquity To Artificial Intelligence. 2021. pp. 673-692. DOI: 10.26650/B/ET07.2021.003.33

[7] Gökbay İZ, Karman Ş, Yarman S, Yarman BS. An intelligent decision support tool for early diagnosis of functional pituitary adenomas. TWMS Journal of Applied and Engineering Mathematics. 2015;**5**(2):169-187

[8] Nursal AF, Aydın PÇ, Uysal MA, Pehlivan M, Oyacı Y, Pehlivan S. PER3 VNTR variant and susceptibility to smoking status/substance use disorder in a Turkish population. Arch Clin Psychiatry. 2020;**47**(3):71-74. DOI: 10.1590/0101-60830000000235

[9] Oyaci Y, Aytac HM, Pasin O, Cetinay Aydin P, Pehlivan S. Detection of altered methylation of MB-COMT promotor and DRD2 gene in cannabinoid or synthetic cannabinoid use disorder regarding gene variants and clinical parameters. Journal of Addictive Diseases. 2021;**39**(4):526-536. DOI: 10.1080/10550887.2021.1906618

[10] Pehlivan S, Aydin PC, Nursal AF, Pehlivan M, Oyaci Y, Yazici AB. A relationship between endothelial nitric oxide synthetase gene variants and substance use disorder. Endocrine, Metabolic & Immune Disorders Drug Targets. 2021;**21**(9):1679-1684. DOI: 10.2174/1871530320666201013154917

[11] Shankarishan P, Borah PK, Ahmed G, Mahanta J. Endothelial nitric oxide synthase gene polymorphisms and the risk of hypertension in an Indian population. BioMed Research International. 2014; **2014**:793040. DOI: 10.1155/2014/793040

[12] Isir AB, Nacak M, Balci SO, Aynacioglu AS, Pehlivan S. Genetic contributing factors to substance abuse: An association study between eNOS gene polymorphisms and cannabis addiction in a Turkish population. Australian Journal of Forensic Sciences. 2016;**48**(6):676-683. DOI: 10.1080/00450618.2015.1112428

[13] Guess J, Burch JB, Ogoussan K, et al. Circadian disruption, Per3, and human cytokine secretion. Integrative Cancer Therapies. 2009;**8**(4):329-336. DOI: 10.1177/1534735409352029

[14] Zhabenko O, Austic E, Conroy DA, et al. Substance use as a risk factor for

sleep problems among adolescents presenting to the emergency department. Journal of Addiction Medicine. 2016;**10**(5):331-338. DOI: 10.1097/ADM.0000000000000243

[15] Blum K, Noble EP, Sheridan PJ, et al. Association of the A1 allele of the D2 dopamine receptor gene with severe alcoholism. Alcohol. 1991;**8**(5):409-416. DOI: 10.1016/0741-8329(91)90693-q

[16] Uhl G, Blum K, Noble E, Smith S. Substance abuse vulnerability and D2 receptor genes. Trends in Neurosciences. 1993;**16**(3):83-88. DOI: 10.1016/0166-2236(93)90128-9

[17] Pickens RW, Svikis DS, McGue M, Lykken DT, Heston LL, Clayton PJ. Heterogeneity in the inheritance of alcoholism. A study of male and female twins. Archives of General Psychiatry. 1991;**48**(1):19-28. DOI: 10.1001/archpsyc.1991.01810250021002

[18] DeLisi M, Vaughn MG. Foundation for a temperament-based theory of antisocial behavior and criminal justice system involvement. Journal of Criminal Justice. 2014;**42**(1):10-25

[19] Ishiguro H, Horiuchi Y, Tabata K, Liu QR, Arinami T, Onaivi ES. Cannabinoid CB2 receptor gene and environmental interaction in the development of psychiatric disorders. Molecules. 2018;**23**(8):18360

[20] Park S, Hong JP, Lee JK, et al. Associations between the neuron-specific glucocorticoid receptor (NR3C1) Bcl-1 polymorphisms and suicide in cancer patients within the first year of diagnosis. Behaviour Brain Function. 2016;**12**(1):22

[21] Schote AB, Jäger K, Kroll SL, et al. Glucocorticoid receptor gene variants and lower expression of NR3C1 are associated with cocaine use. Addiction Biology. 2019;**24**(4):730-742. DOI: 10.1111/adb.12632

[22] TUBİM. 2019 Türkiye Uyuşturucu Raporu, Türkiye Uyuşturucu ve Uyuşturucu Bağımlılığı İzleme Merkezi (Internet) 2020. Available from: http://www.narkotik.pol.tr/kurumlar/narkotik.pol.tr/TUB%C4%B0M/Ulusal%20Yay%C4%B1nlar/2019-TURKIYE-UYUSTURUCU-RAPORU.pdf

[23] Bilici R, Karakaş UG, Tufan E, Güven T, Uğurlu M. Bir bağımlılık merkezinde yatarak tedavi gören hastaların sosyo demografik özellikleri. Fırat Tıp Dergisi. 2012;**17**:223-227

[24] Liu L, Li J, Qing L, et al. Glucocorticoid receptor gene (NR3C1) is hypermethylated in adult males with aggressive behaviour. International Journal of Legal Medicine. 2021;**135**(1):43-51. DOI: 10.1007/s00414-020-02328-7

[25] Yang Y, Li J, Yang Y. The Research of the Fast SVM Classifier Method. In: 2015 12th International Computer Conference on Wavelet Active Media Technology and Information Processing (ICCWAMTIP), IEEE. 2015. pp. 121-124

[26] Berke JD, Hyman SE. Addiction, dopamine, and the molecular mechanisms of memory. Neuron. 2000;**25**(3):515-532. DOI: 10.1016/s0896-6273(00)81056-9

[27] Vandenbergh DJ, Rodriguez LA, Miller IT, Uhl GR, Lachman HM. High-activity catechol-O-methyltransferase allele is more prevalent in polysubstance abusers. American Journal of Medical Genetics. 1997;**74**(4):439-442

Chapter 11

Mathematical Modeling of a Porous Medium in Diesel Engines

Arash Mohammadi

Abstract

Direct injection diesel engines have high power density with low exhaust emission but suffer from particulate matter (PM). Some new technologies were applied to reduce emissions, but they have not solved the emission problem of diesel engines altogether. The main problem of emissions from diesel engines is the simultaneous process of fuel injection and combustion, so non-homogeneous mixture formation occurs in cylinder space, and non-homogeneity is the main reason for emission generation. The solution to this problem is the separation of injection fuel and combustion processes for homogeneous mixture formation in diesel engines. An applicable practical solution for homogeneous mixture formation is the application of porous media (PM) in diesel engine combustion chambers. PM develops stable ultra-lean combustion and decreases emissions. This chapter has three parts for the mathematical modeling of PM diesel engines. The first part is thermodynamically modeling in a closed cycle. The second is zero-dimensional modeling with the chemical kinetics of PM diesel engines, and the third is three-dimensional CFD modeling with the chemical kinetics of PM diesel engines in open or closed cycle. So, mathematical modeling of PM diesel engines, from simple thermodynamically modeling to complicated 3D modeling, is described in this chapter.

Keywords: direct injection diesel engine, porous medium, thermodynamic modeling, zero-dimensional modeling, CFD modeling

1. Introduction

The target of current diesel engines is low fuel consumption with near-zero emission levels for gaseous and particulate matter components at all operational conditions (engine speed and load). Thus, diesel engines require new concepts for the combustion process. The heterogeneous combustion in diesel engines causes non-uniform heat release in the combustion chamber and in the following: NOx, soot, CO, and UHC formation. Homogeneous combustion can solve this problem. Homogeneous combustion in diesel engines is described as a process of homogeneous mixture formation accompanied by volumetric heat release in the total combustion chamber space with a low-temperature gradient inside the chamber. A possible solution to achieving homogeneous combustion is applying PM inside the combustion chamber. One of the different combustion technologies is inside PM combustion. It is a flameless heat release inside an operating PM followed by homogeneous combustion with near-zero emission [1–3]. This process is stable combustion with a high-power density in an extensive dynamic range. In a burner with an injection of liquid fuel, high-temperature PM acts like an effective evaporator. The large specific surface area with high heat transfer between fluid and solid phases of PM causes fast vaporization of fuel droplets. The large heat capacity of the PM leads to homogeneous combustion with approximately constant temperature.

Some remarkable features of the PM which attracts its application for combustion chamber of diesel engine, display in **Figure 1** [1].

Many experimental and numerical research has verified the combination of PM burners with flame stability and low emissions. Such exciting features of combustion inside PM make it plausible for application in diesel engines. However, mixture

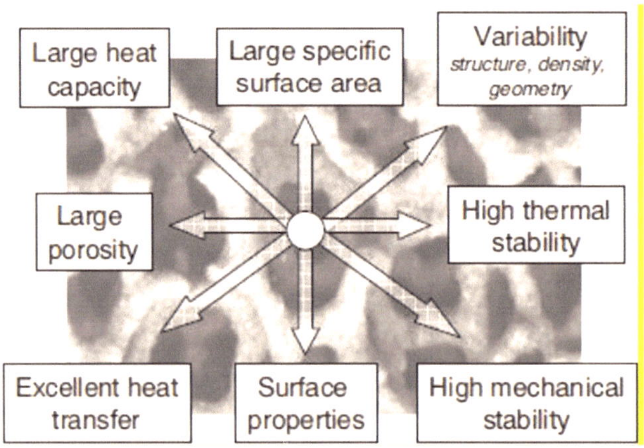

Figure 1.
Remarkable features of PM for application in combustion chamber of diesel engine [1].

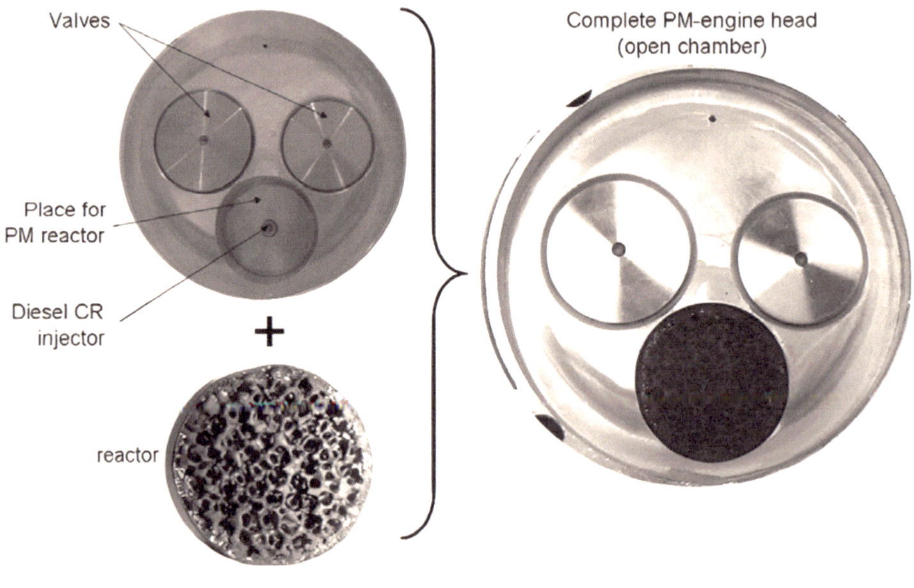

Figure 2.
Diesel engine cylinder head of a with inserted PM [2].

formation and combustion in a diesel engine are complicated nonstationary high-pressure processes in with a direct spray inside the combustion chamber. So, PM can be applied to solve this problem. Ultra-lean burn combustion can be used in diesel engines with stable flame due to the high volumetric heat transfer coefficient between fluid and solid phases of PM. The large surface area of PM and the high heat capacity of PM can absorb some of the heat released during combustion and transfer it to fresh air during the end of the compression process. Hence, the temperature reduction decreases nitrogen oxides. Also, a reduction in the temperature gradient inside the cylinder leads to a decrease in carbon monoxide formation. **Figure 2** displays a view of the PM diesel engine. PM inserts in the cylinder head of a diesel engine, and permanent contact between the in-cylinder mixture and PM (open PM concept) exists [3–6].

2. Diesel engine concept with inserted PM in head of combustion chamber

Durst and Weclas described the diesel engine concept with new mixture formation and combustion processes in a PM reactor. Application of PM in diesel engines generates a homogeneous and flameless combustion process accompanied by a near-zero emission level. Heat recovery from the last combustion process in PM increases the temperature end of the compression process, resulting in raised thermal efficiency and reduced fuel consumption. Heat absorption in PM leads to a reduction in combustion temperature and near-zero NOx. There is difference between the combustion processes in conventional diesel engines and PM-inserted diesel engines. These processes are liquid fuel injection directly into PM, causing multi-jet splitting for fuel distribution throughout PM volume, fuel vaporization, and mixing with air, accomplished by thermal ignition and heat release [1, 6–9].

2.1 Diesel PM-engine with concept of an open PM chamber

A diesel PM engine describes as a diesel engine with a homogeneous combustion process in a PM volume. PM engines recognize these processes in PM volume: energy recovery in the cycle, fuel injection in PM, fuel vaporization for liquid fuels, mixing with air, homogenization of air-fuel mixture, self-ignition of the mixture, and homogeneous combustion.

This mathematical modeling of permanent contact between working fluid and PM is schematically studied, as illustrated in **Figure 3**. The PM-combustion chamber is supposed to be inserted in the cylinder head space, and the PM chamber wall is thermally isolated. During the intake process, the PM-heat capacitor has an ignorable effect on the in-cylinder air conditions. Also, during the start of the compression process, a small amount of air is in contact with hot PM. Before TDC, the fuel is injected into PM space, and very fast fuel vaporization for liquid fuels and mixing with air happen in the PM structure. Hence, the fuel is injected close to TDC of the compression process due to high energy storage in PM volume. There is a very complex process during fuel injection, mixture formation, and combustion initiation in the PM structure. The high initial PM temperature (solid-phase temperature of the PM and gas temperature trapped inside PM volume) with mixture formation inside the PM reactor causes self-ignition and volumetric heat release in the combustion process. The reactor heat capacity, pore density, and pore structure can affect the combustion process [2–7].

Solid phase PM has higher heat capacity than fluid flow and high energy storage capability. Correspondingly, its high surface-to-volume ratio leads to considerable

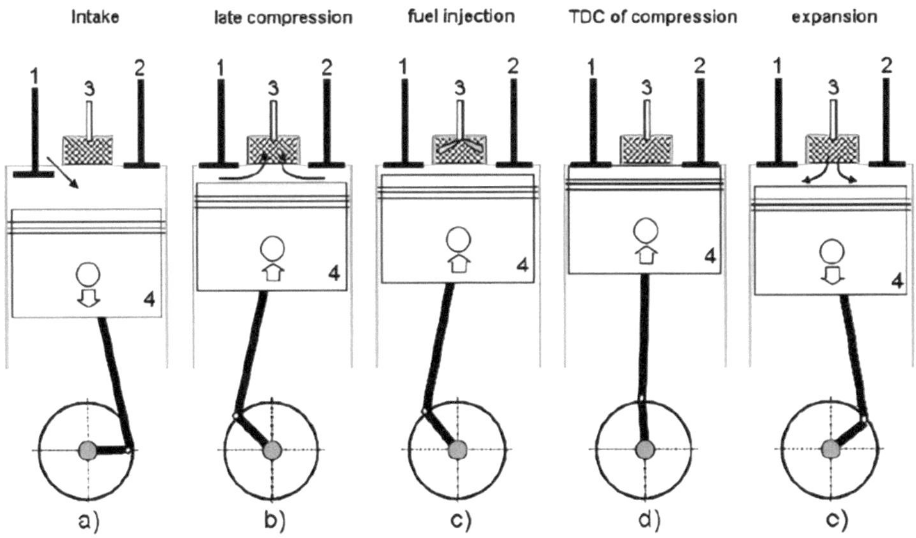

Figure 3.
Schematic of a permanent contact PM- diesel engine [8].

heat exchange. The solid high temperature of PM inside the combustion chamber is a source of the high rate of liquid fuel evaporation and fast mixing with air and self-ignition of the mixture. In diesel engines, mixture formation and combustion simultaneously happen, but in PM engines, these processes occur separately.

The ideal thermodynamic model for combustion in diesel engines is an isobar process, but combustion in PM happens inside diesel engines very fast. Combustion in PM intensifies the reaction rate in a short time scale. Hence, volume change can be considered approximately by a combination of isochoric and isothermal processes.

The last case is illustrated in **Figure 3**. During the intake process (**Figure 3a**), PM has not considerably affected the in-cylinder pressure and temperature. Also, during the early compression process, a low quantity of air is in contact with hot PM. In the following compression process, the heat exchange process increases with the motion of the piston to the TDC (**Figure 3b**). At the TDC, total air is collected in the PM volume. Near the TDC of the compression process, the fuel is injected into the PM volume (**Figure 3c**), and vaporization of liquid fuel and mixing with air occur very fast in the PM. A volumetric self-ignition of the fuel-air mixture follows flameless combustion by uniform temperature distribution in the PM chamber (**Figure 3d**). Fuel injection controls combustion initiation timing in the PM volume. The PM structure creates conditions for a homogeneous combustion process and converts the heat into work (**Figure 3e**). The combustion in a PM can be carried out in the PM volume that cannot occur in the free flame combustion process [1–7].

2.2 Mathematical thermodynamic modeling of PM diesel engine

For an ideal thermodynamic cycle of conventional and PM diesel engines, a closed cycle is assumed, with working fluid as air with no exhaust gases to the environment. The heat capacity of PM is considerably further than that of fluid. Hence, the solid phase temperature of PM is considered constant during the cycle, and heat exchange between the PM and the working fluid does not affect it. Heat losses of the piston, liner wall, and PM chamber to the environment are ignored, and compression and expansion processes have adiabatically happened.

In the ideal closed cycle energy of a diesel engine (compression ignition engine), combustion occurs at the constant pressure assumed. Heat losses through the combustion chamber to the environment are neglected, and compression and expansion (work) processes happen isentropically. **Figure 4a** and **b** shows the P-V (pressure versus volume) and T-S (temperature versus entropy) diagrams of diesel engine closed cycle analysis. The four processes are:

1. process 1 → 2 isentropic compression

2. process 2 → 3 isobar heat addition

3. process 3 → 4 isentropic expansion

4. process 4 → 1 isochoric heat rejection

PM diesel engine in the ideal closed cycle energy of the fuel is added to the air in a combination of isochoric and constant temperature. Heat losses through the PM combustion chamber to the environment are ignored, and compression and expansion (work) processes occur isentropically. **Figure 4a** and **b** shows a closed cycle's P-V and T-S diagrams for permanent contact PM diesel-engine analysis. The five processes are:

1. process 1 → 2 isentropic compression

2. process 2 → 3` isochoric heat addition

3. process 3` → 3 isothermal heat addition

4. process 3 → 4 isentropic expansion

5. process 4 → 1 isochoric heat rejection

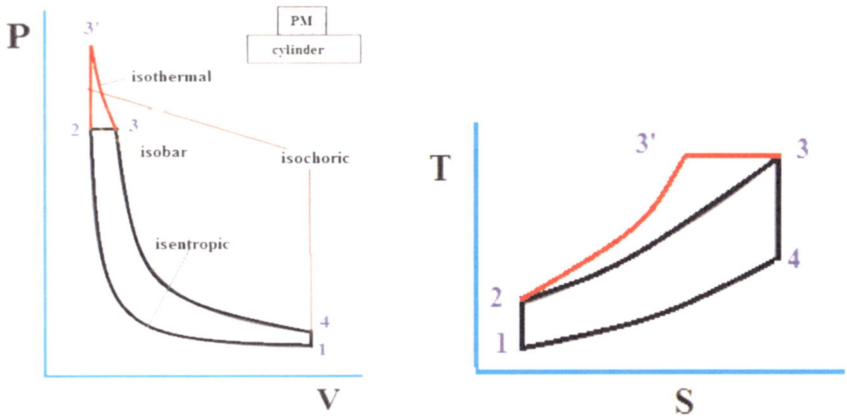

Figure 4.
(a) P-V diagram of closed cycle of conventional diesel engine and PM engine. (b) T-S diagram of closed cycle of conventional diesel engine and PM engine.

Compression ratio is defined as r_c, and ρ is the compression ratio during constant temperature heat addition:

$$r_c = v_1/v_2 \tag{1}$$

$$\rho = v_3/v_{3'} \tag{2}$$

Process 1–2 is isentropic, so:

$$T_2 = T_1(r_c)^2 \tag{3}$$

For PM diesel engines, the energy of fuel is added to air through two process: isochoric process $C_v(T_{3'} - T_2)$ plus isothermal process $RT_3 \ln\left(\frac{v_3}{v_{3'}}\right)$. Heat loss to environment according to conventional diesel engine is $C_v(T_4 - T_2)$.

$$q_{in} = C_v(T_{3'} - T_2) + RT_3 \ln\left(\frac{v_3}{v_{3'}}\right) \tag{4}$$

Heat rejection occurs in a constant volume process:

$$q_{out} = C_v(T_4 - T_1) \tag{5}$$

Engine thermal efficiency is defined according to Eq. (6):

$$\eta = \frac{w}{q_{in}} = 1 - \frac{q_{out}}{q_{in}} \tag{6}$$

Assuming constant specific heat, the thermal efficiency of diesel engine is according to Eq. (7):

$$\eta_{diesel} = 1 - \frac{C_v(T_4 - T_1)}{C_p(T_3 - T_2)} \tag{7}$$

Hence, thermal efficiency of PM diesel engine is according to Eq. (8).

$$\eta_{PM\ diesel} = 1 - \frac{C_v(T_4 - T_1)}{C_v(T_{3'} - T_2) + RT_3 \ln\left(\frac{v_3}{v_{3'}}\right)} \tag{8}$$

2.3 Mathematical zero-dimensional modeling with chemical kinetics of PM diesel engine

Figure 5 illustrates schematically permanent contact PM diesel engine. It is supposed that PM is inserted inside the cylinder head. All necessary conditions for homogeneous combustion are carried out in the PM combustion chamber [6, 7]. During the intake and early compression process, the PM-heat capacitor has a low effect on the in-cylinder thermodynamic pressure and temperature. Near TDC, the fuel is injected into the PM structure.

Furthermore, the fuel is vaporized fast and mixed with air inside the PM. Because of the instant evaporation and combustion of liquid fuel after injection, it is a logical assumption that combustion occurs in the isochoric process. Then, all the combustion

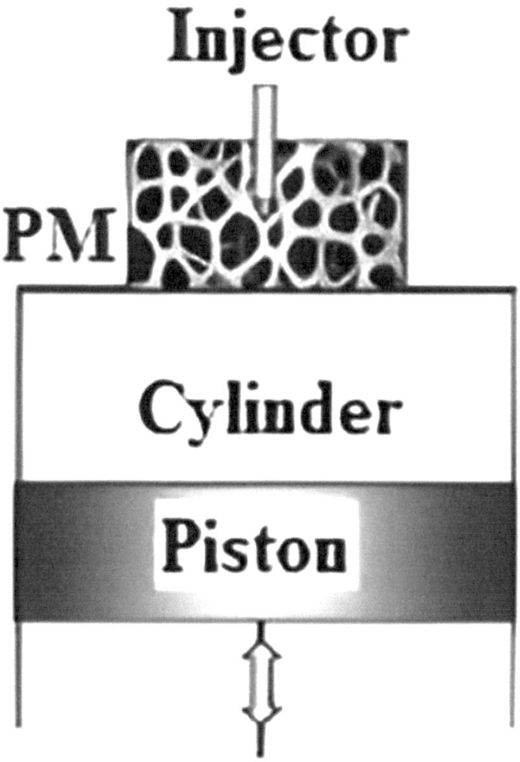

Figure 5.
Schematic illustration of a permanent contact PM diesel engine [6].

products enter the cylinder space instantaneously. So, the energy of the fluid phase of PM is transferred to the in-cylinder volume [1, 6–9].

For thermodynamic modeling of PM in the combustion chamber, two energy equations are used, a modified gas energy equation and a solid phase energy equation. These assumptions are considered for the modeling process [1]:

1. Combustion process is adiabatic.

2. PM only recuperates the energy of combustion and does not participate in combustion.

3. A single-step or multi-step reaction(s) is considered for modeling of combustion process.

4. Radiation heat loss computes only in solid phase.

5. There is permanent contact between PM and inside mixture.

6. Air-fuel mixture inside PM is premixed in specified temperature and pressure.

7. Heat capacity for both phases of PM depends on temperature.

8. Energy equations for both phases of PM are solved at each crank angle.

9. Species of combustion products with fluid phase temperature inside PM are transferred to in-cylinder volume at each crank angle.

Mixture formation process assumes a homogenous before combustion due to PM space. Combustion happens as a result of self-ignition due to the high initial temperature.

3. PM modeling

3.1 Gas phase energy equation

Combustion inside a chamber is considered as thermodynamically modeling. Hence, gas phase energy equation and reaction rate are solved simultaneously. The closed system first law of thermodynamics (energy equation) for a differential crank angle, $d\theta$, is [10, 11]:

$$\frac{dQ}{d\theta} - \frac{dV}{d\theta} = m\,\frac{du}{dt} \tag{9}$$

The definition of enthalpy $h = u + pv$ and differentiating in constant pressure process gives:

$$\frac{dQ}{d\theta} = m\,\frac{dh}{d\theta} \tag{10}$$

The enthalpy in terms of chemical compositions of gases is:

$$h = \frac{\sum_{i=1}^{N} N_i \overline{h_i}}{m} \tag{11}$$

where N_i and $\overline{h_i}$ are the moles and molar enthalpy of species i, so:

$$\frac{dh}{d\theta} = \frac{1}{m}\left[\sum_i \left(\overline{h_i}\,\frac{dN_i}{d\theta}\right) + \sum_i \left(N_i\,\frac{d\overline{h_i}}{d\theta}\right)\right] \tag{12}$$

Assuming ideal gas relation:

$$\frac{d\overline{h_i}}{d\theta} = \frac{\partial \overline{u_i}}{\partial T} \bullet \frac{dT}{d\theta} = \overline{c}_{p,i}\,\frac{dT}{d\theta} \tag{13}$$

where $\overline{c}_{p,i}$ is the constant molar pressure specific heat of species i. Concentration of species i is obtained as:

$$[X_i] = \frac{\varkappa_i\,P}{R\,T_g} \tag{14}$$

The porosity of PM is φ (0 <φ<1), so the volume of fluid in PM is φV, and the volume of solid is (1-φ) V. So $N_i = \varphi V\,[X_i]$, with differentiating of this equation:

$$\frac{dN_i}{d\theta} = \varphi V \dot{\omega}_i \tag{15}$$

The $\dot{\omega}_i$ is the rate of reaction species i.

Gas-phase energy equation by applying PM is determined based on a relation of convective heat transfer between the gas and solid phases of PM [10, 11]:

$$\frac{dT_g}{d\theta} = \frac{-h_v(T_g - T_s) - \varphi \sum_i(\dot{\omega}_i h_i) + \varphi R T_g \sum_i \dot{\omega}_i}{\sum_i [[X_i](c_{p.i} - R)]} \tag{16}$$

$(\dot{Q}/_V)$ is heat loss to the environment.

The empirical relation Eq. (17) is applied to compute the volumetric heat transfer coefficient [1]:

$$h_v = \frac{k_g Nu_v}{d^2} \tag{17}$$

Volumetric Nusselt number is calculated by relation Eq. (18) [10, 11]:

$$Nu_v = 2 + 1.11\ Re^{0.60} Pr^{0.33} \tag{18}$$

Reynolds and Prandtl's numbers are computed by time-varying conditions of the fluid inside the combustion chamber. Reynolds was calculated according to injection velocity. Eq. (19) is used for the calculation of the Reynolds number [10, 11]:

$$Re = \frac{\rho_g u_c\, d}{\mu} \tag{19}$$

The density of gases from the ideal gas equation:

$$\rho = \frac{P}{T\ (R/_{MW_{mix}})} \tag{20}$$

$$MW_{mix} = \sum_i^N x_i MW_i$$

Relation Eq. (21) applies for calculating velocity where ρ_f is the fuel density, the chamber pressure is p_c, and injection pressure is clarified by p_i [10, 11].

$$u_c = \sqrt{\frac{2\ (p_i - p_c)}{\rho_f}} \tag{21}$$

Eq. (22) shows the calculation of chamber pressure [10, 11]:

$$P = \sum_i [X_i]\ R\ T_g \tag{22}$$

The modified Arrhenius equation is used for modeling the fuel oxidation as shown in Eq. (23) [1].

$$\dot{\omega}_{Fuel} = \frac{d[X_{Fuel}]}{d\theta} = -A\ exp\left(-E_a/_{RT}\right)[X_{Fuel}]^m \left[X_{Oxygen}\right]^n \tag{23}$$

A, m, and n are constant coefficients of single-step fuel oxidation Eq. (23), selected as **Table 1**. After solving coupling Eqs. (16) and (23), the gas temperature, species, and fuel consumption rates are determined. After updating the temperature and concentration of species, the pressure was calculated in Eq. (24) [1].

$$P = \sum_i [X_i]\, R\, T \tag{24}$$

NASA seven-term polynomials are applied to compute the thermodynamic properties. For the calculation of constant pressure of specific heat, NASA polynomial can be used, which depends on temperature:

$$c_p = R\left(a_1 + a_2T + a_3T^2 + a_4T^3 + a_5T^4\right) \tag{25}$$

Enthalpy of gas can be determined from [10, 11]:

$$\boldsymbol{h} = \boldsymbol{R}\left(a_1 + \frac{a_2\boldsymbol{T}}{2} + \frac{a_3\boldsymbol{T}^2}{3} + \frac{a_4\boldsymbol{T}^3}{4} + \frac{a_5\boldsymbol{T}^4}{5} + \frac{a_6}{\boldsymbol{T}}\right) \tag{26}$$

$$\boldsymbol{h}(\boldsymbol{T}) = \Delta \boldsymbol{hf}(298) + [\boldsymbol{h}(\boldsymbol{T}) - \boldsymbol{h}(298)]$$

3.2 Solid phase energy equation

Energy equation for the solid phase of PM is:

$$\dot{Q} = m\,\frac{dh}{d\theta} \tag{27}$$

By chain rule differentiable:

$$\frac{dh}{d\theta} = \frac{\partial h}{\partial T_s}\,\frac{dT_s}{dt} = c_s\,\frac{dT}{d\theta} \tag{28}$$

Solid phase volume is (1-φ) V, so:

$$\frac{\dot{Q}}{(1-\varphi)\,V\,\rho_s} = c_s\,\frac{dT_s}{d\theta} \tag{29}$$

where ρ_s is the density of solid phase, and c_s is the specific heat of solid phase.

Fuel	A	E_a/R (K)	m	n
H_2	$1.8 \cdot 10^{13}$	17,614	1.00	0.50
C_3H_8	$8.6 \cdot 10^{11}$	15,098	0.10	1.65
C_8H_{18}	$4.6 \cdot 10^{11}$	15,098	0.25	1.50
$C_{10}H_{22}$	$3.8 \cdot 10^{11}$	15,098	0.25	1.50
CH_3OH	$3.2 \cdot 10^{12}$	15,098	0.25	1.50
C_2H_5OH	$1.5 \cdot 10^{12}$	15,098	0.15	1.60

Table 1.
Single-step oxidation of several fuels [10, 11].

Due to heat transfer between energy gas and solid phase, $h_v(T_g - T_s)$ is the heat transfer from the gas phase to solid phase. Also, $\dot{q}_r$ is radiation from solid phase to environment. Therefore, Eq. (30) calculates solid temperature.

$$\frac{dT_s}{d\theta} = \frac{h_v(T_g - T_s) - \dot{q}_r}{(1-\varphi)\rho_s C_s} \tag{30}$$

The solid phase energy is computed based on Eq. (31) [10, 11]:

$$\frac{dT_s}{d\theta} = \frac{h_v(T_g - T_s) - \varepsilon\sigma\frac{A}{V}\left(T_s^4 - T_0^4\right)}{(1-\varphi)\rho_s C_s} \tag{31}$$

where $\left(\frac{A}{V}\right)_s$ is the surface area to volume ratio of the PM, and T_0 is the environment's temperature where radiation loss occurs.

3.3 Solution method of the equations

Three coupled Eqs. (32)–(34) that are relevant to solid and gaseous phase temperature and species concentrations are solved.

$$\frac{dT_g}{d\theta} = f(\theta, [X_i], T_g, T_s) \tag{32}$$

$$\frac{dT_s}{d\theta} = g(\theta, [X_i], T_g, T_s) \tag{33}$$

$$\frac{d[X_i]}{d\theta} = h(\theta, [X_i], T_g) \tag{34}$$

Runge–Kutta fourth-order method is applied to solve the fluid and solid phase energy equations in PM and compute the species concentrations. Due to the high rate of combustion, the small crank angle (Δθ) should be considered. The step size for solving Runge–Kutta method was assumed 10^{-6} s or 0.1 crank angle that depend on engine speed [10, 11].

$$k_{1,g} = hf(\theta_n, T_g, T_s)$$

$$k_{1,s} = hg(\theta_n, T_g, T_s)$$

$$k_{2,g} = hf\left(\theta_n + \frac{h}{2}, T_g + \frac{k_{1,g}}{2}, T_s + \frac{k_{1,s}}{2}\right)$$

$$k_{2,s} = hg\left(\theta_n + \frac{h}{2}, T_g + \frac{k_{1,g}}{2}, T_s + \frac{k_{1,s}}{2}\right)$$

$$k_{3,g} = hf\left(\theta_n + \frac{h}{2}, T_g + \frac{k_{2,g}}{2}, T_s + \frac{k_{2,s}}{2}\right)$$

$$k_{3,s} = hg\left(\theta_n + \frac{h}{2}, T_g + \frac{k_{2,g}}{2}, T_s + \frac{k_{2,s}}{2}\right)$$

$$k_{4,g} = hf(\theta_n + h, T_g + k_{3,g}, T_s + k_{3,s})$$

$$k_{4,s} = hg(\theta_n + h, T_g + k_{3,g}, T_s + k_{3,s})$$

Updated gas and solid phase temperatures of PM can be obtained from Eqs. (35) and (36). An update of species concentration is computed from Eq. (37):

$$\theta_{n+1} = \theta_n + \Delta\theta$$

$$T_{g,n+1} = T_{g,n} + \frac{1}{6}\left(k_{1,g} + 2k_{2,g} + 2k_{3,g} + k_{4,g}\right) \tag{35}$$

$$T_{s,n+1} = T_{s,n} + \frac{1}{6}\left(k_{1,s} + 2k_{2,s} + 2k_{3,s} + k_{4,s}\right) \tag{36}$$

$$[X_i]_{n+1} = [X_i]_n + h\left(\theta_n, [X_i]_n, T_{g,n+1}\right) \tag{37}$$

4. In-cylinder modeling

4.1 Mass conservation

The total mass of the cylinder is a combination of the energy of the in-cylinder and fluid phase volume of PM:

$$(\mathrm{m}\ )_{cylinder,n} + (\mathrm{m}\ )_{PM\ fluid,n} = (\mathrm{m}\ )_{cylinder,n+1} \tag{38}$$

4.2 Energy equation

The total energy of the cylinder is a combination of the energy of the in-cylinder and the fluid phase volume of PM:

$$(\mathrm{m}\ c_v\ \mathrm{T})_{cylinder,n} + (\mathrm{m}\ c_v\ \mathrm{T})_{PM\ fluid,n} = (\mathrm{m}\ c_v\ \mathrm{T})_{cylinder,n+1} \tag{39}$$

$$\mathrm{m}_{cyl,n} c_{v,cyl,n} T_{cyl,n} + m_{f_{PM},n} c_{v,f_{PM},n} T_{f_{PM},n} = \mathrm{m}_{cyl,n+1} c_{v,cyl,n+1} T_{cyl,n+1} \tag{40}$$

So finally, in-cylinder temperature and pressure were updated according to Eqs. (41) and (42):

$$T_{cyl,n+1} = \frac{\mathrm{m}_{cyl,n} c_{v,cyl,n} T_{cyl,n} + m_{f_{PM},n} c_{v,f_{PM},n} T_{f_{PM},n}}{\left((\mathrm{m}\ )_{cylinder,n} + (\mathrm{m}\ )_{PM\ fluid,n}\right) c_{v,cyl,n}} \tag{41}$$

$$P_{cyl,n+1} = \sum_i [X_i]_{n+1}\, R\, T_{cyl,n+1} \tag{42}$$

5. Mathematical three-dimensional CFD modeling with chemical kinetics of PM diesel engine

The 3D computational domain is composed of structured, unstructured, or hybrid meshes. The computational domain of the PM engine is a combination of in-cylinder volume and PM reactor. For the in-cylinder space, original governing equations are applied, but the governing equations need to be modified to simulate PM volume. Momentum, gas phase energy, and chemical species continuity equation are modified. Also, a new equation for solid phase energy equation with a radiation model is derived [8–10, 12–14]. For modeling the PM reactor, some assumptions were considered:

1. There is non-equilibrium thermal energy between gas and solid phases of PM.

2. The solid phase of PM is homogeneous and isotropic; has a variable property with temperature; and has no catalyst effects.

3. Radiation heat transfer to the environment is considered for the solid phase. Moreover, the effect of its radiation on the evaporation of droplets is neglected.

Considering to the above assumptions, modified governing equations are [8–10, 12–14]:

5.1 Continuity equation for species i is

$$\frac{\partial(\rho_i\varphi)}{\partial t} + \nabla.(\rho_i u\varphi) = \nabla.\left[\rho\varphi D_{im}\nabla\left(\frac{\rho_i}{\rho}\right)\right] + \varphi\,\dot{\rho}_i^c + \dot{\rho}^s\delta_{i1} \tag{43}$$

where the diffusion coefficient D_{im} is based on kinetic theory of gases. u is the velocity vector, ρ_i is the density of species i, and ρ is the density of mixture. $\dot{\rho}_i^c$ is the density of species generation or destruction during combustion, and $\dot{\rho}^s$ is the density of spray.

5.2 Gas phase momentum equation

$$\frac{\partial\left(\rho_g u\right)}{\partial t} + \nabla.\left(\rho_g uu\right) = -\nabla P - \nabla\left(\frac{2}{3}\rho_g k\right) + \nabla.\sigma + F^s - \left(\frac{\Delta P}{\Delta L}\right) \tag{44}$$

F^s is the momentum source of the liquid fuel injection term, and the term $\left(\frac{\Delta P}{\Delta L}\right)$ on the right-hand side of Eq. (43) is the pressure drop source by PM where Ergan equation is used [8–10, 12–14].

$$\left(\frac{\Delta P}{\Delta L}\right) = \left(\frac{\mu}{\alpha}u + c_2\frac{1}{2}\rho_g|u|u\right) \tag{45}$$

$$\alpha = \frac{{d_p}^2}{150}\frac{\varepsilon^3}{(1-\epsilon)^2}$$

$$c_2 = \frac{3.5}{d_p}\frac{(1-\varepsilon)}{\varepsilon^3}$$

5.3 Gas phase energy equation with gaseous fuel injection

$$\frac{\partial}{\partial t}\left(\varphi\rho c_p T_g\right) + \nabla.\left(\varphi\rho c_p T_g u\right) + \varphi\sum_i \dot{\omega}_i h_i W_i = -\varphi P\nabla.u + \varphi A_0\rho\epsilon + (1-A_0)\sigma : \nabla u + \varphi\nabla.$$
$$\left(\left(k_g + \rho_g c_g D_{||}^d\right)\nabla T_g\right) - h_v\left(T_g - T_s\right) + \dot{Q}^s \tag{46}$$

where T_g is the temperature of gas, φ is the PM porosity, u is the velocity vector, h_i is the enthalpy of species i, k_g is the fluid thermal conductivity, ${D_{//}}^d$ is the thermal dispersion coefficient along the length of the PM, and h_v is the volumetric heat transfer coefficient. The term $(\varphi\nabla.\left(\left(k_g + \rho_g c_g D_{\parallel}^d\right)\nabla T_g\right))$ is added to the energy equation of conduction heat transfer in the fluid phase of PM and longitudinal dispersion of mixture in PM. The term $(h_v(T_g - T_s))$ is added to clarify convective heat transfer between gas and solid phases of PM. Heat exchange between solid and gas phases is computed according to convective heat transfer derived by Wakao and Kaguei to estimate heat transfer between packed beds and fluid [10–14].

5.4 Gas phase energy equation with liquid fuel injection

$$\frac{\partial}{\partial t}(\varphi\rho c_p T_g) + \nabla.(\varphi\rho c_p T_g u) + \varphi\sum_i \dot{\omega}_i h_i W_i = -\varphi P\nabla.u + \varphi A_0\rho\epsilon + (1 - A_0)\sigma : \nabla u + \varphi\nabla.$$
$$\left(\left(k_g + \rho_g c_g D_{\parallel}^d\right)\nabla T_g\right) - h_{gs}(T_g - T_s) + (1-\delta)h_{gl}A_p(T_g - T_l) - \delta\,\dot{m}_p H_{gl} \tag{47}$$

$$\delta = \begin{cases} 0 & T_g < T_{sat} \\ 1 & T_g = T_{sat} \end{cases}$$

The term $h_{gs}(T_g - T_s)$ is added to represent volumetric convective heat transfer between gas and solid phases of PM. The term $(1-\delta)h_{gl}A_p(T_g - T_l)$ is the heat transfer among the gas phase and liquid fuel droplets where liquid droplets are lower than saturation temperature of liquid fuel. δ is the Kronecker delta function for sensible energy of fuel droplets and latent heat of vaporization. A_p is the droplet surface area, and T_l is the liquid droplet temperature. **Figure 6** illustrates schematic heat transfer between gas, liquid and solid phases in PM space.

The heat transfer between the gas and solid phases is calculated according to Eq. (47). Wakao and Kaguei derived that heat transfer between solid phase and hot gas inside PM [10–14].

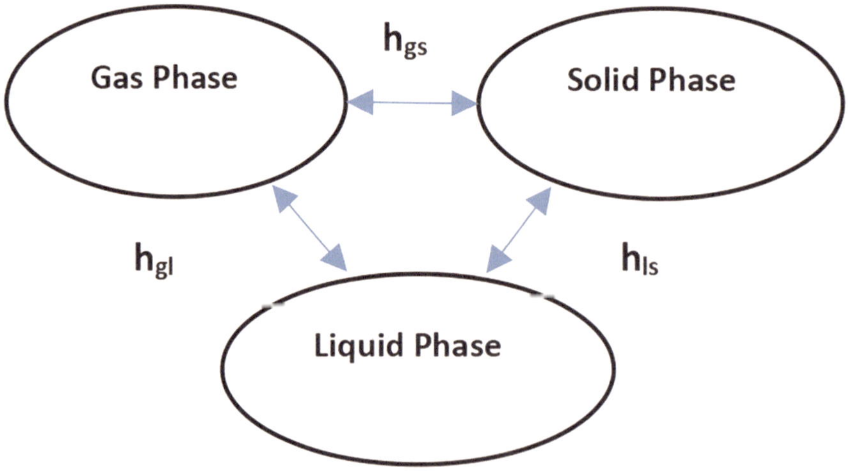

Figure 6.
Schematic heat transfer between gas, liquid and solid phases in PM space.

$$h_{gs} = \frac{6\varphi}{{d_p}^2} k_g \left(2.0 + 1.1\ Re^{0.6} Pr^{0.33}\right) \tag{48}$$

The heat transfer among the gas phase and liquid droplets is computed according to Eq. (48). Correlation (48) was derived by Ranz and Marshal for heat transfer among liquid droplets and gas phase during spray [15–17].

$$h_{gl} = \frac{k_g}{d_p} \left(2.0 + 0.6\ {Re_p}^{0.5} Pr^{0.33}\right) \tag{49}$$

5.5 Solid phase energy equation

$$\frac{\partial}{\partial t}((1-\varphi)\rho_s c_s T_s) = \nabla \bullet [k_s(1-\varphi)\nabla T_s] + h_{gs}(T_g - T_s) - \nabla . q_r \tag{50}$$

The term (1-ϕ) is due to the solid volume of PM. T_s is the solid phase temperature, k_s is the thermal conductivity of solid phase of PM, ρ_s is the density, and c_s is the specific heat of solid phase of PM. q_r is the radiation heat loss of solid. Considering the high capacity of the solid phase of PM and the low volume of liquid droplets, the effect of heat transfer between the solid phase and liquid droplets on the energy equation of the solid phase is neglected.

5.6 Turbulence model

Standard κ- ε equations without modifications were used.
The transport equation for κ turbulent kinetic energy [8–10, 12–14]:

$$\frac{\partial(\rho k)}{\partial t} + \nabla.(\rho u\ k) = -\frac{2}{3}\ \rho\ \kappa \nabla.u + \sigma : \nabla u + \nabla.\left[\left(\frac{\mu}{Pr_k}\right)\nabla k\right] - \rho\varepsilon + \dot{W}^s \tag{51}$$

with a similar one for the dissipation rate ε:

$$\frac{\partial(\rho\varepsilon)}{\partial t} + \nabla.(\rho u\ \varepsilon) - \left(\frac{2}{3}c_{\epsilon_1} - c_{\epsilon_3}\right)\rho\ \varepsilon\nabla.u + \nabla.\left[\left(\frac{\mu}{Pr_\varepsilon}\right)\nabla\epsilon\right] + \frac{\epsilon}{k}\left[c_{\epsilon_1}\sigma : \nabla u - c_{\epsilon_2}\rho\varepsilon + c_s \dot{W}^s\right] \tag{52}$$

6. Equation of state

$$P = \rho_g R\ T/\overline{W} \tag{53}$$

6.1 Combustion model

Chemical mechanism for oxidation of Decane or other hydrocarbon fuels is considered. $\dot{\omega}_i$ chemical production rate:

$$\dot{\omega}_i = \sum_{i=1}^{NR} (v''_{k,i} - v'_{k,i}) R_i \tag{54}$$

$v''_{k,i}$ and $v'_{k,i}$ are stoichiometric coefficients. The Arrhenius model calculates reaction rates. For other equations, the reaction rate is very quickly relative to the main equations; equilibrium reactions are considered. In order to calculate the effects of turbulence on combustion, Spalding's eddy-breakup model is considered. The idea of the eddy-breakup model is that the combustion rate is computed by the rate at which large parcels of unburned gas are broken down into smaller particles. The turbulence length scale is significant in determining turbulent burning rates [8–10, 12–14].

6.2 Radiation model

Because of the high temperature of the combustion zone and solid phase, radiation heat loss should be considered. Absorption coefficient of the solid phase of PM is higher than that of the gas phase; hence, gas phase radiation loss in analogy with solid phase radiation loss can be ignored. Several relations for modeling of radiation intensity were presented in the literatures. The heat source term $\nabla .q_r$, due to radiation in the solid phase of PM in Eq. (54), is computed by the Rosseland method [18].

$$q_r = -\frac{16}{3}\frac{\sigma_b T_s^{\ 3}}{\beta}\nabla T_s \tag{55}$$

6.3 Gas injection model

Gaseous fuel (methane, propane, hydrogen) is directly injected into high-temperature of the PM volume. The gaseous fuel injection model's detail can be found in Ref [5]. The model is for simulating transient direct injection of gaseous fuel into the combustion chamber using a logically refined computational grid [8–10, 12–14].

6.4 Liquid fuel injection

The essential dynamics of a fuel spray and its interactions with an in-cylinder flow are very complex problems. To compute the mass, momentum, and energy exchange among spray and gas, the distribution of drop sizes, velocities, and temperatures should be determined. In many sprays, drop Weber numbers are more significant than unity, and drop oscillations, distortions, and breakup must be calculated. Drop collisions and coalescence in a diesel engine can be significant in many engine sprays [19–22]. Jet interaction with the PM has four phases [3]:

Phase A: free jet formation from outlet of the nozzle to PM surface.

Phase B: multi-jet splitting as a result of jet interaction with PM surface.

Phase C: liquid fuel distribution in the PM space.

Phase D: liquid fuel passes through the PM space.

Therefore, due to four phase of interaction, particles motion and energy equation need to be modified. **Figure 7** displays schematic modeling of liquid-fuel jet impingement with PM.

6.5 Particles motion equation

With the injection of liquid droplets, drag force is applied to particles. By impingement of liquid droplets on PM, more drag force is applied to droplets. Hence, the drag coefficient should be modified with available correlation for the

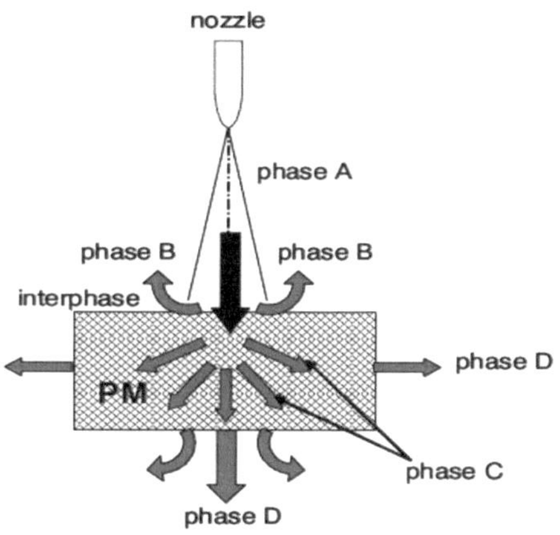

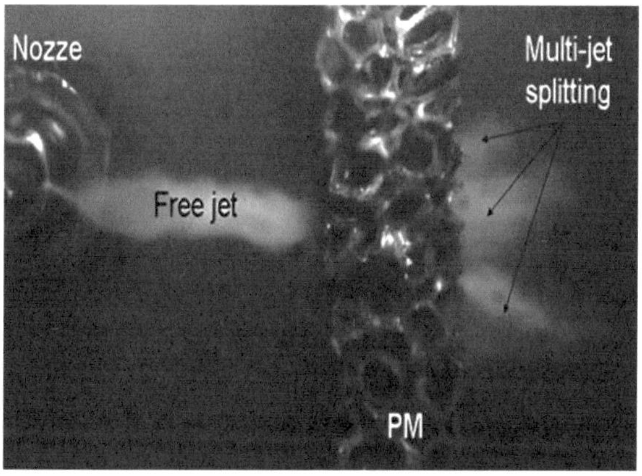

Figure 7.
Displays schematic modeling of liquid-fuel jet impingement with PM [1].

impingement of liquid droplets on a single cylinder. Eq. (55) shows the equation of droplets' motion. μ is the viscosity, ρ_p is the density of liquid fuel, and Re_p is the Reynolds number of droplets based on velocity difference among droplets and in-cylinder fluid. C_d is the modified drag coefficient of the surface of droplets [23–25].

$$\frac{du_p}{dt} = \frac{18\,\mu}{\rho_p d_p^2} \frac{C_D\, Re_p}{24} (u - u_p) \tag{56}$$

$$C_D = \begin{cases} -0.44 + 0.001\ Re_p + \dfrac{5.64}{Re_p} + \dfrac{5.04}{{Re_p}^{0.28}} & Re_p < 400 \\ 1.1 & Re_p \geq 400 \end{cases}$$

6.6 Energy equation for particle (heating-evaporation equation)

The droplet temperature is computed according to a convective heat transfer between the gas and solid phases of PM and latent heat transfer among the droplets with solid and fluid phases for PM. The total convective heat transfer coefficient from in-cylinder gas to liquid droplet is calculated by Eq. (56). Heat transfer coefficient among liquid droplets and solid phase of PM is computed by consideration of two phenomena. The first term is heat exchange among multi-jet splitting with the solid phase of PM modeling by jet impingement of liquid spray on a hot wall. This heat transfer coefficient was calculated by Eq. (57) and was derived by Rosenow [15]. The second term is heat exchange among the solid phase of PM and gas phase flow in PM that is modeled by heat transfer hot wall to the vaporized fuel-air mixture. This heat transfer coefficient was computed by Eq. (58) and derived by McAdams [11, 20]. The heat exchange process of liquid droplets and gas phase is highly complex in PM-volume and has not been clearly understood. That which correlation (57) or (58) is dominated during heat transfer, and what is a portion of Eqs. (57) and (58) in heat transfer from the solid phase of PM to the gas phase of PM and liquid droplets. Hence, the random number α, where α∈0,1, is inserted in Eq. (59), which is generated by the programming language in each time step. This random number determined the portion of each term in Eqs. (57) and (58) on the total heat transfer coefficient (Eq. (59)). The effect of radiation heat transfer through gas and solid phases of PM on the temperature of liquid droplets has been ignored due to a low volume of liquid droplets [10–16].

$$m_p C_p \frac{dT_p}{dt} = (1-\delta) h_{gl} A_p \left(T_g - T_l\right) + (1-\delta) h_{sl} A_p (T_s - T_l) + \delta \dot{m}_p H_{gl} \tag{57}$$

$$h_{sl-1} = \frac{v_0 H_{lg} \ \rho_l}{T_s - T_l} \ exp\left[1 - \left(\frac{T_s}{T_{sat}}\right)^2\right] \tag{58}$$

$$h_{sl-2} = 0.023 \ \frac{k_g}{d} \ Re^{\,0.8} Pr^{0.4} \tag{59}$$

$$h_{sl} = \alpha \, h_{sl-1} + (1-\alpha) h_{sl-2} \tag{60}$$

7. Conclusions

This chapter illustrates the mathematical modeling of PM diesel engines: thermodynamically modeling, zero-dimensional closed cycle modeling with chemical kinetics of PM diesel engine, and three-dimensional CFD modeling with PM diesel engine chemical kinetics. So, mathematical modeling of PM diesel engines, from simple thermodynamically modeling to complicated 3D modeling, has been described.

Author details

Arash Mohammadi
Shahid Rajaee Teacher Training University, Tehran, Iran

*Address all correspondence to: amohammadi@sru.ac.ir

References

[1] Weclas M. Potential of porous-media combustion technology as applied to internal combustion engines. Journal of Thermodynamics. 2010;**2010**(39): 789262. DOI: 10.1155/2010/789262

[2] M. Weclas, Porous Media in Internal Combustion Engines, 2005, Cellular Ceramics: Structure, Manufacturing, Properties and Applications, Editor(s): Michael Scheffler, Ing. Paolo Colombo, Weinheim: WILEY-VCH Verlag GmbH & Co. KGaA; DOI: 10.1002/3527606696.ch5j

[3] Abdul Mujeebu M, Mohamad AA, Abdullah MZ. Chapter 24-applications of porous media combustion technology. In: Fanun M, editor. The Role of Colloidal Systems in Environmental Protection. USA: Elsevier B.V; 2014. pp. 615-633

[4] Chalia S, Kumar Bharti M, Thakur P, Thakur A, Sridhara SN. An overview of ceramic materials and their composites in porous media burner applications. Ceramics International. 2021;**47**(8): 10426-10441

[5] Hongsheng L, Maozhao X, Dan W. Simulation of a porous medium (PM) engine using a two-zone combustion model. Applied Thermal Engineering. 2009;**29**(14–15):3189-3197

[6] Hongsheng L, Maozhao X, Dan W. Thermodynamic analysis of the heat regenerative cycle in porous medium engine. Energy Conversion and Management. 2009;**50**(2):297-303

[7] Abhisek B, Diplina P. Developments and applications of porous medium combustion: A recent review. Energy. 2021;**221**:119868

[8] Weclas M, Cypris J, Maksoud TMA. Thermodynamic properties of real porous combustion reactor under diesel engine-like conditions. Journal of Thermodynamics. 2012;**2012**:798104, 11 pages

[9] Mohammadi A, Jazayeri A, Ziabasharhagh M. Numerical simulation of porous medium internal combustion engine. In: Proceedings of the ASME-JSME-KSME 2011 Joint Fluids Engineering Conference. Vol. 1. Hamamatsu, Japan: Symposia – Parts A, B, C, and D; July 24–29, 2011. pp. 1521-1529. ASME. DOI: 10.1115/AJK2011-03079

[10] Saghaei M, Mohammadi A. Thermodynamic simulation of porous-medium combustion chamber under diesel engine-like conditions. Applied Thermal Engineering. 2019;**153**:306-315

[11] Mohammadi A, Benhari M, Nouri Khajavi M. Numerical study of combustion in a porous medium with liquid fuel injection. In: Proceedings of the ASME 2017 International Mechanical Engineering Congress and Exposition. Vol. 8. Tampa, Florida, USA: Heat Transfer and Thermal Engineering; November 3–9, 2017. V008T10A033. ASME. DOI: 10.1115/IMECE2017-72483

[12] Mohammadi A, Varmazyar M, Hamzeloo R. Simulation of combustion in a porous-medium diesel engine. Journal of Mechanical Science and Technology. 2018;**32**:2327-2337. DOI: 10.1007/s12206-018-0444-x

[13] Mohammadi A, Benhari M. Simulation of combustion in a porous reactor. In: Proceedings of the ASME 2014 12th Biennial Conference on Engineering Systems Design and Analysis. Vol. 2. Copenhagen, Denmark: Dynamics, Vibration and Control; Energy; Fluids Engineering; Micro and Nano Manufacturing; July 25–27, 2014.

V002T11A014. ASME. DOI: 10.1115/ESDA2014-20205

[14] Mohammadi A, Jazayeri A, Ziabasharhagh M. Numerical simulation of direct injection engine with using porous medium. In: Proceedings of the ASME 2012 Internal Combustion Engine Division Spring Technical Conference. ASME 2012 Internal Combustion Engine Division Spring Technical Conference. Torino, Piemonte, Italy: ASME; May 6–9, 2012. pp. 785-795. DOI: 10.1115/ICES2012-81150

[15] Ganich EN, Rohsenow WM. Dispersed flow heat transfer. International Journal of Heat and Mass transfer. 1990;**33**:847-857

[16] Senecal PK, Schmidt DP, Nouar I, Rutland CJ, Reitz RD, Corradini ML. Modeling high-speed viscous liquid sheet atomization. International Journal of Multiphase Flow. 1999;**25**:1073-1097

[17] Ranz WE, Marshall WR. Evaporation from drops. Journal of Chemical Engineering Progress. 1952;**48**: 141-146

[18] Modest MF. Radiative Heat Transfer. California USA: Academic Press; 2013

[19] Cypris J, Schlier L, Travitzky N, Greil P, Weclas M. Heat release process in three-dimensional macro-cellular SiC reactor under diesel-like conditions. Journal of Fuel. 2012;**102**:115-128

[20] Martynenko VV, Echigo R, Youshida H. Mathematical model of self-sustaining combustion in inert porous medium in inert porous media with phase change under complex heat transfer. Journal International Journal Heat and Mass Transfer. 1998;**41**(1):117-126

[21] Periasamy C, Saboonchi A, Gollahalli SR. Numerical Prediction Evaporation Processes in Porous Media Combustor. USA: DETC conference; 2007

[22] Periasamy C, Chinthamony SK, Gollahalli SR. Experimental evaluation of evaporation enhancement with porous media in liquid-fuel burners. Journal of Porous Media. 2007;**10**(2):137-150

[23] O'Rourke PJ, Amsden AA. A Spray/Wall Interaction Submodel for the KIVA-3 Wall Film Model. SAE Technical Paper Series; 2000

[24] Diaz A, Ortega A, Anderson R. Numerical investigation of a liquid droplet impinging on a heated surface. In: Proceedings of the ASME 2009 International Mechanical Engineering Congress and Exposition. Vol. 9. Lake Buena Vista, Florida, USA: Heat Transfer, Fluid Flows, and Thermal Systems, Parts A, B and C; November 13–19, 2009. pp. 1769-1775 ASME

[25] Soriano G, Alvarado L, Lin P. Experimental characterization of single and multiple droplet impingement on surfaces subject to constant heat flux conditions. In: Proceedings of the 2010 14th International Heat Transfer Conference. Vol. 6. Washington, DC, USA: 2010 14th International Heat Transfer Conference; August 8–13, 2010. pp. 707-715 ASME

Chapter 12

Modeling of Thermal Conductivity in Gas Field Rocks

Chis Timur, Jugastreanu Cristina, Tabatabai Seyed Mehdi and Renata Radulescu

Abstract

The thermal conductivity of rocks is a property necessary to be determined at the beginning of the exploitation of oil and gas deposits, both for the design of secondary extraction (hot water injection, steam) and for the development of tertiary extraction technologies (CO_2 injection, injection flue gas, and initiation of underground combustion). In this chapter, we present a new method for determining the thermal conductivity of rocks and we also analyzed the relationships between this parameter and the properties of oil and gas collector rocks (density, porosity).

Keywords: thermal, conductivity, oil, gas, fields, rocks, properties

1. Introduction

The need to discover new deposits of useful mineral substances led to the study of the physicochemical properties of the soil and subsoil constituents (geological layers).

From the beginning of the geophysical research of the subsoil, the knowledge of the temperature of rocks and constituent fluids was an absolutely necessary step to establish working conditions in tunnels and mine shafts (to prevent the formation of explosive mixtures and explosions) [1].

Subsequently, the geothermal phenomena (geothermal gradient and geothermal stage) were analyzed, in order to improve the exploitation of oil and gas deposits [2]

Knowing the geothermal flow of geological layers was useful for determining the temperature of the crust and the structure of the lithosphere and also for understanding how to form oil and gas deposits.

The exploitation of crude oil and gas from the Moesic Platform in Romania has created a database regarding the understanding of the thermals of geological strata and especially the formation of oil deposits in magmatic fields [3, 4].

Theoretical and practical aspects of the use of geothermal steps in the analysis of deposits of useful mineral substances were made by Dowle and Cobb [5].

The interest in determining the geothermal of the subsoil and in particular the geothermal of the oil and gas deposits were due to the following [6–8]:

a. Research into the evolution of the temperature of geological structures in order to recover heat and produce renewable energy,

b. Determining the mode of secondary and tertiary recovery of oil for the application of methods for the injection of flue gas, steam, hot water, and CO_2 into the field,

c. The need to determine the geological period for the formation of oil and gas deposits and the optimal way to extract them,

d. Analysis of the evolution of the waterfront in oil and gas fields,

e. Research on ways to isolate geological layers by cementation.

The calculation of the thermal flow was performed by determining the thermal gradients and thermal conductivity in measurement points and subsequently by determining in the laboratory the physical properties of the rocks (cores) collected from geological research drilling [9].

The determined values had an estimative character (being often point values), but they were useful in determining the thermal structure of the geological layers [10].

Thus, following the measurements of the temperature of the oil fluid extraction wells, a low thermal flux was determined (45–57 mW/m^2 and 33–58 mW/m^2) in the areas rich in gas deposits and quite high in the areas with coal deposits (200 mW/m^2).

But the most important physical property of rocks and the constituent fluids of oil and gas deposits, namely thermal conductivity, is very useful in establishing the tertiary oil recovery system and in determining the flow of fluids through rock pores.

That is why this property is determined in the laboratory, by analyzing the heat transfer that passes through the rocks collected from the oil fields.

Thermal conductivity (k) is the property (ability) of rocks and continuous media to transmit, to a greater or lesser degree, thermal energy (relation 1) [11].

$$k = \frac{Q}{(t_2 - t_1) \bullet \left(\frac{s}{l}\right) \bullet \tau} [\mathrm{w/m\ ^\circ C}] \tag{1}$$

In Eq. (1), Q is the amount of heat that passes through rock with cross section s, in a time τ, and with a length l.

Factors influencing rock conductivity to take into account the structural-textural peculiarities of rocks (composition, size and orientation of rock granules, porosity, and fluid content), the temperature measured at the faces of the analyzed rocks (t_2, t_1), and the pressure to which the structure is subjected: geological analysis [12].

The conductivity of the rocks decreases as the size of the rock granules decreases because the number of contacts between the granules and the heat flux increases in the flow of fluids through the rock pores.

Also, the orientation of the rock granules influences their thermal conductivity, schistosity, stratification, and fracturing, reducing their values.

The thermal conductivity of dry porous rocks is lower than that of compact rocks (air with a thermal conductivity value of 0.55 mcal/cm°C s).

At the same time, the flow of fluids or the fluid content of rocks changes the value of thermal conductivity, and rocks with water have higher conductivity than those containing oil or natural gas.

The pressure in the geological layers not influences the values of thermal conductivity.

But the increase of the pressure increasing the conductivity due to the friction of the rock particles.

The increase in temperature leads to a decrease in conductivity due to the increase in the interaction speed of the particles of the crystal lattice.

The determination of the thermal conductivity of the rocks is necessary for the evaluation of the thermal flow of the deposit and especially for the choice of the most useful technology in increasing the recovery factor of crude oil and petroleum gases.

2. Measurement of rocks conductivity

The method of determining the thermal conductivity of rocks and constituent fluids is to measure the amount of heat that passes through a system consisting of two discs of known rocks (quartz) and which includes the rock disc under analysis (**Figure 1**).

The system is placed in a thermostatic bath, measuring the temperature difference between the three discs.

After reaching the thermal equilibrium, the values Q_1, Q_2, and Q_3 are almost equal and the heat transfer relation becomes [13]:

$$2Q_2 \cong Q_1 + Q_3 \tag{2}$$

As explained in relation 1, we obtain the value of thermal conductivity:

$$k_r = \frac{k_q \frac{\Delta t_1}{z_1} S_1 + k_q \frac{\Delta t_3}{z} S_3}{2 \frac{\Delta t_2}{z_2} S_2} \tag{3}$$

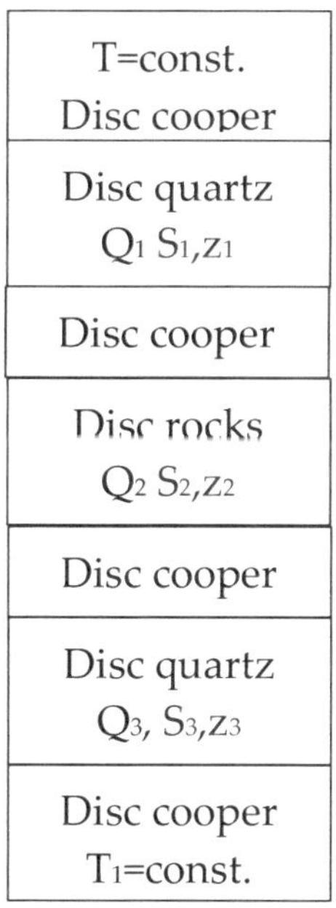

Figure 1.
Thermal conductivity measurement system (split bar method) [1, 9, 13].

where k_r is the thermal conductivity of the rock sample with cross section S_2 and thickness z_2, k_q is the thermal conductivity of quartz discs with cross section S_1, S_3 and thickness z_1, z_3.

The thermal conductivity of crystalline quartz can be determined by the relation [10, 12–14]:

$$k_q = \frac{1}{60{,}7 + 0{,}242t} \tag{4}$$

Figure 2 shows a device for determining the thermal conductivity by the plate method, with a single rock sample.

The mathematical model for determining the thermal conductivity of rocks in the exploitation areas of oil and gas deposits starts from the equation:

$$k_r = k_q \left(\frac{\Delta t_1 + \Delta t_3}{2\Delta t_2} \right) \left(\frac{z_2}{z} \frac{s}{s_2} \right) \tag{5}$$

where k_r is the thermal conductivity of the rock sample with cross section S_2 and thickness z_2, and k_q is the thermal conductivity of quartz discs with cross section S_1, S_3 and thickness z_1, z_3.

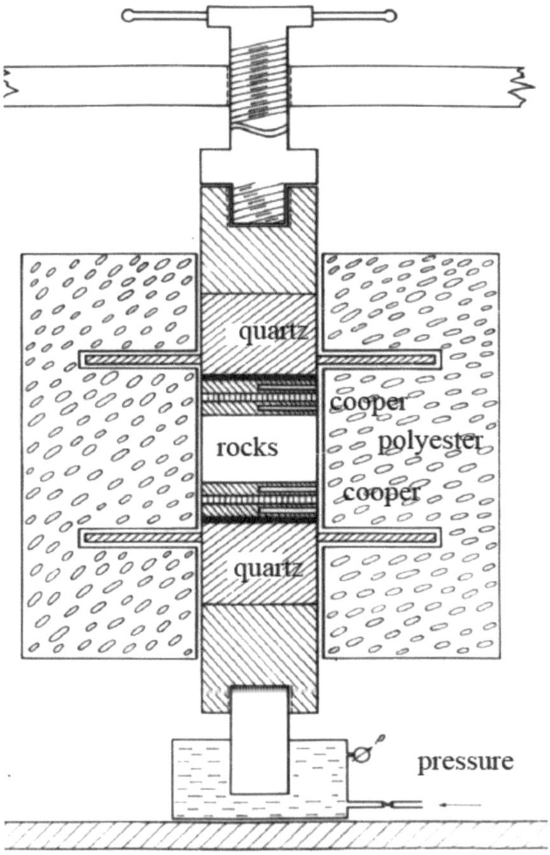

Figure 2.
Device for determining the thermal conductivity by the plate method, with a single specimen [9, 13, 14].

For ease of determinations, identical quartz dikes were made, and then:

$$S_1 = S_3 = S \tag{6}$$

$$z_1 = z_3 = z \tag{7}$$

Logarithming Eq. (5) we get:

$$log\,k_r = log\,k_q + \log(\Delta t_1 + \Delta t_3) - log\,2 - log\,\Delta t_2 + log\,z_2 - logz + logs - log\,s_2 \tag{8}$$

To reduce the thermal resistance to the contact between the crucible and the analyzed rock, a thin layer of vaseline is used.

Another method for determining the thermal conductivity starts from the knowledge of the mineralogical composition and porosity.

Thus, knowing that rocks from oil and gas deposits are characterized by a liquid phase (the pore space saturation fluid) and a solid phase (the mineral skeleton), the thermal conductivity of the analyzed rock depends on the thermal conductivity of the two constituent phases.

If the two phases are oriented parallel to each other (a maximum thermal conductivity value is obtained), then:

$$k_{max} = pk_f + (1 - p)k_m \tag{9}$$

where k_{max} is the maximum conductivity of the analyzed rock, p represents the porosity of the rock, k_f tests the conductivity of the fluid phase, k_m represents the conductivity of the matrix (mineral skeleton).

If the two phases are oriented in series (which gives a minimum value of the thermal conductivity), we can write the equation of the conductivity of the deposit swarm, as:

$$\frac{1}{k_{min}} = \frac{p}{k_f} + \frac{1-p}{k_m} \tag{10}$$

For an average thermal conductivity value, the following relationship is used:

$$k = k_f^p \bullet k_m^{1-p} \tag{11}$$

Logarithmizing Eq. (11), we can obtain a linear relationship between the effective conductivity of the porous medium and the effective conductivity of the constituent fluid, namely:

$$\log k = p\ log\,k_f + (1-p)\,log\,k_m \tag{12}$$

For rocks with low porosity and complex mineralogical composition, their conductivity can be obtained based on the relations:

$$\frac{1}{k_{min}} = \frac{V_1}{k_1} + \frac{V_2}{k_2} + \ldots + \frac{V_n}{k_n} \tag{13}$$

$$k_{max} = V_1k_1 + V_2k_1 + \ldots + V_nk_n \tag{14}$$

where $V_1, V_2 \ldots, V_n$, are the volumes of the mineral fractions, 1, 2 , ... , n, and $k_1, k_2, \ldots, k_n$, are the thermal conductivity of minerals.

3. Mathematical modeling

The equation that best describes thermal conductivity is given by the relationship:

$$\frac{\partial t}{\partial \tau} = a\left(\frac{\partial^2 t}{\partial x^2} + \frac{\partial^2 t}{\partial y^2} + \frac{\partial^2 t}{\partial z^2}\right) \tag{15}$$

where x, y, and z are the heat dissipation directions, and t is the heat flow temperature, measured after time τ.

Sometimes it is used as a way of calculation, writing the thermal conductivity in spherical coordinates:

$$\frac{\partial t}{\partial \tau} = a\left(\frac{\partial^2 t}{\partial r^2} + \frac{1}{r}\frac{\partial t}{\partial r} + \frac{1}{r^2}\frac{\partial^2 t}{\partial \theta^2} + \frac{\partial^2 t}{\partial z^2}\right) \tag{16}$$

And from Eq. (16) it is written:

$$\nabla^2 t = -\frac{1}{a}\frac{\partial t}{\partial \tau} = -\frac{A}{k} \tag{17}$$

Given that in the stabilized regime $\frac{\partial t}{\partial \tau} = 0$, ecuația 17 se Eq. (17) turns into Poisson's equation, namely:

$$\nabla^2 t = -\frac{A(x, y, z)}{k} \tag{18}$$

where $A(x,y,z)$ is the amount of heat that dissipates in the analyzed rock volume, and k represents the thermal conductivity of the rocks.

Oil drill	Depth	Rocks structures	Geological layers	Density g/cm³	Porosity cm	Thermal conductivity (W/m K)
Valea Rației	2334–2335	Clay	Ponțian	2,46	4,20	0,57
Valea Rației	2530–2531	Hone	Ponțian	2,56	4,20	1,24
Valea Rației	2802–2810	Hone	Ponțian	2,80	4,40	1,37
Căldărușanca	2608–2611	Hone	Meoțian	2,10	3,92	0,79
Căldărușanca	3356–3364	Sandstone marl	Meoțian	2,28	4,05	0,42
Boldești	2942–2943,5	Hone	Helvețian	2,30	4,45	0,55
Boldești	4177–4179	Marl	Helvețian	2,52	4,20	1,27
Boldești	4468–4469	Marl	Helvețian	2,65	4,00	1,28
Izvoare	1216–1217	Clay	Eocen	2,46	4,20	1,30
Izvoare	2243–2245	Hone	Eocen	2,50	3,95	0,56
Izvoare	2841–2844	Hone	Eocen	2,60	4,05	1,01
Moreni	884–887	Hone	Helvețian	2,16	4,9	0,98
Moreni	1627–1630	Hone	Helvețian	2,54	4,5	1,08
Moreni	1730–1740	Hone	Helvețian	3,75	4	1,12

Table 1.
Analysis of the productive states of the studied deposits (oil and gas) (Moesic platform) [9, 13, 14].

Geological structures	Thermal conductivity equation (y) as a function of density (x)	Thermal conductivity equation (y) as a function of porosity (x)	Density equation (y) as a function of porosity (x)
Valea Rațe	$y = -18{,}113x^2 + 97{,}626x - 129{,}98$	$y = -68333x^2 + 55{,}417x - 110{,}97$	$y = 46569x^2 - 25{,}377x + 38{,}647$
Căldărușanca	$y = 53704x^2 - 22{,}8x + 24{,}987$	$y = -0{,}539x^2 + 34{,}402x - 42{,}826$	$y = -14{,}444x^2 + 62{,}433x - 63{,}34$
Boldești	$y = -91309x^2 + 47{,}284x - 59{,}9$	$y = -62889x^2 + 51{,}519x - 104{,}17$	$y = -11489x^2 + 44{,}011x + 0{,}4049$
Moreni	$y = -0{,}2447x^2 + 0{,}9433x - 0{,}3824$	$y = -0{,}1889x^2 + 15{,}256 - 1{,}96$	$y = 0{,}4021x^2 - 29{,}427x + 9{,}38$
Izvoare	$y = -59{,}643x^2 + 307{,}08x - 393{,}92$	$y = -40{,}476x^2 + 316{,}83 - 618{,}7$	$y = 39286x^2 - 0{,}736x + 31{,}286$

Table 2.
Equations for simulating thermal conductivity as a function of porosity and density and density as a function of porosity.

Rocks	Thermal conductivity, (determinată) λ, (W/m°C)	Density, value of literatures [14] ρ, (kg/m^3)	Thermal conductivity value of literature [14] k_r (W/m°C)	Absolute error, thermal conductivity	Density determinations, ρ, (kg/m^3)	Absolute error, density
clay	0,57	0,55	0,03508772	2,46	2,5	0,0162602
tiles	1,24	1,5	0,20,967,742	2,8	2,57	0,0821429
Sandy marl	0,42	0,66	0,57,142,857	2,28	11,455	0,4,975,877

Table 3.
Differences between the thermal conductivity of rocks drilling and density (determined values and values in the literature).

The cores (rocks) collected from the oil deposits in the Moesic platform of Romania were subjected to a heat transfer by measuring the temperature variations on each side of the rocks and crystals.

The density was achieved by weighing and measuring the volume, the determination error being 0.1% g/cm^3.

The porosity was determined by drying the cores and filling them with helium (according to Boyle-Marriote's law pV = const.) in a controlled closed tank.

The data measured in the laboratory are shown in **Tables 1** and **2**.

In order to model the geothermal structure of the analyzed deposit, we created numerical equations that best describe the variation of thermal conductivity depending on the density and porosity of the constituent rocks.

The equation form is:

$$y = ax^2 + bx + c \tag{19}$$

where a, b, and c are experimentally determined coefficients.

The absolute error of the determined values are compared to the values from the specialized literature we determined with the relation (**Table 3**):

$$AE = \frac{x_{models} - x_{international\ papers}}{x_{models}} \tag{20}$$

The error is a maximum of 0,5 in conductivity, in the case of shale clay, due to the fact that it was not pure.

The density errors are large because the chosen cores were not pure, being impure with other materials.

4. Conclusion

Research on the fluidity of oil and gas deposits also led to the understanding of the role of thermal conductivity in studying the phenomena of hydrocarbon migration and their formation.

Also, the use of thermal conductivity in determining the secondary (injection of gases and hot water into the deposit) and tertiary (underground combustion, steam injection into the deposit) recovery mode created the need for the experimental determination of this property of the constituent rocks.

Tikhomirov scientifically established based on the determination of the thermal conductivity of the rocks constituting the oil and gas deposits that the calculation relationship of this property (thermal conductivity) is [1, 14]:

$$k_{sat,T} = 26{,}31 e^{0{,}6\ \rho + 0{,}6\ S_A} T^{-0{,}55} \tag{21}$$

where $k_{sat,T}$ is the thermal conductivity of the rock saturated with the constituent fluids, ρ is density (g/cm^3), T is temperature of determination (°C).

The author estimates an error in the thermal conductivity determination of 16%.

Cermak also used a relationship to determine conductivity as a function of density (for calcareous rocks) [1, 14]:

$$k_{sat} = 2{,}15 \bullet 10^{-3} \rho - 3{,}16 \tag{22}$$

For carbonate rocks, the above relation can be written [1, 14]:

$$k_{sat} = 4{,}18 \bullet 10^{-3} \rho - 7{,}97 \tag{23}$$

In relations (22) and (23) the density is expressed in kg/m^3.

But even these equations give errors of over 15%.

But our calculation method, namely the statistical determination of thermal conductivity as a function of porosity and density, indicated a very small error (maximum 0.5%).

In this chapter, the conductivity of fluid-saturated rocks can also be determined.

As can be seen, the data in the literature are very close to the determined values.

Also in this book chapter, we managed to model the thermal conductivity depending on the density and porosity of the rocks.

The equations that best describe this behavior are polynomial, the values of the coefficients being a function of the porosity and density of the rocks.

Author details

Chis Timur[1*], Jugastreanu Cristina[2], Tabatabai Seyed Mehdi[2] and Renata Radulescu[2]

1 Ovidius University Constanta, Romania

2 Oil-Gas University, Ploiesti, Romania

*Address all correspondence to: tchis@univ-ovidius.ro

References

[1] Jugastreanu C, Petrache S, Chis T. Thermal regime analysis in areas with hydrocarbon potential. In: 21st International Scientific Multidisciplinary Conference, on Earth and Planetary Sciences SGEM 2021, Conference. DOI: 10.5593/sgem2021/1.1/s.06126, ISBN: 978-619-7603-20-0

[2] Cristescu T, Vasiliu EV. Evaluation of the thermal resistance of certain earth crust layers based on geophysical properties obtained from laboratory analysis of drilling samples, Revista de chimie (Bucharest). 2014;**65**(3):350-353. Available from: www.revistadechimie.ro

[3] Demetrescu C, Ene M, Andreescu M. New heat flow data for the Romanian territory. Anuarul Institutului de Geologie Geofizica. 1983;**6**:45-57

[4] Demetrescu C, Veliciu S, Burst D. Heat flow map of Romania. In: Hurtig E, Cermak V, Hanel R, Zui V, editors. Geothermal Atlas of Europe. Gotha: VEB Hermann Haack; 1991

[5] Dowle WL, Cobb WM. Static formation temperature from well logs–An empirical method. Journal of Petroleum Technology. 1975;**27**:1326-1330

[6] Hall NH. Compressibility of reservoir rocks. Transactions of AIME. 1953;**5**:198

[7] Kappelmayer O, Haenel R. Geothermics with special reference to application. Gebruder Borntraeger, Berlin-Stuttgart. 1974;**4**:238

[8] Monicard RP. Properties of reservoir rocks: Core analysis. Editions Technip. 1980:21-221

[9] Negoiță V. Etude sur la distribution des temperatures en Roumanie. Rev. Roum.Geol.Geophys., Geogr.Se. Geophysique. 1970;**14**(1):25-30

[10] Popov Y, Spasennykh M, Shakirov A, Chekhonin E, Romushkevich R, Savelev E, et al. Advanced determination of heat flow density on an example of a West Russian Oil Field. Geosciences. 2021;**11**:346. DOI: 10.3390/geosciences 11080346

[11] Satter A, Thakur G. Integrated Petroleum Reservoir. Tulsa, Oklahoma: Pennwell Books; 1994. pp. 22-254

[12] Sonney R. Groundwater flow, heat and mass transport in geothermal systems of a Central Alpine Massif. In: The Cases of Lavey-les-Bains, Saint-Gervais-les-Bains and Val d'Illiez. Geochemistry. Université de Neuchâtel; 2010. Available from: https://tel.archives-ouvertes.fr/tel-00923368/file/These-Sonney-2010.pdf

[13] Jugastreanu C, Tabatabai SM, Chis T. Thermal properties of oil and gas reservoires rocks modeling. International Journal of Research – GRANTHAALAYAH. 2022;**10**(2):125-144. ISSN (Online): 2350-0530

[14] Tabatabai SM, Jugastreanu C, Chis T. Mathematical modeling of the geothermal gradient of oil and gas deposits. International Journal of Engineering Research and Applications. 2022;**12**(2):21-27. Available from: www.ijera.com. ISSN: 2248-9622

Section 3

The Role of Computational Modeling in Numerical Simulation

Chapter 13

Simulation Study of Microwave Heating of Hematite and Coal Mixture

Prasenjit Singha, Sunil Yadav, Soumya Ranjan Mohanty, Abhishek Tiwari and Ajay Kumar Shukla

Abstract

Temperature distribution in hematite ore mixed with 7.5% coal was predicted by solving a 1-D heat conduction equation using an implicit finite difference approach. In this work, a square slab of 20 cm x 20 cm was considered, which assumed the coal to be uniformly mixed with hematite ore. MATLAB 2018a software was used to solve the equations. Heat transfer effects in this one dimensional slab having convective and radiative boundary conditions are also considered in this study. Temperature distribution is obtained inside the hematite slab by considering microwave heating time, thermal conductivity, heat capacity, carbon percentage, sample dimensions, and many other factors, such as penetration depth, permittivity, and permeability of coal and hematite ore mixtures. The resulting temperature profile used as a guiding tool for optimizing the microwave-assisted carbothermal reduction process of hematite slab which was extended to other dimensions as well, viz., 1 cm x 1 cm, 5 cm x 5 cm, 10 cm x 10 cm, and 20 cm x 20 cm. The model predictions are in good agreement with experimental results.

Keywords: Hematite ore, coal, microwave processing, heat transfer, implicit method, temperature distribution

1. Introduction

Current iron and steel-making industries are experiencing challenges, such as low-grade ore (<35% iron), poor sintering performance, and CO_2 gas emissions. FASTMET, ITmk3, and Hi-QIP processes can be adopted to increase the iron concentration of low-grade ores. For such processes, a mixture of powdery iron ore and carbonaceous material is used as raw materials, whereas the conductive heat flow from the surface to the interior of materials is the rate-controlling step, resulting in decreased productivity. The demand for microwave heating is escalating day by day in iron-making industries owing to its faster heating rate, volumetric heat generation for specific materials, energy savings, and less processing time [1–7]. Hematite can be converted to magnetite for iron ore beneficiation purposes. It can be achieved by heating the hematite. There are two ways of heating: one, conventional heating, and second, microwave heating. In any conventional heating processes, along with hematite, other impurity oxides will be heated, which is not beneficial. This unnecessary heating can be bypassed by adopting microwave heating. In microwave heating, high efficiency in heating can be achieved because not all impurity oxides are heated, such as SiO_2 and Al_2O_3 [8]. Hayashi et al. [7] experimentally generated temperature versus time profile and predicted the effect of graphite content on the temperature characteristic of a mixture of hematite and graphite powders. Standish and Huang [9] investigated the reduction behaviors of hematite fines by mixing carbon and suggested microwave reduction in a non-isothermal process, and the temperature had a significant impact on the reduction of hematite fines. Agrawal and Dhawan [10] evaluated the reduction behavior of low iron hematite ore containing coke and charcoal and reported coke to be a better reductant than charcoal. Mishra and Roy [11] introduced carbon content that had an important effect on the reduction efficiency of the iron ore–coal composite pellet. They [11] obtained a 70% degree of reduction at 1250°C for 20 minutes when the C to Fe_2O_3 molar ratio is three. Zhulin et al. [12] presented the effect of carbon content on the reduction efficiency of iron ore–coal composite pellets at 1200°C for 15 minutes and predicted that higher carbon-containing pellets reduced faster than lower carbon containing pellets. Mourao et al. [13] have demonstrated the reduction rate and maximum temperature of process depending upon the fraction of carbonaceous material. The authors found that if the fraction of carbonaceous material increases in the mixture of hematite ore and carbonaceous material, the maximum temperature increases in the process, which increases the reduction reaction rate of hematite ore. Mishra et al. [14] developed a thermal profile considering powder size, emissivity, and susceptibility using microwave heating, and reported that the model predictions were in good agreement with experimental results. The previous studies [7–14] experimentally demonstrated coke or coal to have a significant effect on temperature and reduction rate of hematite ore. However, their work lacks temperature distribution during the process. Shukla et al. [15] simulated temperature distribution in 2D cylinders of varying radii and physical properties using an explicit infinite method and found that the efficacy of temperature distribution in cylinders depends on the sample size and its thermal conductivity. They considered constant thermal conductivity and heat capacity throughout the process. Peng et al. [16] have demonstrated the heat transfer numerically during microwave heating of magnetite 1D slab. They solved the heat equation using an explicit infinite method based on fewer dimensions. Peng et al. [17] predicted the dielectric and magnetic behavior of the nonstoichiometric ferrous oxide at 823 K and 1373 K, respectively, and evaluated the temperature dependence of the microwave absorption capability of the ferrous oxide by considering the phase transformation during heating. Leo et al. [18] have shown effective permittivity of soils using Lichtenecker's mixing model. Although microwaves have superiority in materials heating, a major drawback known as nonuniform temperature distribution inside materials has also been observed [5, 6]. To address this problem, accurate temperature determination inside the materials under microwave irradiation is necessary. For last 30 years, microwave heating has been utilized extensively in the food processing industry. Most of those works considered heat diffusion and/or convection to predict the temperature distribution in the material.

Although a lot of work has been done on microwave heating, the present study investigates the heat transfer process in microwave heating to predict the temperature distribution inside a 1D hematite slab using an implicit finite-difference approach by considering heat diffusion convection and radiation effects. The objective of the current study is to design and produce a well-defined heating profile in a 1D rectangular slab of hematite ore using microwave heating routes. The resulting temperature profile can be used as a guiding tool to optimize the carbothermal reduction process of hematite in industry, where we studied on large slab (20 cm x 20 cm) as well as laboratory approach small slab (1 cm x 1 cm).

2. Mathematical formulation

The microwave heating process is modeled using a 1D heat transfer equation given below.

$$\frac{\partial T}{\partial t} = \frac{1}{\rho c_p}\frac{\partial k}{\partial x}\frac{\partial T}{\partial x} + \frac{k}{\rho c_p}\frac{\delta T^2}{\delta x^2} + \frac{P(x)}{\rho c_p} \tag{1}$$

$$P(x) = \frac{P_0}{D_p} \times \left(\exp\left(X - x\right)/D_p\right) \tag{2}$$

$$\mathrm{D_p}\ (\text{depth of penetration}) = \frac{\lambda_0}{2 * \pi\sqrt{(2 * \mu_r')}} * \left\{\left[1 + \left(\frac{\varepsilon_r''}{\varepsilon_r'}\right)^2\right]^{1/2} - 1\right\}^{-1/2} \tag{3}$$

Eq. (1) was discretized to study the heat distribution in each node with respect to time. The slab was discretized into 50 × 50 nodes using the equations given below.

$$\frac{\partial T}{\partial x} = \frac{T_{i+1}^{n+1} - T_{i-1}^{n+1}}{2\delta x} \tag{4}$$

$$\frac{\partial k}{\partial x} = \frac{k_{i+1}^{n+1} - k_{i-1}^{n+1}}{2\delta k} \tag{5}$$

$$\frac{\delta T^2}{\delta x^2} = \frac{T_{i+1}^{n+1} - 2\,T_i^{n+1} + T_{i-1}^{n+1}}{\delta x^2} \tag{6}$$

The schematic representation of a square slab with node points is shown in **Figure 1**. Here, i represents the horizontal domain, and i+1 and i−1 represent forward and backward nodes, respectively. Where i, varying from 1 to M, $i = 1$ represents the left boundary, and $i = M$ represents the right boundary. The time-domain [0, t] was divided into n segments, each of duration $\delta t = t/n$.

The implicit finite difference approximation method was used in this study, where the governing equations were discretized (**Figure 1**) into different domains in the form of node points.

Model assumptions are as follows:

1. Heat source is uniformly distributed.

2. No change in properties after mixing ore and coal

3. Ash and volatile matter does not affect the loss factor and dielectric constant of the mixture (**Figure 2**)

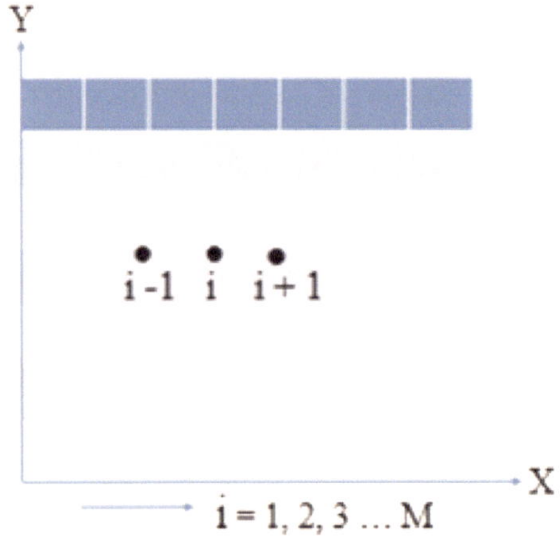

Figure 1.
Schematic representation of a square slab with node points i representing the horizontal domain, and i+1 and i−1 represent forward and backward nodes, respectively.

2.1 Modeling and Process Parameter

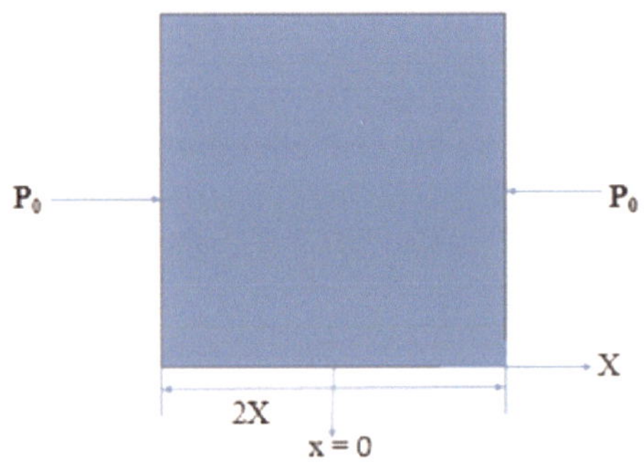

Figure 2.
Schematic representation of a square slab where P_0 is microwave power flux (MW/m^2) and 2X is the length of the slab.

2.2 Simulation parameter calculation process

Proximate analyses and ultimate analyses of coal were taken from Ref. [19]. Now, the amount of carbon monoxide required to convert hematite to magnetite was estimated by using equations given below (**Tables 1** and **2**) [21].

$$2\,C + O_2 = 2\,CO \tag{7}$$

$$\Delta G_f^0 = -111700 - 87.65T \tag{8}$$

$$3\,Fe_2O_3 + CO = 2\,Fe_3O_4 + CO_2 \tag{9}$$

$$\Delta G_f^0 = -44300 - 39.89\mathrm{T} \tag{10}$$

The fraction of coal needed is estimated using proximate and ultimate analyses, and we found that 7.5% of coal is required. Iron ore and coal powder were mixed

Parameter	Value (Wt. % dry basis)
Ash	9.89
Volatile	30.25
Fixed Carbon	59.86
Sulfur	1.03

**Value calculated based on the data reported in Ref. [20].*

Table 1.
*Proximate analyses of the coal sample.**

Parameter	Value (Wt.%, dry, Ash free)
Carbon	82.71
Hydrogen	5.39
Nitrogen	1.57
Sulfur	1.49
Oxygen	8.73

**Value calculated based on the data reported in Ref. [20].*

Table 2.
*Ultimate analyses of the coal sample.**

properly according to a stoichiometric calculation for the reduction of hematite into a desirable amount of magnetite.

Initial and final boundary conditions are given below:

$$t = 0, T = T_0, 0 \leq x \leq X \tag{11}$$

$$x = 0, -k \frac{\partial T}{\partial x} = 0, t > 0 \tag{12}$$

$$x = X, -k \frac{\partial T}{\partial x} = h\,(T - T_\infty) + \mathcal{E}\,\sigma\,\left(T^4 - {T_\infty}^4\right) \tag{13}$$

3. Results and discussion

The heating profile characteristic of hematite ore mixed with coal was modeled using simulation parameters of **Table 3**. Variations of thermal conductivity, heat capacity and thermal diffusivity as a function of temperature are shown in **Figure 3**.

3.1 Temperature estimation

When solving the 1D heat equation, convection and radiation conditions were also considered. It is observed that there is an increase in temperature till 100 s, and beyond this time, the increase in temperature is negligible. The temperature of the slab at the center (x = 0 m) is considered to be 30°C. After heating for 100 s, it gave an indication that the thermal runaway may occur during the microwave heating.

Parameter name	Value	Unit
k^*	$4.4072 - 32^*(10^{-4})\ ^*T$	W/m/K
ρ	3866	kg/m^3
h^{**}	10	W/m^2 °C
ε^{***}	0.96	None
T_0	25	°C
T_∞	25	°C
C_p^{****}	$211.122 - 0.10993{*}T - 0.0003{*}T^2 - 0.000001{*}T^3$	J/kg/K
α	$k/\ (\rho\ {*}\ C_p)$	m^2/s
D_p (at 915Hz)*****	$0.4474 - 0.034 \times 10^{-3}{*}T + 2 \times 10^{-5}\ {*}T^2 - 1 \times 10^{-7}\ {*}T^3 - 6 \times 10^{-9}{*}T^4 + 9 \times 10^{-13}\ {*}T^5 - 4 \times 10^{-16}\ {*}T^6$	mm

**Value calculated based on the data reported in Refs. [19, 22]*
***Value calculated based on the data reported in Ref. [16].*
*****Value calculated based on the data reported in Ref. [19, 22].*
****Value calculated based on the data reported in Ref. [16].*
******Value calculated based on the data reported in Ref. [19, 22–24].*

Table 3.
Thermophysical properties and modeling parameters were used in the simulation, and k, ρ, C_p, and D_p calculation processes are shown in Appendices.

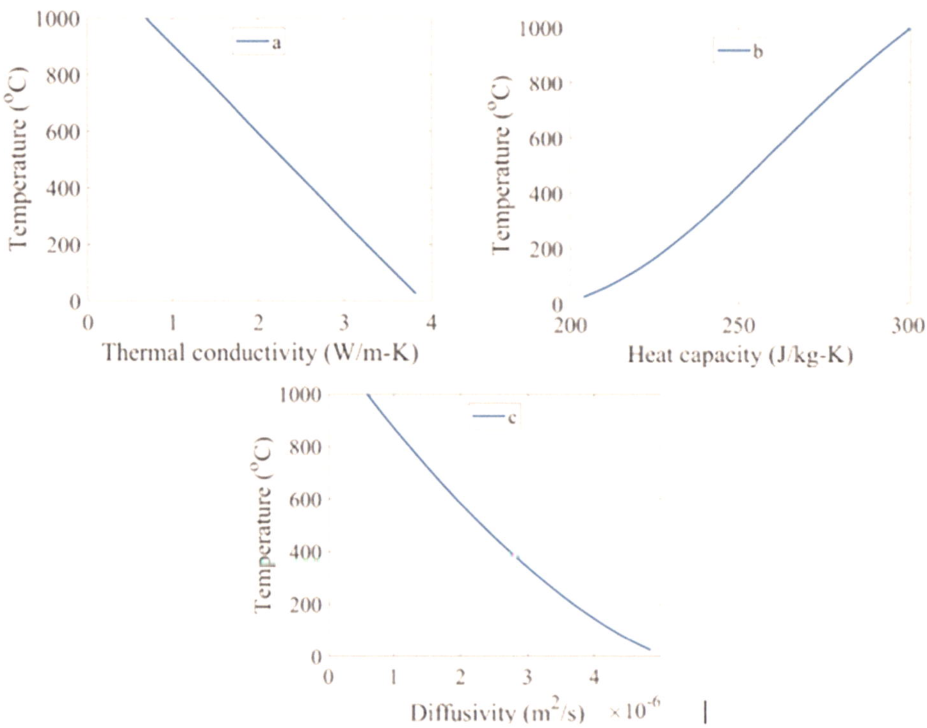

Figure 3.
Temperature dependence on thermal conductivity, heat capacity, and thermal diffusivity.

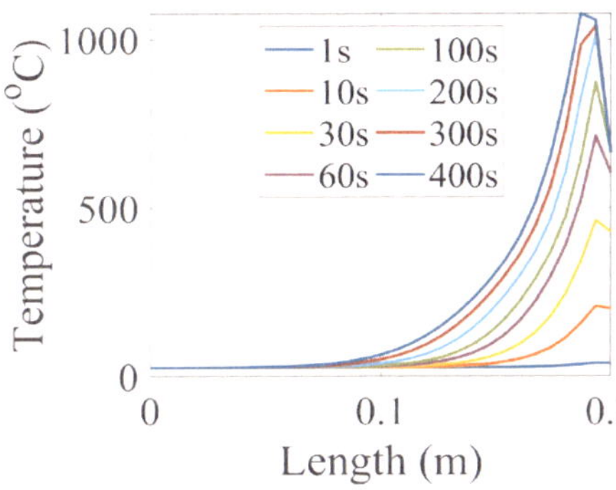

Figure 4.
Temperature distribution in a 20 cm × 20 cm slab for different heating periods, viz., 1 s, 30 s, 60 s, 100 s, 200 s, 300 s, and 400 s, respectively, at a constant power.

Temperature profiles for different heating time periods 1s, 30 s, 60 s, 100 s, 200 s, 300 s, and 400 s (at 915 MHz and 1 MW) are shown in **Figure 4**. The highest temperatures inside the object are around 33°C, 220°C, 415°C, 719°C, 912°C, 1016°C, and 1042°C for 1 s, 10 s, 30 s, 60 s, 100 s, 200 s, 300 s, and 400 s respectively. Within the first 100 s of heating, the temperature in the 1D slab increases rapidly from 25 to 912°C. The reason for this rapid increase is due to the increase in thermal energy supplied by microwave radiation [8]. From 100 s to 400 s, the temperature increases from 912 to 1042°C. This increase is slower as compared to heating from 0 s to 100 s. It is mainly attributed to the initial heating (0–100 s). The thermal contribution from microwave heat generation dominates the temperature rise in the sample due to the weak thermal radiation effect and the relatively low temperature of the object. As the heating continues, the temperature of the object increases, which leads to a high radiation effect. At higher temperatures, thermal conductivity reduces, and heat capacity increases. The heat diffusivity is found in **Figure 3c** to be in the order of 4 x 10^{-6} m^2 /s and decreases with increasing temperature. Heat capacity is found in **Figure 3b** to be 200 J/kg/K and increases with increasing temperature.

3.2 Temperature distribution across length for different powers of microwave

The 1D heat equation was solved by considering different values of power flux. Convection and radiation conditions were also considered. It was observed that there was an increase in temperature till a power flux of 2.5 MW/m^2, and beyond this power flux, the increase in temperature was negligible. Temperature profiles are plotted for a 1D slab with dimensions 0.20 m x 0.20 m. In this slab, different microwave powers (P_0) of 0.5, 1.5, 2.5, and 3.5 MW/m^2 were considered as shown in **Figure 5**. The temperature of the object increases with increasing microwave power. The highest temperatures attained in microwave heating for 60 s using different microwave powers of 0.5 MW/m^2, 1.5 MW/m^2, 2.5 MW/m^2, and 3.5 MW/m^2 are 370°C, 723.6°C, 1062°C, and 1112°C, respectively. Similarly, for times of 1 s, 120 s, and 180 s, temperatures attained are shown in **Table 4**. It is clear that if the source power increases, the microwave heating rate increases rapidly within 60 s. However, after 60 s, the temperature does not increase rapidly, even for higher microwave powers because of radiation and convection losses at high temperatures. It demonstrates that an appropriate power applied in microwave heating is influential in obtaining a high heating

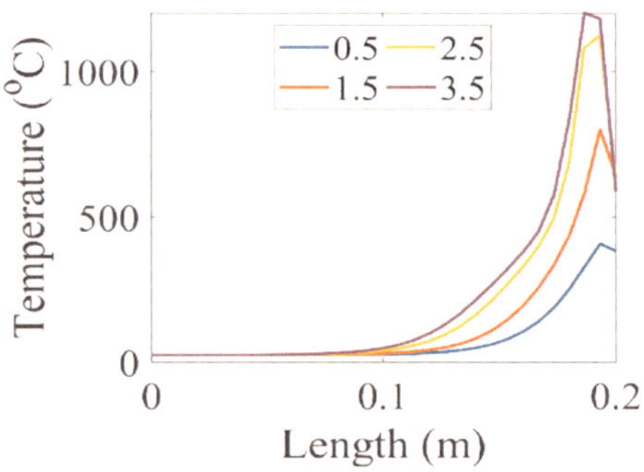

Figure 5.
Temperature distribution in a 20 cm × 20 cm slab by varying power flux, viz., 0.5 MW/m², 1.5 MW/m², 2.5 MW/m², and 3.5 MW/m², respectively, at a constant time of 50 s.

S.no.	Power flux (MW/m^2)	Time (s)	Temperature (°C)
1	0.5	1, 60, 120, 180	27.8, 210.34, 441.63, 706.38
2	1.5	1, 60, 120, 180	30.58, 410.65, 757.98, 807.6
3	2.5	1, 60, 120, 180	39.67, 796, 850.4, 1025.6
4	3.5	1, 60, 120, 180	44.33, 1063.3, 1110.2, 1162.3

Table 4.
Temperature distribution at various times and various power flux.

rate in a short time. On account of more radiation and convection heat losses to the environment at high temperatures, the peak migrates inward to keep heat balance between the object and surroundings.

3.3 Object Dimensions

Volumetric heating also depends on the object dimensions; it is shown in **Figure 6**. Temperature distribution in the slab with different dimensions of 0.01 m × 0.01 m, 0.05 m × 0.05 m, and 0.1 m × 0.1 m, respectively, were observed. More homogeneous temperature is achieved in samples of lesser dimensions than in larger dimensions. The highest temperature peak is obtained in a very short duration, which is an indication that more homogeneity was also observed. The temperature at the slab center increases from 25 to 84°C as the dimension decreases from 0.15 to 0.01 m. This indicates an optimal dimension of the material is required to obtain the minimum temperature nonuniformity and high heating performance. Here the highest temperature peaks for sample dimensions 0.01 m × 0.01 m, 0.05 m × 0.05 m, and 0.1 m × 0.1 m are 1000°C, 912°C, and 850°C, respectively.

3.4 Temperature distribution across length for different coal percentages

Microwave heating depends upon coal percentage in samples, as shown in **Figure 7**. The maximum peak temperatures of the slab for different coal percentages of 6.5%, 7.5%, and 8.5% are 736°C, 743°C, and 753°C, respectively. Coal, a carbonaceous material, absorbs microwave energy, resulting in an increase in temperature.

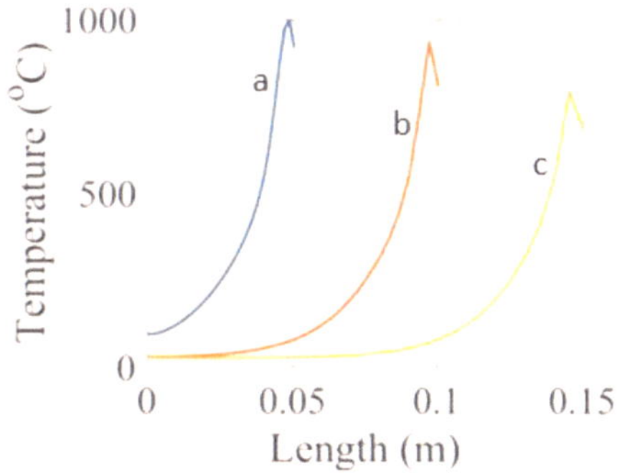

Figure 6.
Temperature distribution in the hematite slab of dimensions, 0.01 m × 0.01 m (a), 0.05 m × 0.05 m (b), and 0.1 m × 0.1 m (c) for a constant heating period of 60 s and power of 1 MW/m².

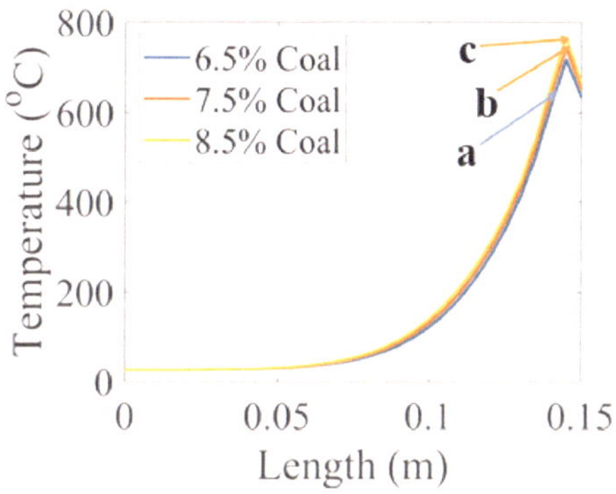

Figure 7.
Temperature distribution in the hematite slab of dimensions, 0.15 m × 0.15 m and coal percentage (a), 6.5 (b), 7.5, and (c) 8.5 for a constant heating period of 180 s and power of 1 MW/m².

Xu et al. [24] reported that coal could enhance permeability, loss factor, and dielectric constant of mixture samples and ensure the optimal heating effect. D_p (depth of penetration) of microwave heating is directly proportional to $\sqrt{\frac{\varepsilon''}{\varepsilon'}}$, The loss tangent is defined as $\tan\delta = \frac{\varepsilon'}{\varepsilon''}$. The attenuation constant, $\alpha = \frac{2\pi}{\lambda}\left[\frac{\varepsilon'}{2}\left(\sqrt{1+(\tan\delta)^2}-1\right)\right]^{1/2}$ represents the rate of absorption of the wave into the sample. Finally, the output of penetration depth is achieved by $D_p = \frac{1}{2\alpha}$, so the depth of penetration is inversely related to the attenuation. So, it means that coal increases the loss factor and the dielectric constant of the samples increases. For that, penetration depth of the samples, will be enhanced which also respond to temperature increases. The temperature peak (**Figure 7**) is seen at similar locations in all three samples because the D_p expression accounts for the loss factor and permeability also. During microwave heating, for lean iron ore (Fe – 58%, Si – 42%, NMDC, Hyderabad) beneficial (Fe_2O_3 to Fe_3O_4) purposes, coal percentage should be limited (7.5%) when coal composition is as per **Table 2**. Otherwise, the temperature will increase, and a non-magnetic phase (FeO) will be produced. Because above 800°C temperature, formation of FeO and Fe is dominant due to the high rate of carbon gasification. Therefore, more than 7.5% carbon is not a desirable carbon percentage for the reduction study in the microwave furnace for such type of lean iron ores

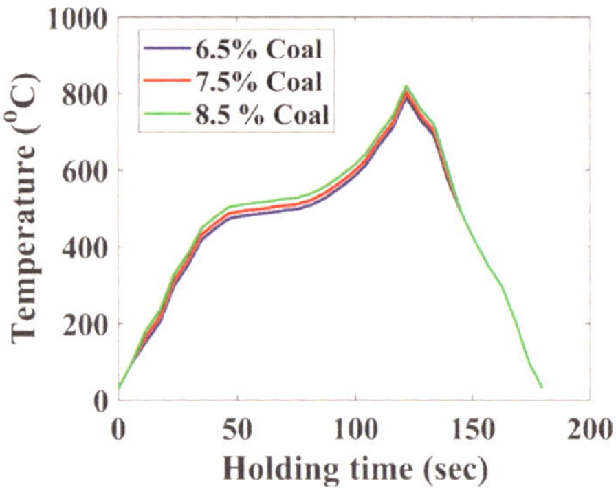

Figure 8.
Variations of temperature with blowing time-varying carbon 6.5%, 7.5%, and 8.5%, respectively.

3.5 Verification with experimental results

Temperature variations of the samples for 6.5, 7.5, and 8.5 coal is shown in **Figure 8**. Our model predictions are consistent with experimental results, which found temperature increases from 791°C to 821°C when coal content in the samples increases from 6.5% to 8.5%.

3.6 Sample preparation

Received iron ore is pulverized in a ball mill for 30 min with 10 mm diameter steel balls. The milled powder is further sieved to below 100 μm (-100 mesh size). The composition of iron ore samples is (65 % Fe_2O_3 and 35% SiO_2). The proximate analysis result of coal is given below in **Table 1**. 100 grams of iron ore and the stoichiometric amount of reducing agent like 7.5 gram coal are entirely mixed. Then, using this mixture, pellets are made with dimensions of 15 mm × 15 mm by pressing the machine (5 T pressing m/c, VB Ceramic). The pressing condition is 2-ton load and dwells time, 2 min. Each pellet weighs about 6 gms. Three samples were made by changing carbon percentages to 6.5, 7.5, and 8.5.

3.7 Microwave treatment

A power (1 MW) and 2.45 GHz hybrid microwave oven (VB Ceramic, Chennai, India) was used for microwave heating. Samples were placed in the oven in a microwave in an alumina crucible. The crucible was located in the central position of the microwave chamber to minimize the effect of the field pattern variations in the oven. Then heating is performed at 800°C for 3 minutes.

4. Conclusions

In this study, microwave heating effects were discussed in detail, which can be very useful in verifying the microwave-assisted carbothermal reduction process of hematite slab in iron-making plants. Even though the findings are valuable in optimizing microwave heating parameters for reduction, further verification is needed before implementing them in plants. Numerical simulations of heat transfer in one-dimensional hematite 20 cm x 20 cm slab under microwave irradiation by considering conduction, convection, and radiation effect, concludes the following

1. The temperature distribution inside the object is nonuniform.
2. The temperature in the object increases rapidly from 0 s to 100 s due to the quick transfer of thermal energy to the object by microwave irradiation. However, after 100 s of heating, the temperature does not increase rapidly due to higher thermal radiation and convection losses.
3. A pre-determined microwave power source is required to attain the highest temperature of the hematite slab within 100 s.
4. Highest peak temperature of the slab increased from 736°C to 753°C using model prediction and 791°C to 821°C using experiments respectively due to increased coal percentage from 6.5% to 8.5%.

Acknowledgements

The authors would like to thank Prince Gollapalli for his valuable discussions.

Conflicts of interest

The authors declare no conflict of interests.

Nomenclatures

P_0	Microwave power flux (MW/m^2)
T	Temperature (°C)
ρ	Density (kg/m^3)
C_p	Specific heat (J/kg-K)
k	Thermal conductivity (W/m-K)
D_p	Depth of penetration (m)
t	time (s)
T_0	Initial temperature (°C)
T_∞	Environmental temperature (°C)
h	Heat transfer coefficient (W/m^2K)
ε	Emissivity (W/m^2)
σ	Stefan-Boltzmann constant ($W{\cdot}m^{-2}.K^{-4}$)
μ_r'	The real part of permeability (-)
μ_r''	The complex part of permeability (-)
X	Half-length of slab (m)
ε_r'	The real part of permittivity
ε_r''	The complex part of permittivity (-)
ε'	dielectric constant (-)
ε''	loss factor (-)

A. Appendices

Density, permittivity, and permeability, specific heat calculation process after mixing

The density of mixture = $\frac{M_t}{\frac{M_a}{\rho_a}+\frac{M_b}{\rho_b}}$

M_t = total mass of mixture
M_a = mass of component a
ρ_a = density of component a
M_b = mass of component b
ρ_b = density of component b
$\ln \varepsilon = V_{iron}\, \varepsilon_{iron} + V_{coal} \ln\varepsilon_{coal}$
V_{iron} = Volume fraction of iron ore
V_{coal} = Volume fraction of coal
$\ln\varepsilon' = V_{iron}\, \varepsilon'_{iron} + V_{coal} \ln\varepsilon'_{coal}$
$\ln\varepsilon'' = V_{iron}\, \varepsilon''_{iron} + V_{coal} \ln\varepsilon''_{coal}$
V_{iron} = volume fraction of iron ore
V_{coal} = volume fraction of coal
C_p of mixture = $\frac{M_a}{M_{mix}} C_{p_a} + \frac{M_b}{M_{mix}} C_{p_b}$
M_a = mass of component a
M_b = mass of component b
M_{mix} = total mass of components a and b
C_{p_a} = Specific heat of component a
C_{p_b} = Specific heat of component b

Author details

Prasenjit Singha[1*], Sunil Yadav[1†], Soumya Ranjan Mohanty[1†], Abhishek Tiwari[2] and Ajay Kumar Shukla[1]

1 Department of Metallurgical and Materials Engineering, IIT Madras, Chennai, Tamil Nadu, India

2 Department of Metallurgical and Materials Engineering, Indian Institute of Technology Kharagpur, West Bengal, India

*Address all correspondence to: psinghanifft@gmail.com

† These authors contributed equally.

References

[1] Kusuma J, Ribal A, Mahie AG. Influence of advection in box models describing thermohaline circulation. International Journal of Differential Equations and Applications. 2018;**2018**:1

[2] Yoshikawa N, Ishizuka E, Ashiko KM, Chen Y, Taniguchi S. Brief review on microwave (MW) heating, Its application to iron & steel industry and to the relevant environmental techniques. ISIJ International. 2007;**47**:523

[3] Malmberg D, Hahlin P, Nilsson E. Microwave technology in steel and metal industry, an overview. ISIJ International. 2007;**47**:533

[4] Makino Y. Characteristics of millimeter-wave heating and smart materials synthesis. ISIJ International. 2007;**47**:539

[5] Peng Z. Michigan Technological University, Michigan. Available from: https://digitalcommons.mtu.edu./etds/17. [Accessed 22 Feb 2021]

[6] Vadivambal R, Jayas DS. Non-uniform temperature distribution during microwave heating of food materials—a review. Journal of Food and Nutrition Research. 2010;**3**:161

[7] Hayashi M, Takeda K, Kashimura K, Watanabe T, Nagata K. Carbothermic reduction of hematite powders by microwave heating. ISIJ International. 2013;**53**:1125

[8] Barani K, Javad Koleini SM, Rezaei B. Magnetic properties of an iron ore sample after microwave heating. Separation and Purification Technology. 2011;**76**:331

[9] Standish N, Huang W. Microwave application in carbothermic reduction of iron ores. ISIJ International. 1991;**31**:241

[10] Agrawal S, Dhawan N. Microwave carbothermic reduction of low-grade iron ore. Metal Material Transaction B. 2020;**51**:1576

[11] Mishra S, Roy GG. Effect of amount of carbon on the reduction efficiency of iron ore-coal composite pellets in multi-layer bed rotary hearth furnace (RHF). Metallurgical and Materials Transactions B. 2016;**47**:2347

[12] Zhulin L, Xuegong B, Zeping G, Wei L. Carbothermal reduction of iron ore in its concentrate-agricultural waste pellets. Advanced Materials Science and Engineering. 2018;**2018**:1

[13] Nishioka K, Taniguchi T, Ueki Y, Ohno K, Maeda T, Shimizu M. Gasification and reduction behavior of plastics and iron ore mixtures by microwave heating. ISIJ International. 2007;**47**:602

[14] Mishra P, Upadhyaya A, Sethi G. Modeling of microwave heating of particulate metals. Metals, Metallurgical and Materials Transactions B: Process Metallurgy and Materials Processing Science. 2006;**37**:839

[15] Shukla AK, Mondal A, Upadyaya A. Effect of porosity and particle size on microwave heating of copper. Science of Sintering. 2010;**42**:169

[16] Peng Z, Hwang JY, Mouris J, Hutcheon R, Huang X. Microwave penetration depth in materials with non-zero magnetic susceptibility. ISIJ International. 2010;**50**:1590

[17] Zhiwei P, Hwang J-Y, Andriese M, Bell W, Huang X, Wang X. Numerical simulation of heat transfer during microwave heating of magnetite. ISIJ International. 2011;**51**:884

[18] Peng Z, Yang Hwang J, Gon Kim B, Mouris J, Hutcheon R. Microwave absorption capability of high volatile bituminous coal during pyrolysis. Energy & Fuels. 2012;**26**:178

[19] Molgaard J, Seltzer WW. Thermal conductivity of magnetite and hematite. Journal of Applied Physics. 2003;**42**: 3644

[20] Peng Z, Hwang JY, Mouris J, Hutcheon R, Sun X. Microwave absorption characteristics conventionally heated nonstoichiometric ferrous oxide. Metals, Metallurgical and Materials Transactions. 2011;**42**:2259

[21] Ghosh A, Chatterjee A. Ironmaking and Steelmaking Theory and Practice. New Delhi: PHI Learning Private Limited; 2012. pp. 55-87

[22] Hu W, Deng J, Li Q-W, Yang X, Shu C-M, Zhang Y-N. Predictive models for thermal diffusivity and specific heat capacity of coals in Huainan mining area, China. Thermochemical Acta. 2017;**656**: 101

[23] Leão TP, Perfect E, Tyner JS. Evaluation of Lichtenecker's mixing model for predi effective permittivitty of soils at 50 MHz. Transactions of the ASABE. 2015;**58**:83

[24] Xu G, Huang J, Hu G, Yang N, Zhu J, Chang P. Experimental study on effective microwave heating/fracturing of coal with various dielectric property and water saturation. Fuel Processing Technology. 2020;**202**:1

Chapter 14

Perspective Chapter: Computational Modeling for Predicting the Optical Distortions through the Hypersonic Flow Fields

Tao Wang

Abstract

Optical aberrations caused by supersonic/hypersonic flow files can lower the guidance accuracy of high-speed flying interceptors with onboard infrared guider. This chapter mainly summarizes the related research achievements on these issues based on the past works. First, the current developments on this important topic are discussed comprehensively. And secondly, the basic theories for predicting the aero-optical distortions used in this chapter, that is the computational flow field dynamics and its characteristics used for establishing the relationships with the flow fields and the optical light, are carefully provided. And then the density field of the flow field acquired from the large-eddy simulation (LES) can be transformed into the refractive index field in terms of the Gladstone-Dale relation. Recursive ray tracing method of optical propagation through the flow of fluids is given out. In the following, the chapter discusses the information optical modeling approach for the solutions to this issue. In the flow fields, every CFD grid is thought of as a uniform and isotropic cell. This chapter utilized the angular spectrum propagation theory for considering the optical waves propagating cell by cell. The suggested method can give out the optical transformation function (OTF), which can be directly used for modeling the aero-optical image. In the end, this chapter concludes the research works and points out the future development and potential applications of these presented research works.

Keywords: aero-optical effects, hypersonic flow fields, computational model, ray tracing, angular spectrum

1. Introduction

The interceptor mounted with inferred detector moves at speed of the supersonic, and an aerodynamic window will be formed at the face of the detector. Supersonic flows produce time- and position-dependent density fields, which directly lead to changes in optical properties dominated by the refractive index. When light passes through a field of varying refractive index, the initial optical path changes, causing distortions and phase errors in the light. This causes optical distortions such as blurring, shifting, jitter, loss of brightness, and loss of resolution. These image distortions are often referred to as the aero-optical effect (AOE). The aberration will affect the image quality of the aeronautical optical sensor, seriously affect the guidance accuracy, and even cause the interceptor to fail. The study of the principle of aero-optical effects and the measurement of aero-optical aberrations are of great importance for the endo-atmospheric aircraft. Consequently, it is necessary to study the influence of the supersonic flow fields on optical propagation and imaging in order to acquire higher guidance accuracy. The research on the aero-optical transmission not only has theoretical merit, but also has important merits in the instruction of the optical system design and restoration of the turbulence-degraded images.

Unlike atmospheric optics, aero-optics is near-field optics [1], which includes turbulent boundary layers, wake layers, and shear layers. Sutton [2] carried out detailed studies of the fundamentals and applications of aero-optics. Aero-optics is a phenomenon of fluid-optic interactions. The refractive index of air and many other fluids is linearly related to the fluid's density through the Gladstone-Dale relationship. In general, supersonic flows are turbulent. Density fluctuations are the root cause of optical aberrations. Liepmann first studied the aero-optical effects on turbulence in 1952 [3]. After this, methods for simulating and measuring aero-optical effects have been widely developed, and research in aero-optics has a history of almost 60 years. However, there are many difficult problems in numerical modeling of aero-optical images based on computational simulation of flow fields and optic transmission, which can be used to adjust imaging sensors' measurement, predict potential distortion, and improve guidance accuracy. It is worth our attention that this effectively reduces experimental costs and helps guide wavefront sensor design in the field of adaptive optics. There is still a difficult problem in aero-optical research, and a lot of researchers around the world have fared better in such field.

In the 1990s, researchers improved dynamic measurement and analysis of aero-optical interactions to obtain wavefront phase variance, Strehl ratio, and optical transmission function (OTF) to compensate for images degraded by turbulence. Shack-Hartmann wavefront sensors [4] have been used to measure wavefront distortion for many years. The sensor frequency has recently reached 1Mz [5, 6], almost meeting the requirements of dynamic wavefront phase measurement. Jumper [4] provided a brief perspective on traditional approaches to measure and quantify aerial-optical interactions. Meanwhile, the theory of numerical analysis from aero-optics is integrated into the CFD codes. Sutton [3, 7–9] pointed out that the procedure of aero-optic, and he devoted efforts to aero-optical performance predictions and analyzed the effect of nonuniform turbulence on the point-spread function (PSF) for imaging through turbulent flow fields. Clark and Farris [10] employed CFD codes and wave optics to provide a numerical method for calculating the aero-optical performance of a hypersonic flow field. Lockheed Martin Aeronautics has published a CFD-based aero-optical analysis of unstable aerodynamic flow fields [11, 12] that has been successfully applied to programs such as ARROW, THAAD, and ENDO LEAP. Catrakis et al. [13] studied aero-optical interactions along the propagation path in shear layers of turbulent compressible separation through direct imaging experiments of refractive-index fields, and the amount and RMS values of the differences in the optical path of interaction, that are a function of the distance traveled in the direction of the beam and a function of the laser aperture size. Roberto et al. [14] described an experimental imaging technique in which the refractive index field and the propagation optical wavefront can be measured simultaneously and based on the results of quantitative image analysis of the refractive index field and the calculated optical wavefront. Frumker et al. [15] proposed a general method to calculate the average MTF flux for a supersonic flying spherical dome using Code V and FLUENT. Monteiro and Jarem [16] studied the mutual interference function in the theory of strong fluctuations when light passes through a nonuniform layer of optical turbulence of gas, and deduced the point scattering function, optical transfer function, and related imaging equations. Michael [17] solved the Laplacian and Runge-Kutta integral parabolic beam equations along the beam path in aero-optics using higher-order compressed differentials. Zhang and Fan [18] and Wang et al. [19–21] used a grid-based model to study aero-thermal optical effect and aerodynamic optical effect near side-mounted optical windows. Juan and David et al. [22] performed a 1:1 scale validation study of a computational fluid dynamics-based aero-optics model in a wind tunnel experiment and found that the overall performance of the CFD-based predictive model was better. These studies facilitated the study of aero-optics. Numerical simulation of aero-optics propagation and imaging is an important topic in the experimental study of aero-optics physics in the wind conditioning process, and the two are considered complementary to each other.

This chapter makes use of the CFD grid model respectively with geometrical optics and information optics in order to describe a computation model of the light propagation through the supersonic flow field. The CFD grid model is thought of as the foundation of the computational simulation. The first method is based on geometrical optical so as to build up a ray tracing model for optic transmission through the supersonic flow fields. By tracking the ray path in the turbulent flow field, the wavefront aberrations can be calculated and the aero-optical performances were predicted. The algorithms in the cases of the normal incidence and the oblique incidence are worked out, and accurate ray tracing is done well. Provided data from CFD numerical simulations on certain conditions, the optical path differences (OPDs), wavefront phase variances, and the Strehl ratios used for measuring the effect of the high-speed flow fields on the optical intensity are calculated. In addition, the maximum offset angles of the line of sight (LOS) are figured out. The influences of the initial incident angle, the altitude, and the Mach number on the optical transmission through the high-speed flow fields are discussed. The results show the coincident with the prior knowledge on the characteristics of aero-optical phenomena. The second method integrates the CFD grid model with angular spectrum propagation model so as to study the aero-optical imaging through the supersonic flow fields directly. In this point of view, the aero-optical propagation is viewed as the optic angular spectrum of plane wave transmitting grid by grid, and the total optical transfer function of such flow fields can be derived and further digital image processing method is utilized to simulate the aero-optical imaging through supersonic flow fields. Finally, theoretical studies of the side-mounted IR window aero-optical imaging are made and figure out a way to model the imaging through the hypersonic flow fields.

Three kinds of computational simulation methods of aero-optics have been developed: One is to use the ray tracing method, which uses the wave delay phenomenon to measure the change in the direction of the light, but it cannot give the light deviation or the blurring of the uncertain image. One is physical optics, which predicts diffraction caused by interference between light waves; the other is wave optics, which calculates the transmission between wavefronts along the optical path and calculates the complex amplitude distribution on each wavefront. Aero-optics itself studies the interaction of light and fluids, and the application of optics theory is associated with numerical simulation methods of fluids. The density and other related data are obtained through the CFD method to simulate the flow field, and the refractive index field is calculated. Combining geometric optics theory and wave optics to quantitatively study the occurrence of light wavefront through the flow field has always been the focus of aero-optics computational simulation research. The wavefront can accurately compensate for the imaging. In adaptive optics applications, such as the Shack-Hartmann wavefront sensor, the wavefront is directly measured and used to reconstruct the wavefront. The geometry of the turbulent degraded light wavefront accurate prediction of the structure is crucial for inferring and controlling the aero-optical phenomena existing in aerospace applications and assisting in the design of optical systems.

The arrangement of this chapter is described as follows. The first section is the introduction to research on the computational study of aero-optical transmission through supersonic flow fields. In Section 2, the computational fluid dynamics model is analyzed and the Gladstone-Dale relationship used for transforming the density fields into the refractive index fields is figured out. Then, the method based on geometrical optics used for modeling aero-optical transmission is illustrated in detail in Section 3. The corresponding computational results are also given out in Section 3. In Section 4, the method using the angular spectrum propagation model for studying the aero-optical imaging is shown and the simulation results are presented. In the end, the conclusions are described.

2. CFD analysis and Gladstone-Dale relationship

2.1 CFD analysis

A simple flat was used to represent the side-mounted infrared window and CFD grids were constructed to evaluate the CFD/aero-optical analysis method. Solving the dominant flow equations in these CFD meshes is a method for numerically simulating flow fields. The grid is more uniform and rectangular without losing generality. Nonuniform grids used for physical planes must be converted to uniform meshes. If the grid resolutions are good, it can be assumed that the gaseous medium within a single grid is homogeneous and isotropic. Otherwise, the CFD data must be interpolated to increase resolution and obtain approximate streaming data. Ali Mani et al. [23] and Haris et al. [24] have respectively discussed resolution requirement of aero-optical simulation from the theoretical and experimental point of view.

In this chapter, the grids generated from CFD are uniform and hexahedral, the size of which is equal to 1 mm. This chapter considers each CFD hexahedral grid as an index cell with a uniform refractive index, respectively. Each grid is considered as a thin plate glass here. Consequently, the flow field model has cell configuration. **Figure 1** describes the optical transmission through the flow fields. The supersonic flow field data used in this chapter are calculated through large-eddy simulation (LES). **Figures 2** and **3** show samples of the computed density fields. Generally, supersonic flow fields should be completely viewed as turbulent flows. It is known that turbulence shows violent inhomogeneity and anisotropy with time and space changes. And turbulence can be theoretically seen as the flow fields consisting of mean flows and fluctuations. The blur and centroid shift of image degraded by aero-optical effects can be brought about by the mean flows.

The accurate modeling of high temperature, high pressure, and high-speed complex flow fields considering turbulence has always been a scientific problem in fluid mechanics, and until now, there are still some basic problems that have not been solved. This chapter does not involve the mechanism research of fluid mechanics, nor

DOI: http://dx.doi.org/10.5772/intechopen.106591

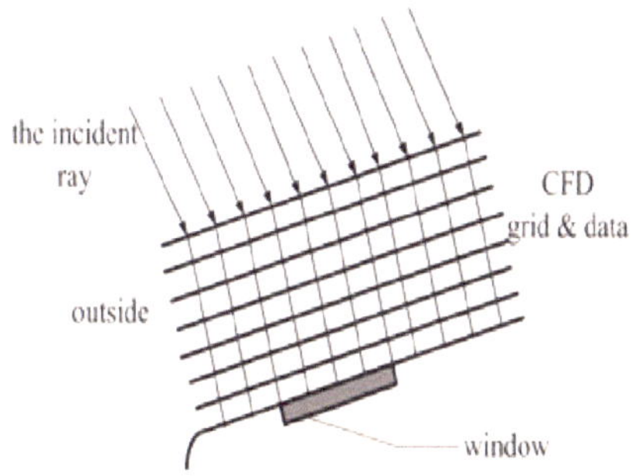

Figure 1.
The relationship between the CFD calculation mesh and the optical transmission.

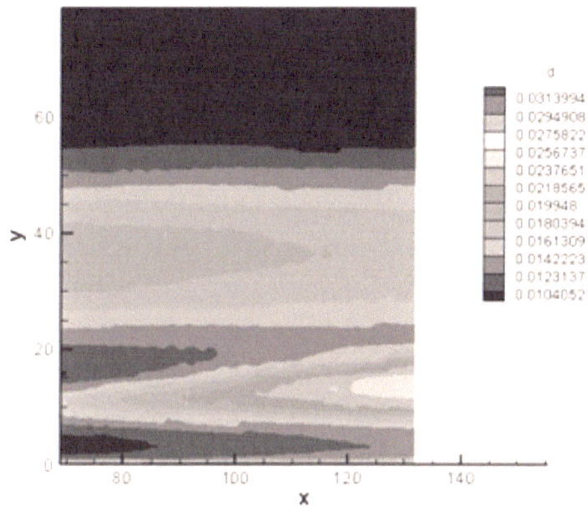

Figure 2.
A sample of the density distributions along the X direction.

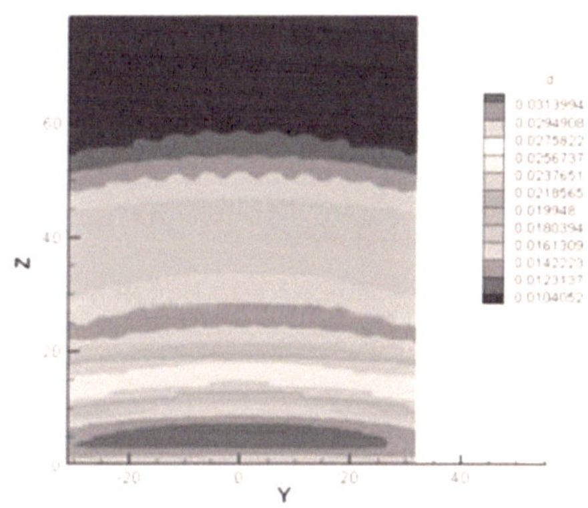

Figure 3.
A sample of the density distributions along the Y direction.

does it pursue the innovation of flow field modeling and solution methods. Instead, it adopts the most mature and reliable calculation model provided by fluid mechanics and widely accepted in the industry to obtain the flow field data that can be used for this chapter to calculate the imaging migration, and then explore the internal relationship between the related physical quantities such as height, line of sight angle,

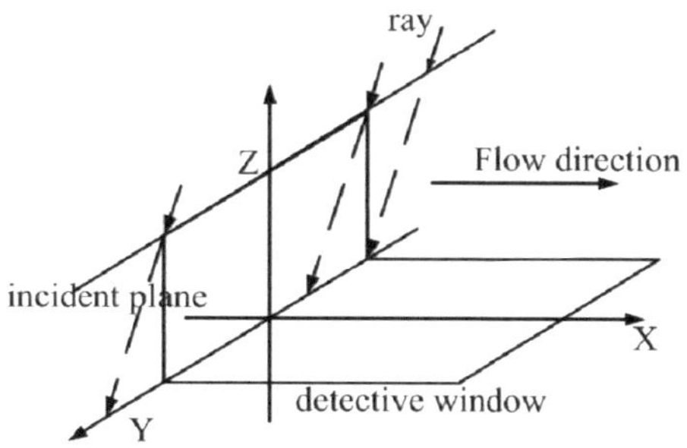

Figure 4.
The reference frame.

optical propagation path in the flow field, and the negative value of imaging migration. There are scientific problems behind the application, such as the internal reasons why the variation law of imaging migration is disturbed at different heights. Although the Navier-Stokes (N-S) equation can be used to describe turbulence, the nonlinearity of the N-S equation makes it extremely difficult to accurately describe all the details related to three-dimensional time with analytical methods. Even if these details can be really obtained, it is not of great significance for solving practical problems. From the point of view of engineering applications, it is important that the change in the average flow field caused by turbulence is the overall effect. In engineering calculation, geometric optics is used to calculate the imaging offset caused by the aero-optical effect, which is exactly the result of the action of the average flow field.

Large eddy simulation divides turbulence into large-scale turbulence and small-scale turbulence. By solving the three-dimensional modified N-S equation, the motion characteristics of large eddies are obtained, and the above model is also used for small eddies. Large eddy simulation has unparalleled advantages in the following aspects: (1) prediction of transition from laminar flow to turbulence; (2) prediction of unsteady turbulence; and (3) prediction of high-speed turbulence. However, it must be emphasized that the application of LES in industrial fluid simulation is still in its infancy.

The reference frame between the computational meshes and the incident rays is shown in **Figure 4**. The computational mesh has $64 \times 64 \times 80$ grid points, ranging from 69 to 132 in the X direction, from -31 to 32 in the Y direction, and from 0 to 79 in the Z direction.

2.2 Gladstone-Dale relationship

The Lorentz-Lorenz formula provides the bridge of linking Maxwell's electromagnetic theory with the micro-substances. The relationship between the flow field density ρ and the refractive index n is modeled by [25]

$$\left(\frac{n^2-1}{n^2+2}\right)\frac{1}{\rho}=\frac{2}{3}K_{GD}. \tag{1}$$

where K_{GD} is the Gladstone-Dale constant. In general, the refractive index of air depends on its density at room temperature. When the air temperature is high, the refractive index mainly depends on the temperature and fluid composition. This chapter ignores the effects of aerodynamic heating and ionization on the index of refraction and only considers the effects of different current densities on the index of refraction. As the constant airflow index is approximately 1, Gladstone-Dale (G-D) relationship can be obtained as

$$n = 1 + K_{GD}\rho, \tag{2}$$

where ρ is the local density of the flow field. In the ideal air, the G-D relation is a universal description of the connection between the light rays and the air. Particularly for the infrared, the Gladstone-Dale coefficient K_{GD} is just dependent on its wavelength. Its values taken from the IR Handbook are fitted with the formula where ρ is the local density of the flow field. In ideal air, the G-D relation is a general description of the connection between light and air. Particularly for infrared, the Gladstone-Dale-K_{GD} coefficient depends only on the wavelength. Its values taken from the IR Handbook are fitted with the formula as follows

$$K_{GD} = 2.24 \times 10^{-4} \times \left(1 + \frac{7.52 \times 10^{-3}}{\lambda^2}\right)\left(\mathrm{m^3/kg}\right) \tag{3}$$

where λ is the wavelength in micron. In this chapter, the wavelength of 8 μm is used for simulations.

3. Geometrical optical method for modeling aero-optical transmission

3.1 Ray tracing model

Geometrical optics is used in this chapter because wavelengths are considered to be negligible. According to the principle of geometrical optics, light exists in the form of straight lines in a uniform medium. When light passes through two homogeneous media with different refractive indices, its behavior can be determined by the laws of refraction and reflection.

According to the Gladstone-Dale relationship above, CFD grids can be converted to indexed grids. The beam axis is oriented parallel to the negative Z direction. As geometric optics show, the refracted/reflected rays are in the same plane as the incident rays. Therefore, the transport of light in a 3D flow field can be seen as consisting of the transport of multiple layers in a 2D cross-section of the flow field. Light incident on CFD grid points now starts at the top of the computer field. **Figure 5** shows how light travels in a plane.

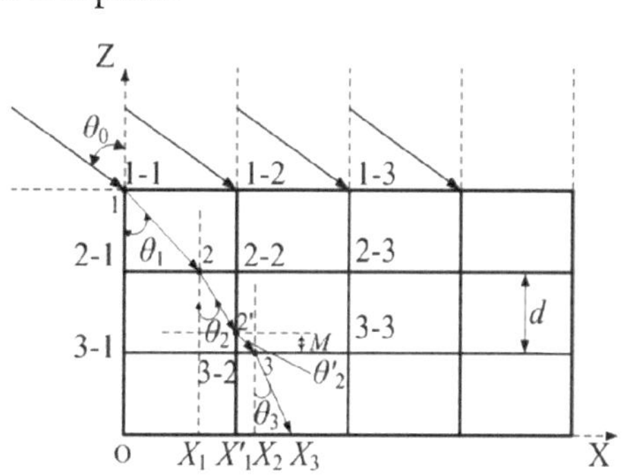

Figure 5.
Geometry of the optical transmission in two dimensions.

The index of refraction at grid point 1–1 is denoted by n_{11} (see **Figure 3**). Likewise, the index of refraction at point i-j is denoted by n_{ij}. Points 1–1 to 1–2 are chosen as n_{11} for the refractive index of the region above the interface, and the index in the next grid of the threshold is selected as n_{21}. d is the size of the square grid taken from a hexahedral grid plane. Assuming that the initial angle of incidence is θ_0, the angle of refraction of the ray as it passes through the interface isθ_1. Thus, when a ray passes through the grid in the negative Z direction, the angle of refraction at point k is written asθ_k. The ray coordinate is (x_k, z_k) and its offset is denoted by Δx_k. The offsets are expressed by$\Delta x_1 = |X_1 O| = x_1 - O$, $\Delta x_2 = |X_2 X_1| = x_2 - x_1$ ... $\Delta x_k = |X_k X_{k-1}| = x_k - x_{k-1}$. Here, the total real offset is marked by$\Sigma \Delta x_k$. Through the geometrical relation, the angle of incidence at point 2 is equivalent to the angle of refraction at point 1. According to the Snell's law, an equation at grid point 1–1 $n_{11} \sin \theta_0 = n_{21} \sin\theta_1$ is acquired. At point 2′, an equation $n_{31} \cos \theta_2 = n_{32} \sin \theta_2'$ is obtained. The point where the transmission is similar to point 2′ is denoted by l'. The optical path length (OPL) from point 1 to point 2 is indicated as OPL_1. Likewise, the length of the path from point k to point k + 1 is denoted byOPL_k. Total reflection can occur when light is transmitted from an optically denser layer to an optically thinner layer due to the introduction of an interface. But interfaces obviously do not apparently exist in real airflow. Total reflection causes light to rotate. This definitely increases the complexity of the algorithm.

Rayleigh pointed out that the wavefront cannot be changed when the wavefront error between the real and reference wavefronts was less than a quarter wavelength. The refractive indices of two adjacent grids are denoted n_i and n_j, respectively. According to the Rayleigh criterion, the wavefront error is negligible if the following equation is satisfied, that is, if the light transmittance is not deformed. That is, the initial light path can be considered unchanged. Rayleigh's criterion is applied to the algorithm. Resort to judging the criterion when total reflection occurs. If yes, the hypothetical interface is considered nonexistent and the light propagates in the same direction. Otherwise, the virtual interface is assumed to exist and full reflection occurs.

$$|n_j - n_i| s < \lambda/4 \tag{4}$$

where s is the geometrical path length and $1 \leq s \leq \sqrt{2}$ (mm) The results show that the approach is useful and enough for simplifying the calculations. Modeling the propagation through the CFD grids, we can derive the recursive algorithm for tracing the light rays. First of all, the relationship at point 1 is expressed by Eq. (5).

$$\begin{cases} \Delta x_1 = d \tan \theta_1 \\ n_{21} \sin \theta_1 = n_{11} \sin \theta_0 \end{cases} \tag{5}$$

At point k, the ideal relation of the light transmission was gained as

$$\begin{cases} \Delta x_k = d \tan \theta_k \\ n_{k+1,l} \sin \theta_k = n_{k,l} \sin \theta_{k-1} \quad (l, k = 1, 2, \ldots) \\ OPL_k = n_{k+1,l} d / \cos \theta_k \end{cases} \tag{6}$$

If $\Sigma \Delta x_k > l \times d$, where $l \in N$and l is counter, we should modify the offset ΔX_k and the OPL_k. Through the triangular relation, Eq. (7) is obtained by

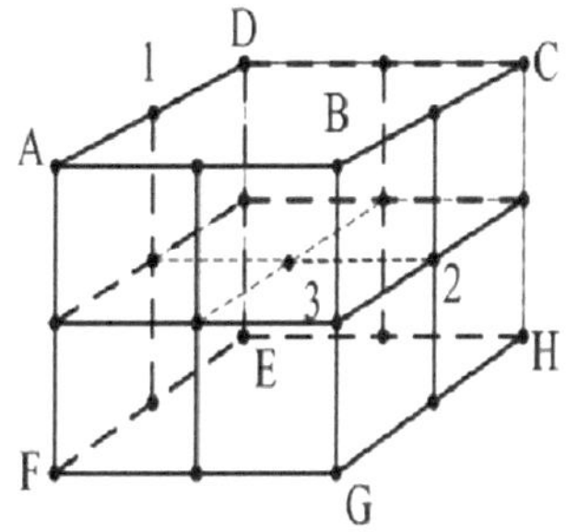

Figure 6.
Interpolated mesh cell.

$$M = (\Sigma\Delta x_k - l \times d)\cot\theta_k \tag{7}$$

The modified offset and OPL would be acquired by

$$\begin{cases} \Delta x_k = M\cot\theta'_k + (d - M)\tan\theta_k \\ OPL_k = n_{k+1,l}(d - M)/\cos\theta_k + n_{k+1,l+1}M/\sin\theta'_k \end{cases} \tag{8}$$

If no reflection over the boundary is produced, the relation between θ_k and θ'_k should be given by

$$n_{k+1,l}\cos\theta_k = n_{k+1,l+1}\sin\theta' \tag{9}$$

$$\theta_{k+1} = 90^\circ - \theta'_k \tag{10}$$

In the case of total reflection, it is assumed that the transmission direction has no changes if Eq. (4) is satisfied. Then, Eq. (9) should be changed into Eq. (11).

$$\theta_k = \theta'_k \tag{11}$$

Integrated with Eqs (4)–(11), the recursive algorithm for tracing the ray based on the CFD grids is derived. For the fine resolution, a method to interpolate the discrete flow field data is shown in **Figure 6**.

The index at point 1 is interpolated by $n_1 = (n_A + n_D)/2$.

The index at point 2 is gained by $n_2 = (n_A + n_B + n_C + n_D)/4$.

The index at point 3 is expressed by $n_3 = (n_A + n_B + n_C + n_D + n_E + n_F + n_G + n_H)/8$.

The rest can be inferred by analogy and higher resolution indexed fields are obtained. **Figure 7b** shows the interpolated index field. This method helps to make the data more contiguous with the initial index data and to adopt the algorithm described above.

3.2 Aero-optical analysis

The aero-optical quantities, measured and calculated, mainly are the wavefronts' distortion, Strehl ratio, and the line of sight error. There is a general assumption that the mean flow fields produce time-averaged blurring and the line of sight error, whereas the turbulence produces jitter and blurring. In the sight of geometrical optics,

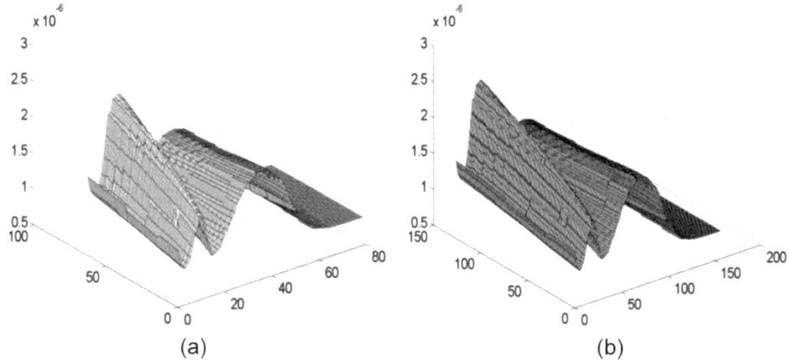

Figure 7.
(a) The initial index data (b) the interpolated index data.

the wavefronts' aberrations arise from the OPL changing. OPL is expressed by Eq. (12).

$$OPL = \int_{ray} ndl \tag{12}$$

As the rays penetrate the disturbed air, they pick up the absolute *OPD* as follows[11]:

$$OPD = \int_0^L (n-1)dl = K_{GD}\int_0^L \rho dl \tag{13}$$

where L is the total geometrical path length, and ρ is the averaged-density value of the local flow. Additionally, if the central ray is considered as the reference ray, whose OPL is marked OPL_{ref}, relative OPD will be obtained by Eq. (14) [26].

$$OPD = OPL - OPL_{ref} \tag{14}$$

Then, OPD data acquired can be transformed to wavefront phase distortion by using the following formula:

$$\Delta\phi\left(\vec{r}, t\right) = kOPD \tag{15}$$

where k is the wave number, and $k = 2\pi/\lambda$. Therefore, optical path differences directly reflecting the variations of the wavefront phase errors. Another quantity of wavefront aberrations is the root-mean-squared (RMS) optical path difference denoted by σ^2, that is, it is wavefront variance gained as follows [27].

$$\sigma^2 = 2{K_{GD}}^2\int_0^L (\rho')^2 l' dl \tag{16}$$

where ρ' is the fluctuation density and l' is the turbulence length scale calculated by CFD. Wavefront variance is a measure of the dispersion along with the mean optical path length of a wavefront and is important for modeling aero-optical parameters

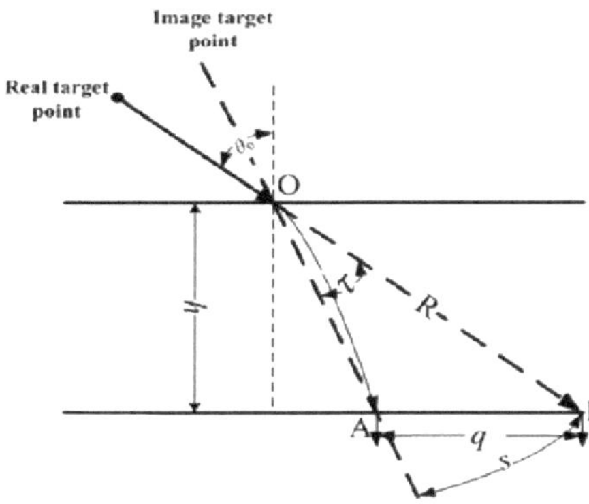

Figure 8.
The LOS deviation angle.

such as Strehl ratios and OTF for the turbulent flows. Only if the RMS optical path difference is not very large, is the time-averaged Strehl ratio for a given wavefront phase shift approximated as [28].

$$SR = \exp\left[-\delta_\phi^2\right] \tag{17}$$

Here, δ_ϕ^2 is denoted the wavefront phase variance and it is obtained by the following relationship [29]

$$\delta_\phi{}^2 = k^2\sigma^2. \tag{18}$$

In **Figure 6**, the line of sight error is described in terms of optical beam path reversibility. The LOS errors result in the coordinate position excursion of the image formed by the light propagation through the supersonic flow fields. Generally, q is of the small scale. So, the arc length s is approximately equal to q. If R is selected as the radius, the LOS deviation angle denoted by τ will be expressed as

$$\tau = q/R. \tag{19}$$

Here, the unit of $\tau$is rad. q is acquired by Eq. (20)

$$q = h\tan\theta - \Sigma\Delta X_i. \tag{20}$$

Here, the maximum displacement angle $\tau_{\max}$ representing the maximum displacement of the image position. In **Figure 8**, when R = |OB|, the value of τ is smaller than the real angle, and when R = |OA|, the value of τ is large. Using each CFD grid as a CCD imaging unit, the detection window is abstracted into an image plane consisting of 64 × 64 pixels. The maximum displacement angle directly reflects the maximum displacement of the intensity distribution, that is, the maximum displacement of the image.

3.3 Simulation results

The supersonic flow field CFD data were calculated using the large-eddy simulation (LES) method. All angles of attack in a flow field simulation are the same. The

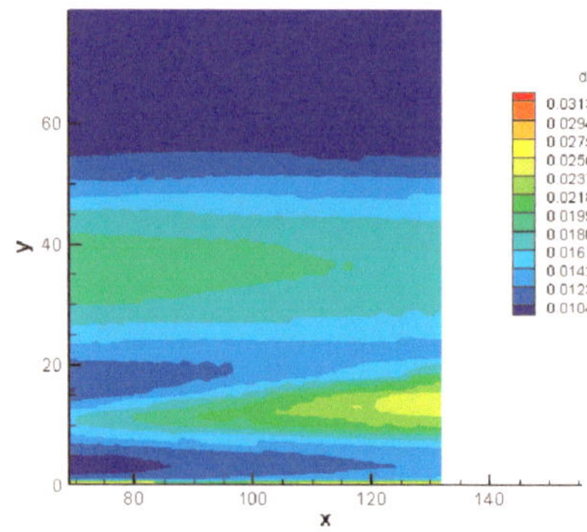

Figure 9.
Solution of density fluctuation along flow direction. Solution at H = 35, and Ma = 7.

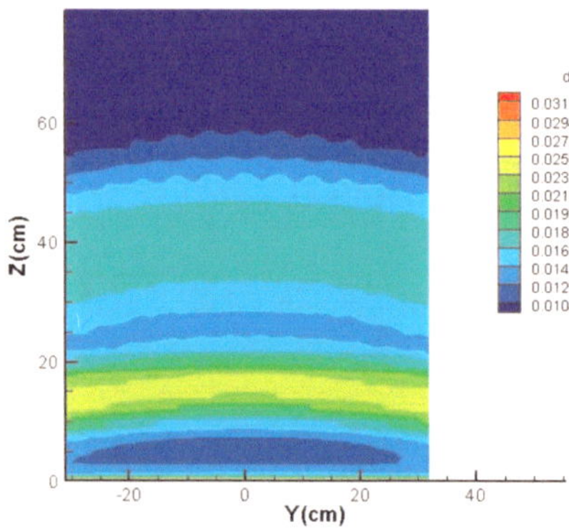

Figure 10.
Solution of density fluctuation in the Y direction. Solution at H = 35, and Ma = 7.

distribution of density fluctuations in the flow field at a height of 35 km and Mach number 7 are shown in **Figures 9** and **10**. In **Figure 9**, density fluctuations near the detection window along the flow direction become bigger and bigger. The distribution of density fluctuations in the positive Y direction is symmetrical. The RMS OPD distribution obtained from Eq. (18) shows the change in phase shift of the wavefront, which in turn shows the characteristics of the flow field in **Figures 11** and **12**. The RMS

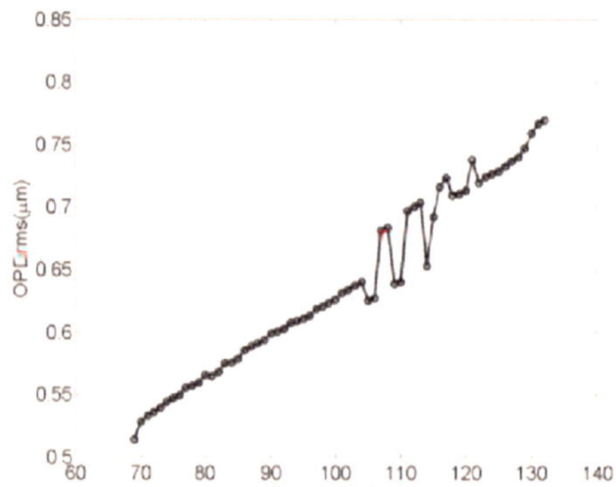

Figure 11.
The RMS wavefront errors in the flow direction (H = 35 km, Ma = 7).

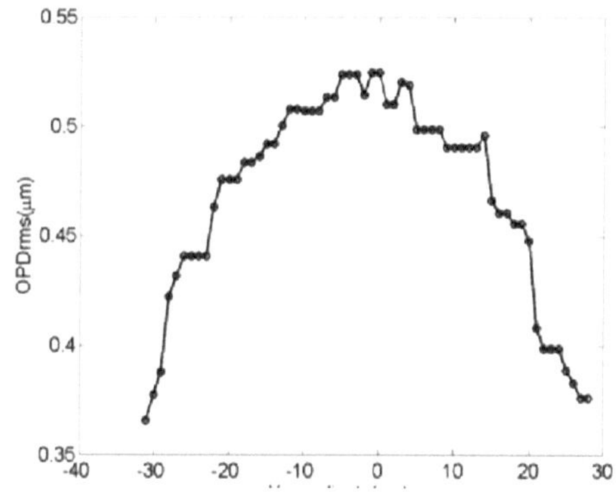

Figure 12.
The RMS wavefront errors in the Y positive direction (H = 35 km, Ma = 7).

optical path difference increases along the flow direction, whereas, in general, in the Y direction, the RMS OPD is the greatest at the center of the window and decreases from the center to the other.

Strehl ratios are shown in **Figures 13** and **14**. On the whole, light intensity is weakened along the flow direction, while in Y positive direction, the light intensity at

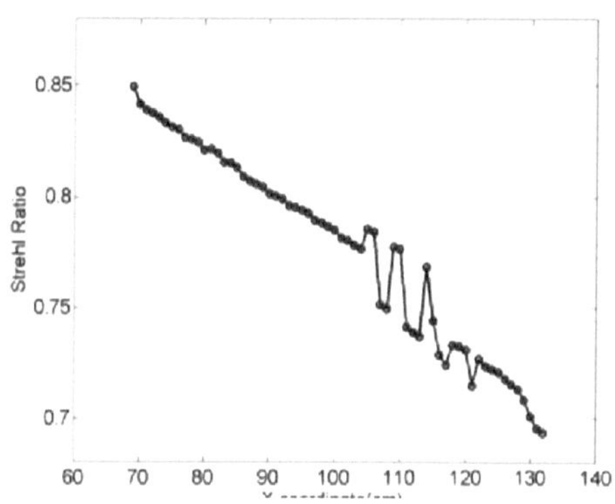

Figure 13.
Strehl ratio in the flow direction (H = 35 km and Ma = 7).

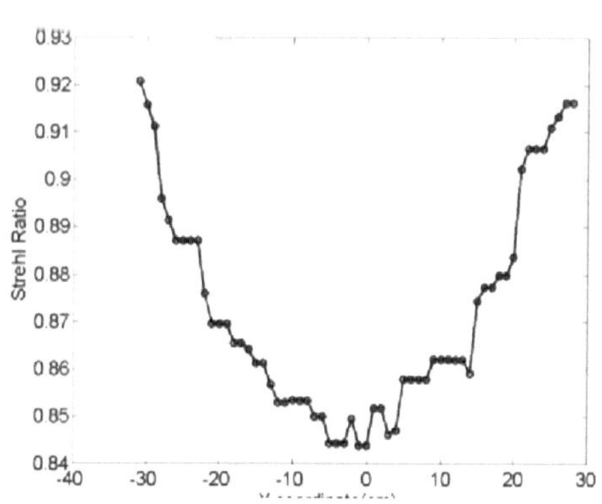

Figure 14.
Strehl ratio in the Y positive direction (H = 35 km and Ma = 7).

the center of the window is reduced to minimum and is decreased from here to other two sides. Apparently, all the calculation results are qualitatively correct.

The absolute differences of the optical path in the positive Y direction are shown in **Figures 15–17**. It can be seen in **Figure 15** that the optical path difference increases as the angle of incidence increases. Therefore, it can be concluded that the sensor within the detection window needs better incident light to reduce wavefront distortion. It can be seen in **Figure 16** that the optical path difference value decreases as the height increases. It can be seen from the standard atmospheric table that the atmospheric density in the range from 0 to 80 km decreases with increasing altitude. Of course, as the free air flow density becomes thinner, the density near the detection window inevitably becomes thinner at higher altitudes. Although the supersonic flux field near the window is compressible, the change in density is small. Therefore, the aberration of the accumulated optical wavefront is still reduced. **Figure 17** shows the optical path differences at different Mach numbers, at the same height and at the same angle of incidence. In aerodynamics, the effect of the compression ratio becomes greater as the velocity of the flow field increases. As the Mach number increases, the density necessarily increases. Therefore, the optical path difference becomes large.

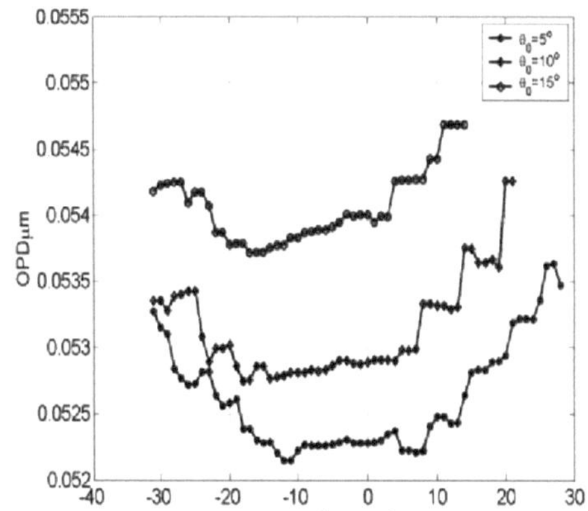

Figure 15.
Comparison of OPD at different angles of incidence (H = 40 km and Ma = 7).

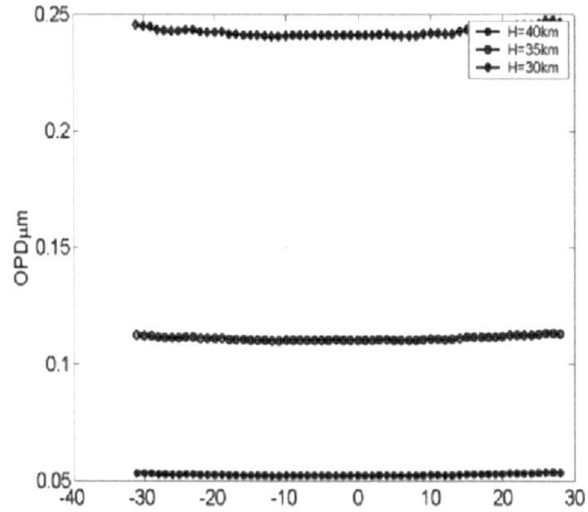

Figure 16.
Comparison of OPD at different altitudes (Ma = 7, and θ_0 = 5°).

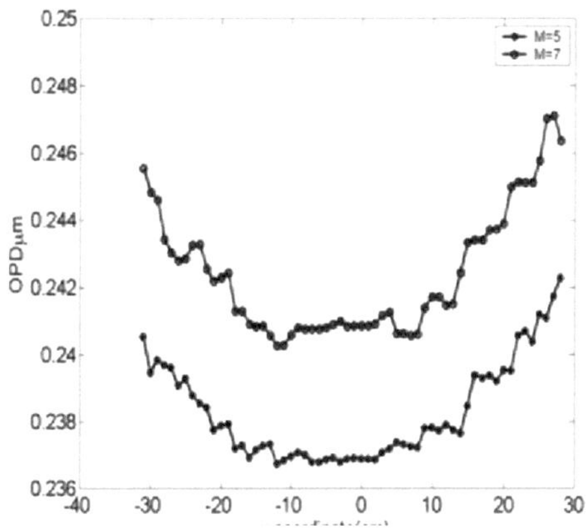

Figure 17.
Comparison of OPD at different Mach Numbers (H = 30 km, and θ_0 = 5°).

	H = 30 km	H = 35 km	H = 40 km
Ma = 7	2.17884e-3/rad	2.17878e-3/rad	2.17865e-3/rad
Ma = 5	2.17880e-3/rad	2.17871e-3/rad	2.17848e-3/rad

Table 1.
The maximum deviation angles.

The maximum deviation angles calculated at different flow fields are given in **Table 1**. It is magnified with the increasing velocity of the flow and reduced with the altitude being higher.

4. Information optical method for modeling aero-optical imaging

4.1 Angular spectrum propagation model

Figure 18 depicts the propagation of the angular frequency spectrum of plane wave. As to the optical wave, the complex amplitude of the plane wave is expressed by

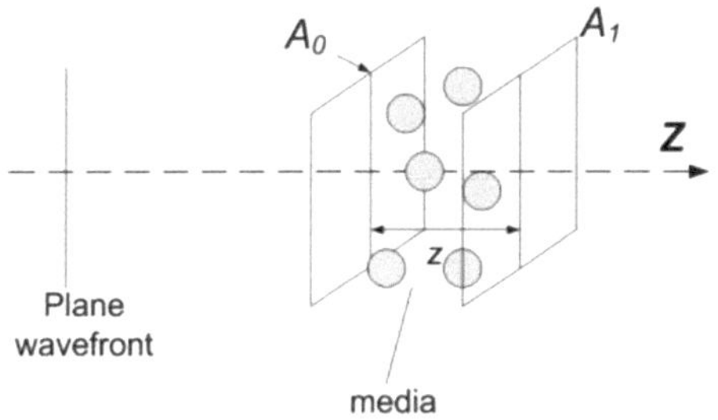

Figure 18.
The angular spectrum propagation of plane wave in uniform optical media.

$$\begin{aligned} U(\mathbf{r}, t) &= a \exp(ik \cdot \mathbf{r}) \\ &= a \exp[ik(x\cos\alpha + y\cos\beta + z\cos\gamma)] \end{aligned} \tag{21}$$

where a is the constant amplitude; $(\cos\alpha, \cos\beta, \cos\gamma)$ denotes the direction cosine; $k = 2\pi/\lambda$ is the wave vector [30].

The angular frequency spectrum is just the 2-D Fourier transformation of the complex amplitude of optical wave. Given that a monochromatic light injects the X-Y plane along with the direction of Z axis, its angular spectrum can be expressed as

$$A\left(f_x, f_y, z\right) = \iint U(x, y, z) \exp\left[-i2\pi\left(xf_x + yf_y\right)\right] dxdy \tag{22}$$

where $U(x, y, z)$ is the complex amplitude distribution. On the other hand, the inverse Fourier transform of the angular spectrum is a complex amplitude distribution. Here, the transmission of aerial optics through the flow fields can be seen as the transmission of each spectrum. Each plane wave spectrum in $(x, y, 0)$ is denoted by $A_0\left(f_x, f_y, 0\right)$, and each spectrum in (x, y, z) is denoted by $A_0\left(f_x, f_y, z\right)$, where $f_x = \frac{\cos\alpha}{\lambda}, f_y = \frac{\cos\beta}{\lambda}$. According to the Helmholtz scalar equation, angular spectrum propagation function can be obtained as

$$A_1\left(\frac{\cos\alpha}{\lambda}, \frac{\cos\beta}{\lambda}, z\right) = A_0\left(\frac{\cos\alpha}{\lambda}, \frac{\cos\beta}{\lambda}, 0\right) \exp\left(jkz\sqrt{1 - \cos^2\alpha - \cos^2\beta}\right) \tag{23}$$

which describes the propagation between the two parallel planes. As a matter of fact, a transfer function of frequency filtering is obtained by

$$\begin{aligned} H_c\left(f_x, f_y\right) &= \frac{A\left(f_x, f_y\right)}{A_0\left(f_x, f_y\right)} \\ &= \begin{cases} \exp\left[ikz\sqrt{1 - \left(\lambda f_x\right)^2 - \left(\lambda f_y\right)^2}\right] & f_x{}^2 + f_y{}^2 < \frac{1}{\lambda^2} \\ 0 & others \end{cases} \end{aligned} \tag{24}$$

The model of aero-optical imaging proposed in this chapter is inspired by the above analysis. Hence, the aero-optical transmission is translated into spatial filtering with limited bandwidth of the lights.

4.2 Linear filter model of aero-optical imaging

The light source described in this chapter may be too far away from the built-in detector, causing the output wave to appear as a plane wave. Therefore, the transmission of light at hypersonic speed can be considered as the transmission of plane waves. The term "diffraction" is conveniently described by Sommerfeld as " any deviation of light rays from rectilinear paths that cannot be interpreted as reflection or refraction.". Furthermore, the results obtained from the scalar diffraction theory approximate the real effect if the wavelength is smaller than the diffraction aperture and the observation point is far from the diffraction aperture [31]. The supersonic flow field transmission process considering the above discussion, aero-optics transmission can be seen as a scalar diffraction problem, so angular spectrum propagation can be used for aero-optics research.

From the point of view of information optics, each cubic grid can be seen as an optical system that composes any optical filtering system that characterizes the flow fields. After these considerations, the supersonic flow fields can be divided into 63 × 63 × 79 optical filtration subsystems, which are serially connected in the negative direction along the Z axis. **Figure 19** maps a sketch of the plane optical wave through CFD cubic grids.

Figure 20 shows the structure of one cubic CFD grid. Each cubic grid has eight points with the determined index of refraction. The following equation gives out the characteristic parameter $n^i{}_{oc}$ of the cubic optical filtering system.

$$n^i{}_{oc} = \frac{(n_1 + n_2 + \ldots + n_8)}{8} \tag{25}$$

where i is the order number of the optical filtering system ($i = 63 \times 63 \times 79$). One of these serial systems is described in **Figure 21**. $H_i\left(f_x, f_y, z_i\right)$ denotes the transfer

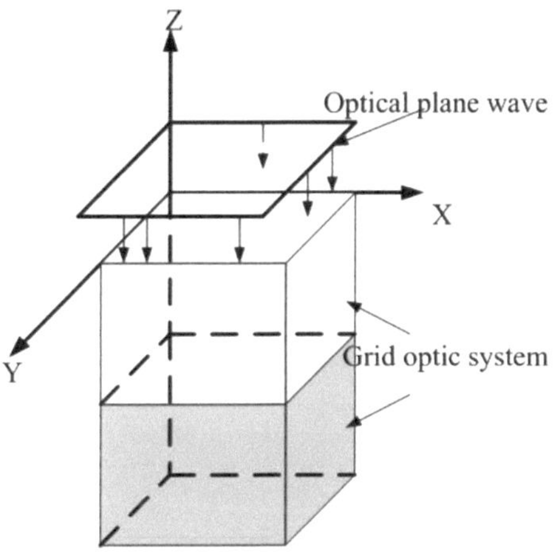

Figure 19.
The sketch map of CFD cubic grids.

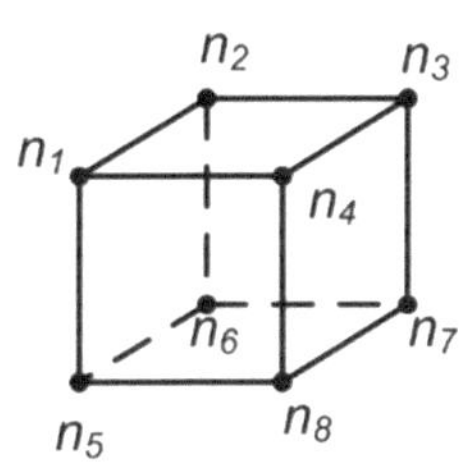

Figure 20.
The distribution of CFD grid points in one cubic optic system.

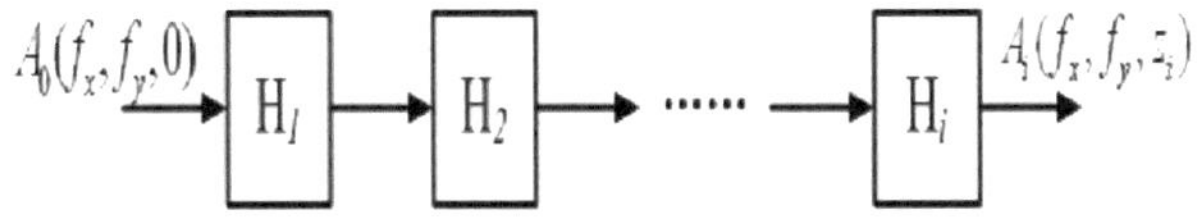

Figure 21.
The diagram of serially connected systems representing the transmission of angle spectrum.

function of the ith optical system, which can be gained through Eq. (23) based on angular spectrum propagation model.

In the frequency domain, the output of such serially linear filtering system can be obtained through

$$A_i\left(f_x, f_y, z_i\right) = A_0\left(f_x, f_y, 0\right)H_1H_2\cdots H_i \tag{26}$$

However, Eq. (23) is actually called as coherence transfer function (CTF). As to aero-optics, it should be viewed as a noncoherent imaging system, which is a linear system concerning on the distribution of light intensity. To the noncoherent imaging system, optical transfer function (OTF) is utilized for the study of light propagation. The OTF can be derived from CTF through the following equation

$$H(u, v) = \frac{H_c(u, v) \otimes H_c(u, v)}{\iint_{-\infty}^{\infty} |H_c(\alpha, \beta)| d\alpha d\beta} \tag{27}$$

where CTF is $\mathbb{F}[h_c(x, y)] = H_c(u, v), u = f_x, v = f_y$ and OTF is $\mathbb{F}[h(x, y)] = H(u, v)$. Thus, OTF of the flow fields is equal to the 2D Fourier transformation of the point-spread function (PSF, $PSF = h(x, y)$) of light intensity distributions. Relationship between the input light intensity distribution $I_o(x, y)$ and output light intensity distribution $I(x, y)$ is obtained by

$$\mathbb{F}[I_o(x, y)] \cdot \mathbb{F}[h(x, y)] = \mathbb{F}[I(x, y)] \tag{28}$$

where $I(x, y) = |U(x, y)|^2$ is the light intensity distribution.

Through the above theoretical analysis, a discrete OTF matrix 63×63 could be acquired. In other words, PSF can be gained for the spatial filtering of image. The relationship between PSF and the light intensity distributions is satisfied by

$$I(x, y) = I_o(x, y) * h(x, y) \tag{29}$$

Here, the transmission of right incident light through the supersonic flow fields is considered. And position of the image centroid can be calculated by

$$x_c = \frac{\sum_i i \cdot I(i, j)}{\sum_i \sum_j I(i, j)}, y_c = \frac{\sum_j j \cdot I(i, j)}{\sum_i \sum_j I(i, j)} \tag{30}$$

where (i, j) is the coordinate of the corresponding image pixel at the image coordinate, and $I(i, j)$ is the corresponding pixel value. Thus, the centroid shift of the degraded image can be evaluated through,

$$\Delta x = x_c' - x_c^o, \Delta y = y_c' - y_c^o \tag{31}$$

where $(x_c', y')_c, (x_c^o, y_c^o)$ are respectively the centroids of the degraded image and original image. To estimate the total aberrations induced by the flow fields, Euclidean distance (ED) of image shift is used for an implicit assessment through

$$r_{ED} = \sqrt{(x' - x)^2 + (y' - y)^2} \tag{32}$$

4.3 Simulation results

Digital images of aircraft obtained from the Internet are used to simulate aero-optical images. To study the shift of image centroid without considering the effect of temporal integration on the image, only snapshots were numerically investigated. **Figure 22** shows the original image. **Figures 23–25** show the degradation results of a supersonic flow fields with the same Mach number (Ma = 7) and height of 30, 35 and 40 km, respectively. **Figures 23** and **26** show the results when the height is 30 km and the Mach numbers are 7 and 5, respectively. **Table 1** shows the results of calculating the shifts of image centroid. Although the evaluation method cannot calculate the total

Figure 22.
The original image.

Figure 23.
The aero-optically degraded image (H = 30 km, Ma = 7).

Figure 24.
The aero-optically degraded image (H = 35 km, Ma = 7).

Figure 25.
The aero-optically degraded image (H = 40 km, Ma = 7).

Figure 26.
The aero-optically degraded image (H = 30 km, Ma = 5).

aero-optical distortion, the validity of the proposed model of the aero-optical imaging method based on optical information can be easily verified.

It must qualitatively satisfy the aero-optical effect of light propagation in a supersonic flow fields. That is, at the same Mach number, the lower the height, the stronger

	Δx	Δy	**Euclidean distance r_{ED}**
H = 30 km Ma = 7	2.260931	0.181078	2.2682
H = 30 km Ma = 5	−0.327743	0.174146	0.3711
H = 35 km Ma = 7	−1.229878	−0.048415	1.2308
H = 40 km Ma = 7	−0.535493	−0.131637	0.5514

Table 2.
Image centroid shifts related to the original image.

the aero-optical effect. The higher the Mach number at the same height, the stronger the optical effect of air. The image is blurred and the centroid shift is larger in the quantitative analysis.

Furthermore, from the results in **Table 2**, it can be seen that the Mach number in the aero-optical images has a greater weight than the flight altitude. That is, the compressive effect of the flow fields by the flight speed compared with atmospheric density at different altitudes is a key factor influencing the change in the density field. Through the above analysis, the simulation results are consistent with the basic facts of the aero-optic effect.

5. Conclusion

This chapter concentrates on the numerical study of modeling aero-optical transmission and imaging through the supersonic flow fields. We have developed two computational models for predicting aero-optical performance of the supersonic flow fields, respectively, using geometrical optical method and information optical method. Firstly, this model combining the CFD model with the geometrical optics is discussed in detail. The calculation results are coincident favorably with prior knowledge from the completed research about aero-optics. The model has been compared with the experimental knowledge about the influences of the supersonic flow fields on optical transmission. Due to test complication and lack of experiment facilities, a complete comparison cannot be allowed. The model has merits not only in predicting optical performance of supersonic flow field but also in understanding the aero-optical characteristic of a particular design.

In addition, this chapter also provides a solution to aero-optical problems in terms of information optics. A computational model for studying aero-optical imaging through the supersonic flow fields is presented. This model integrates the CFD grids with the model of angular spectrum propagation to construct serially connected optical subsystems for representing the supersonic flow fields. The simulation results are qualitatively coincident with prior knowledge about aero-optical effects. The proposed model can be directly helpful in restoring the image degraded by the supersonic flow fields. Compared with the geometrical optical method, the provided approach based on the information optics can overcome the complexity of ray tracing, and give the explicit shifts of the image and describe the blur circle of the degraded image. However, the computational results are discussed qualitatively due to unavailability of the corresponding wind tune experiments. In future, more extensive computational experiments will be done so as to study the imaging under different field of view and do the comparisons under different premises.

The research methods in this chapter are also suitable for the other optical transmitting through the air turbulence and high-speed flow fields. If the issues discussed above can be take consideration into the passive aero-optical system, the other similar problems will be sure to be faced in the active optical systems. As we all know, in the near the future, airborne active laser emitter/laser communication systems will come true. There are optical beams emitting through the high-speed flow fields, and then, airborne lasers are adversely affected by this flow field when operating. In the active system, the flow field causes the projected beam energy to attenuate and deviate from the illumination target, and the laser imaging system will cause image blur and jitter. Similarly, the approaches provided in this chapter will still have the advanced technical merits in the solutions of the aero-optical effects.

Author details

Tao Wang[1,2]

1 School of Intelligent Systems Engineering, Sun Yat-Sen University, Guangzhou, China

2 Guangdong Provincial Key Laboratory of Fire Science and Intelligent Emergency Technology, Guangzhou, China

*Address all correspondence to: philips211@126.com

References

[1] Jumper EJ, Fitzgerald EJ. Recent Advances in Aero-optics. Progress in Aerospace Science. 2001;**37**: 299-339

[2] Sutton GW. Aero-optical foundations and applications. AIAA Journal. 1985; **23**(10):1525-1537

[3] Sutton GW, Pond JE, Snow R, Hwang Y. Hypersonic interceptor performance evaluation center: aero-optics performance predictions. In: Proceedings of the 2nd Annual AIAA SDIO Interceptor Technology Conference. AIAA-93-2675. Washington, D.C.: American Institute of Aeronautics and Astronautics; 1993

[4] Jumper EJ. Recent Advance in the Measurement and Analysis of Dynamic Aero-Optic Interactions (Review Chapter). 28th Plasmadynamics and Lasers Conference. Atlanta, GA, USA: AIAA-97-2350; 23-25 June 1997

[5] Thurow B, Samimy M, Lempert W. Simultaneous MHz rate flow visualization and wavefront sensing for aero-optics. In: proceedings of the 41st AIAA Aerospace Sciences Meeting and Exhibit. AIAA-2003-0684. Washington, D.C.: American Institute of Aeronautics and Astronautics; 2003

[6] Wyckham CM, Zaidi SH, Miles RB, Smiths AJ. Characterization of optical wavefront distortions due to a boundary layer at hypersonic speeds. In: Proceedings of the 34th AIAA plasmadynamics and Lasers Conference. Washington, D.C.: American Institute of Aeronautics and Astronautics; 2003

[7] George W, Pond JE. Hypersonic interceptor aero-optics performance predictions. Journal of Spacecraft and Rockets. 1994;**31**(4):592-599

[8] Sutton GW. Effect of inhomogeneous turbulence on imaging through turbulent layers. Applied Optics. 1994; **33**(18):3972-3976

[9] Michael M, Sutton GW. Beam-jitter measurements of turbulent aero-optical path differences. Applied Optics. 1992; **31**(22):4440-4443

[10] Clark RL, Farris RC. A numerical method to predict aero-optical performance in hypersonic flight. In: Proceedings of the 19th Fluid Dynamics, Plasma Dynamics and Lasers Conference. Washington, D.C.: American Institute of Aeronautics and Astronautics; 1987

[11] Gierloff JJ, Robertson SJ, Bouska DH. Computer analysis of aero-optic effects. In: Proceedings of AIAA SDIO Annual Interceptor Technology Conference. Washington, D.C.: American Institute of Aeronautics and Astronautics; 1992

[12] Mike I, Bender EE. CFD-Based computer simulation of optical turbulence through aircraft flowfields and wakes. In: Proceedings of the 32nd Plasmadynamics and Lasers Conference. Washington, D.C.: American Institute of Aeronautics and Astronautics; 2001

[13] Zubair FR, Catrakis HJ. Aero-optical Interactions Along Laser Beam Propagation Paths in Compressible Turbulence. AIAA Journal. July 2007; **45**(7):1663-1674

[14] Aguirre RC, Nathman JC, Garcia PJ, Catrakis HJ. Imaging of turbulent refractive interfaces and optical wavefronts in aero-optics. In: 36th AIAA Plasmadynamics and Laser Conference. Toronto, Ontario Canada; 2005

[15] Frumker E, Pade O. Generic method for aero-optic evaluations. Applied Optics. June 2004;**43**(16):3224-3228

[16] Monteiro A, Jarem J. Determination of the mutual coherence function and determination of the point-spread function in a transversely and longitudinally inhomogeneous aero-optic turbulence layer. Applied Optics. January 1993;**32**(2):210-224

[17] Michael D. High-order parabolic beam approximation for aero-optics. Journal of Computational Physics. 2010; **229**(15):5465-5485

[18] Zhang Y-P, Fan Z-G. Study on the optical path difference of aero-optical window. Optik. 2007;**118**:557

[19] Wang T, Zhao Y, Xu D, et al. Numerical study of evaluation optical quality of supersonic flow fields. Applied Optics. August 2007;**46**(23):5545-5551

[20] Zhao Yan, Wang Tao, Xu Dong et al.. CFD grids-based transmission model of the rays propagating through the hypersonic flow field. Acta Armamentarii. 2008;**29**(3):282-286

[21] Wang T. A Study of Optical Transmission through the Flow of Fluid-hypersonic. Beijing, China. 2007

[22] Ceniceros JM, Nahrstedt DA, Hsia Y-C. Wind tunnel validation of a CFD-based aero-optics model. In: 38th AIAA Plasmadynamics and Laser Conference. Miami, FL: AIAA; 2007. pp. 2007-4011

[23] Mani A, Wang M, Moin P. Resolution requirements for aero-optical simulations. Journal of Computational Physics. 2008;**227**:9008-9020

[24] Zubair FR, Catrakis HJ. Aero-optical Resolution Robustness in Turbulent Separated Shear Layers at Large Reynolds numbers. AIAA Journal. November 2007;**45**(l1):2721-2728

[25] Born M, Wolf E. Principles of Optics (7th Edition and 60th Anniversary Edition). Cambridge, UK: Cambridge University Press

[26] Chang X, Wang T, Wan S, et al. A method based on 3D ray tracing for aero optical wavefront analysis. Optik—International Journal for Light and Electron Optics. 2015;**126**(23):4392-4396

[27] Baxter MR, Truman CR. Predicting the optical quality of supersonic shear layer. In: Proceedings of AIAA Thermo physics, Plasma dynamics and Lasers Conference. Washington, D.C.: American Institute of Aeronautics and Astronautics; 1988

[28] Joseph W. Introduction to Fourier Optics. Vol. 5.W.H. Freeman; USA. 2017

[29] Peters B, Brown D, Cole T. CFI-aero-optic images: Issues for wave optics modeling. In: Proc. of 18th AIAA Aerospace Ground Testing Conference, Colorado Springs, June 20-23,194. AIAA 94-2622

[30] Wang T, Zhao Y, Dong X. Numerical study using angular Spectrum propagation model for aero optical imaging. Optik. 2013;**124**:411-415

[31] Shifan W. Information Optical Theory and its Applications. Beijing, China: Beijing University of Posts and Telecommunications Press; 2004

Chapter 15

Moving Node Method for Differential Equations

Umurdin Dalabaev and Malika Ikramova

Abstract

The chapter contains information about new approaches to solving boundary value problems for differential equations. It introduces a new method of moving nodes. Based on the approximation of differential equations (by the finite difference method or the control volume method), introducing the concept of a moving node, approximately analytical solutions are obtained. To increase the accuracy of the obtained analytical solutions, multipoint moving nodes are used. The moving node method is used to construct compact circuits. The moving node method allows you to investigate the diskette equation for monotonicity, as well as the approximation error of the differential equation. Various test problems are considered.

Keywords: finite difference, boundary value problem, moving node, approximation, differential equations, difference equation, approximation error, several moving nodes, compact schemes, convective-diffusion, finite volume

1. Introduction

Methods for solving problems of mathematical physics can be divided into the following four classes [1–7].

Analytical methods (the method of separation of variables, the method of characteristics, the method of Green's functions [8], etc.) have a relatively low degree of universality, i.e. focused on solving rather narrow classes of problems. As a result of applying these methods, a solution is obtained in the form of analytical formulas. The use of these formulas for the implementation of the calculation may require the solution of auxiliary computational problems (solution of nonlinear equations, calculation of special functions, numerical integration, summation of an infinite series). Nevertheless, in a number of cases, the application of these methods makes it possible to quickly and with high accuracy calculate the desired solution.

Approximate analytical methods (projection, variational methods, small parameter methods, operational methods, and various iterative methods [4, 9]) are more universal than analytical ones. The use of such methods involves modifying the original problem or changing the problem statement in such a way that the new problem can be solved by the analytical method, and its solution itself differs little enough from the solution of the original problem.

Numerical methods (finite difference method, method of lines, control volume method, finite element method, etc. [1, 2, 5–7, 10–34]) are very universal methods. Often used to solve nonlinear problems of mathematical physics, as well as linear problems with variable operator coefficients.

Probabilistic methods (Monte Carlo methods) are highly versatile. It can be used to calculate discontinuous solutions. However, they require large amounts of calculations and, as a rule, lose with the computational complexity of the above methods when solving such problems to which these methods are applicable.

Comparing methods for solving problems of mathematical physics, it is impossible to give unconditional superiority to any of them. Any of them may be the best for solving problems of a certain class. At the same time, when characterizing a specific method, it is advisable to highlight those features that often determine its advantages or disadvantages in practical application compared to an alternative method.

The advantages of the finite difference method include its high universality, for example, much higher than that of analytical methods. The application of this method is often characterized by the relative simplicity of constructing a decision algorithm and its software implementation. Often it is possible to parallelize the decision algorithm.

The shortcomings of the method include: the problematic nature of its use on irregular grids; a very rapid increase in computational complexity with an increase in the dimension of the problem (an increase in the number of unknown variables); the complexity of the analytical study of the properties of the difference scheme.

The proposed method of moving nodes combines numerical and analytical methods [7, 8, 13, 35–38]. In this case, we can obtain, on the one hand, an approximate analytical solution to the problem, which is not related to the methods listed above. On the other hand, this method allows one to obtain compact discrete approximations of the original problem. Note that obtaining an approximate analytical solution to differential equations is based on numerical methods. The nature of numerical methods also makes it possible to obtain an approximate analytical expression for solving differential equations. For this, a so-called "movable node" is introduced.

The aim of the study is to develop a computing technology based on the proposed method of moving nodes, develop a two-point convective-diffusion problem an analytical method generated by numerical methods based on the method of moving nodes, and give test examples.

2. Derivation of approximate analytical solutions of differential equations by the moving nodes method

This chapter introduces the concept of a roaming node and provides approximate solutions to simple problems using a moving node. We also studied the derivation algorithm for nonstationary and two-dimensional problems.

Note that the concept of a movable node in this context is considered for the first time.

2.1 The concept of a moving node

The solution of differential equations (DE) (ordinary or partial derivatives) by the method of finite differences is based on a finite-difference approximation of derivatives. When applying the finite difference method to the solution of DE, there is a transition from a continuous region to a finite difference one. A grid of "nodal points" is introduced into the solution area. Representing the derivatives in a finite difference

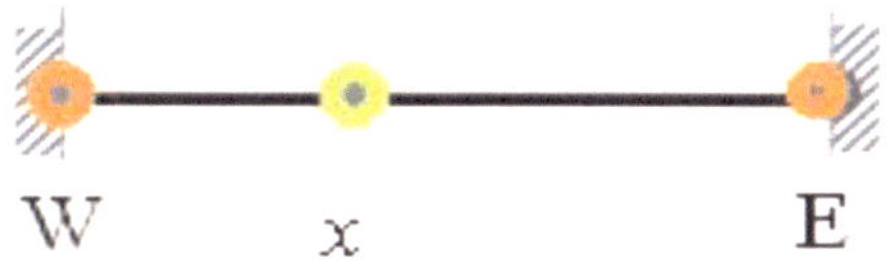

Figure 1.
One moving node.

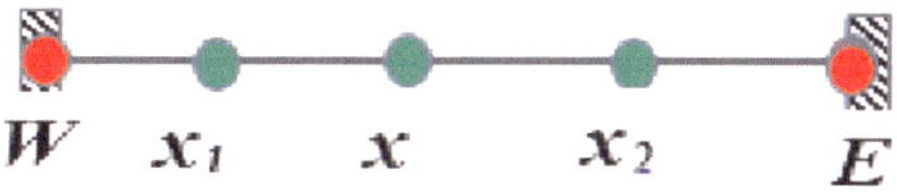

Figure 2.
Three moving node.

form, they bring it to the form of a difference equation. The difference equation is written for all grid nodes and results in a system of algebraic equations [4, 36].

Most of the DEs found in the equations of mathematical physics contain only partial derivatives of the first and second orders, while for the approximation of the derivatives they try to use no more than three nodes of the difference grid (in the case of ordinary DEs) (**Figure 1**). Let the node W and E be considered fixed, and the node x changes into segments (W, E). Then the approximation of the derivatives (first or second order) also changes based on the location of the node. The node x is said to be movable.

You can increase the number of moved nodes. Let us select additional moving nodes as follows: $x_1 = (\mathrm{W} + x)/2$, $x_2 = (\mathrm{W} + x)/2$. When node x changes its position, x_1 and x_2 automatically change their positions (**Figure 2**). In this way, you can increase the number of moved nodes. The increase in the number of moved nodes is related to the accuracy of the difference equations.

The displacement of nodal points is not only related to finite-difference equations, this approach can be successfully applied when discretizing differential equations using the control volume method.

2.2 Obtaining an approximate analytical solution with one moving node

Let, it is necessary to find $\Phi(x)$ a solution to the DE in the region $W \leq x \leq E$ with the corresponding boundary conditions. Let us take an arbitrary point $x \in (W, E)$. We have three nodes: W, E boundary nodes and an internal node x. The position of a point inside the region is determined by the node being moved x. The difference equation is usually written for an arbitrary node, x. When approximating differential operators, the first derivatives on the moving node are approximated by different relations:

$$\frac{d\Phi(x)}{dx} \approx \frac{U(x) - U(W)}{x - W}, \tag{1}$$

$$\frac{d\Phi(x)}{dx} \approx \frac{U(E) - U(x)}{E - x}, \tag{2}$$

$$\frac{d\Phi(x)}{dx} \approx \frac{U(E) - U(W)}{E - W}. \tag{3}$$

The approximation of the derivative by (1) and (2) is called the approximation of this derivative using a one-sided difference, and (3) is the approximation using the central difference.

The second derivative on the moving node is approximated as follows [4] (similarly to the approximation of the second derivative in a non-uniform grid):

$$\frac{d^2\Phi(x)}{dx^2} \approx \frac{2}{E-W}\left(\frac{U(E)-U(x)}{E-x} - \frac{U(x)-U(W)}{x-W}\right) \tag{4}$$

Let us consider some model problems of applying the moving nodes method (MNM) to obtain an analytical solution.

2.2.1 Flow in a flat pipe

The flow of a viscous fluid in a flat pipe in a one-dimensional formulation is described by the equation

$$\frac{d^2U}{dy^2} = -\frac{\Delta p}{\mu\, l} \tag{5}$$

where U is the fluid velocity, is the vertical coordinate perpendicular to the flow, $\Delta p/l$ is the pressure drop (const), μ is the viscosity. Let $y = 0$ and $y = h$ motionless walls.

We average (5) over the liquid volume: $[y/2, (h-y)/2]$, here "y" is a moving node (**Figure 3**). Then we have

$$\int_{y/2}^{(h+y)/2} \frac{d^2U}{dy^2} dy = \int_{y/2}^{(h+y)/2} \left(-\frac{\Delta p}{\mu\, l}\right) dy$$

From here

$$\left.\frac{dU}{dy}\right|_{(h+y)/2} - \left.\frac{dU}{dy}\right|_{y/2} = \left(-\frac{\Delta p}{\mu\, l}\right)\frac{h}{2} \tag{6}$$

We replace the derivatives in (6) with the difference relation:

$$\left.\frac{dU}{dy}\right|_{(h+y)/2} \approx \frac{u(h)-u(y)}{h-y}, \left.\frac{dU}{dy}\right|_{y/2} \approx \frac{u(y)-u(0)}{y}. \tag{7}$$

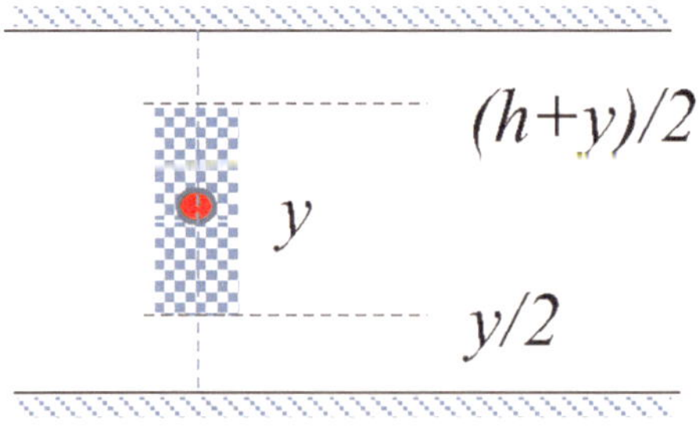

Figure 3
Control volume.

Here $u(y)$ is an approximate value $U(y)$. Thus, approximation (5) with respect to the moving node has the form:

$$\frac{u(h)-u(y)}{h-y}-\frac{u(y)-u(0)}{y}=\left(-\frac{\Delta p}{\mu l}\right)\frac{h}{2}. \quad (8)$$

Hence, taking into account the no-slip condition $(u(h)=u(0)=0)$

$$u(y)=-\frac{\Delta p}{2\mu l}y(h-y).$$

Here $u(y)$ is the average solution. For this problem, the averaged solution coincides with the exact solution.

This means that the approximation (7) for this problem is exact. The reason for the coincidence of the solution obtained with the help of the MNM with one node and the exact solution is explained by the following fact.

Lagrange's mean value theorem states that if a function $f(x)$ is continuous on an interval $[a,b]$ and differentiable on an interval (a,b), then in this interval there is at least one $x=\xi$ point such that

$$\frac{f(b)-f(a)}{b-a}=f'(\xi). \quad (9)$$

It is easy to check that if $f(x)$ represents a parabola, then in (9) $\zeta=(a+b)/2$. The exact solution (5) is a parabola. Integrating (5) over the control volume $[x/2,\ (h+x)/2]$, we obtain

$$\int_{y/2}^{(h+y)/2}\frac{d^2u}{dy^2}dy=\left.\frac{du}{dy}\right|_{(h+y)/2}-\left.\frac{du}{dy}\right|_{y/2}=\int_{y/2}^{(h+y)/2}\frac{1}{\mu}\frac{\Delta p}{l}dy.$$

Since $u(y)$ there is a parabola, therefore

$$\left.\frac{du}{dy}\right|_{(h+y)/2}=\frac{u(h)-u(y)}{h-y},\ \left.\frac{du}{dy}\right|_{y/2}=\frac{u(y)-u(0)}{y-0},$$

and (8) is the exact difference analog of (5).

2.2.2 Heat distribution in the plate

Heat propagation in the plate is described by the equation

$$\frac{d^2T}{dx^2}+\frac{q}{k}=0, \qquad \frac{dT(0)}{dx}=0, \quad T(1)=1 \quad (10)$$

where k is the thermal conductivity and q is the heat release per unit volume (k and q = const). It is assumed that the source does not depend on temperature. Replacing (10) with a difference equation with a moving node, we have

$$\frac{2}{1-0}\left[\frac{T(1)-T(x)}{1-x}-\frac{T(x)-T(0)}{x-0}\right]+\frac{q}{k}=0 \quad (11)$$

Solving Eq. (11), we obtain

$$T(x) = 1 + \frac{q}{2k}\left(1 - x^2\right) \tag{12}$$

Solution (12) coincides with the exact solution. Note that the exact solution is obtained not only for the Dirichlet problem but as for the problem of flow in a flat pipe. Here the boundary conditions are of mixed type.

2.2.3 Magnetohydrodynamic Couette flow

Consider the Couette flow, when a conducting fluid flows in a uniform magnetic field between two plates, one of which is stationary, and the other moves in its own plane at a constant speed. Based on the Navier-Stokes equation, taking into account the magnetic field and taking into account the one-dimensionality of the flow, it can be written in a dimensionless form as follows:

$$\frac{d^2u}{dy^2} - M^2u = P \tag{13}$$

Boundary conditions

$$u(0) = 0, \;\; u(1) = 1 \tag{14}$$

Here, u is the dimensionless flow velocity and y is the dimensionless coordinate. Dimensionless quantities M – Hartmann number, P – pressure coefficient (M and P = const).

Replacing the second-order derivative in (13) with a difference relation similar to (7), and considering the boundary condition (14), we can obtain an approximate solution

$$u_1(y) = \frac{2y - Py(1-y)}{2 + M^2y(1-y)} \tag{15}$$

This solution comes close to the exact solution (**Figure 4**).

2.2.4 The method of moving nodes for the convection-diffusion equation

Consider the transport equation

$$\frac{d\Phi}{dx} = \frac{1}{Pe}\frac{d^2\Phi}{dx^2} + S(x), \tag{16}$$

Here, Φ the unknown function, $S(x)$ the source, Pe is the Peclet number. The equation is considered under appropriate boundary conditions.

The convective term of Eq. (16) is approximated by (1), and the diffusion term by (4). Consider (16) into segments with boundary conditions $\Phi(0) = 0, \Phi(1) = 1$ and $S(x) = 0$. Then, using the upwind scheme, we replace Eq. (16) with a difference equation that looks like this:

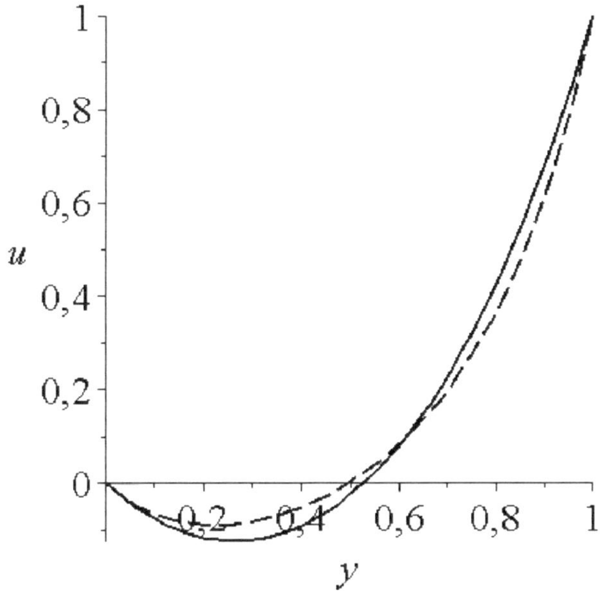

Figure 4.
Comparison of exact and approximate solutions (M = 2, P = 4). The solid line is the exact solution, the dotted line is according to (5).

$$\frac{U(x)}{x} = \frac{2}{Pe}\left(\frac{1-U(x)}{1-x} - \frac{U(x)}{x}\right) \tag{17}$$

From here, we can easily determine $U(x)$:

$$U(x) = \frac{2x}{2 + Pe(1-x)} \tag{18}$$

Figure 5 shows a comparison of the exact and approximate solutions. The solid line corresponds to the exact solution, and the dotted line corresponds to the solution (8). It can be seen from the graph that numerical diffusion takes place.

For $\Phi(0) = 0, \Phi(1) = 1$ and $S(x) = 5\cos 4x\ Pe = 5$, the results of the exact and approximate solutions are shown in **Figure 6**. It can be seen from the graph that there are large errors. Here the Peclet number plays an important role. Indeed, for $\Phi(0) = 0, \Phi(1) = 1$ and $S(x) = 5\cos 4x\ Pe = 0, 1$, we obtain solutions shown in **Figure 7**, which shows that the approximate and exact solutions are close.

2.2.5 Equation with variable coefficient

Consider the equation

$$\varepsilon u''(x) + 2xu'(x) = 0, \tag{19}$$

into segments (−1,1) with boundary $u(-1) = -1,\ u(1) = 2$ conditions $u(-1) = -1,\ u(1) = 2$. The exact solution is determined through the error functions:

$$u(x) = \frac{erf(1/\sqrt{\varepsilon}) + 3erf(x/\sqrt{\varepsilon})}{2erf(1/\sqrt{\varepsilon})}.$$

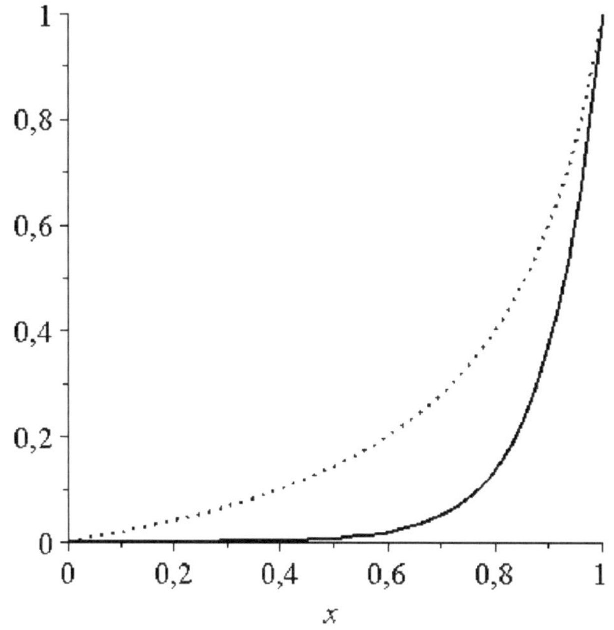

Figure 5.
Comparison of exact and approximate solutions. Pe = 10.

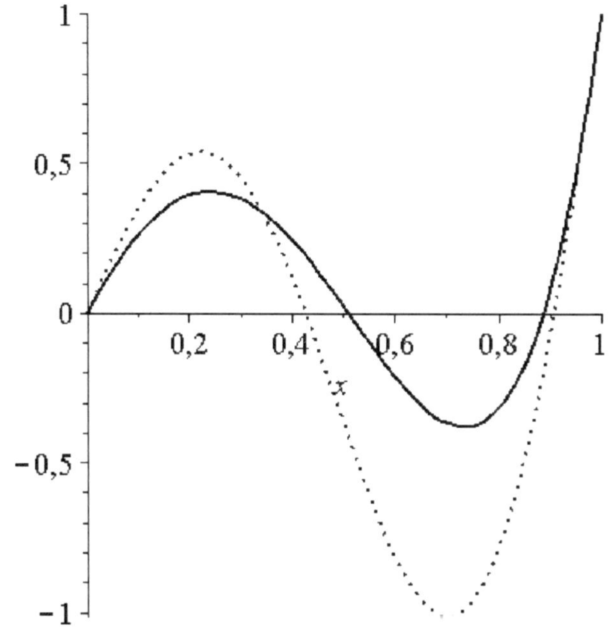

Figure 6.
Comparison of exact and approximate solutions. Pe = 5.

The difference scheme with a moving node for (19) has the form (upwind scheme):

$$\varepsilon\left[\frac{2-U(x)}{1-x}-\frac{U(x)+1}{x+1}\right]+\frac{1}{2}(2x-|2x|)\frac{U(x)+1}{1+x}+\frac{1}{2}(2x+|2x|)\frac{2-U(x)}{1+x}=0.$$

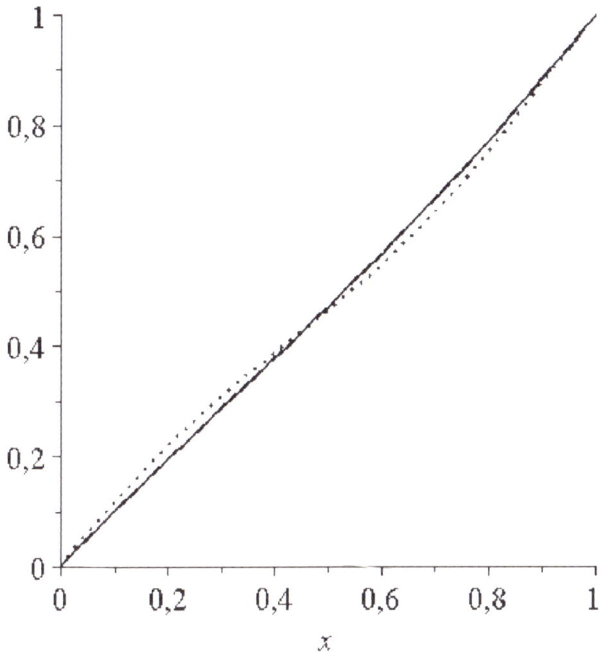

Figure 7.
Comparison of exact and approximate solutions. Pe = 0.1.

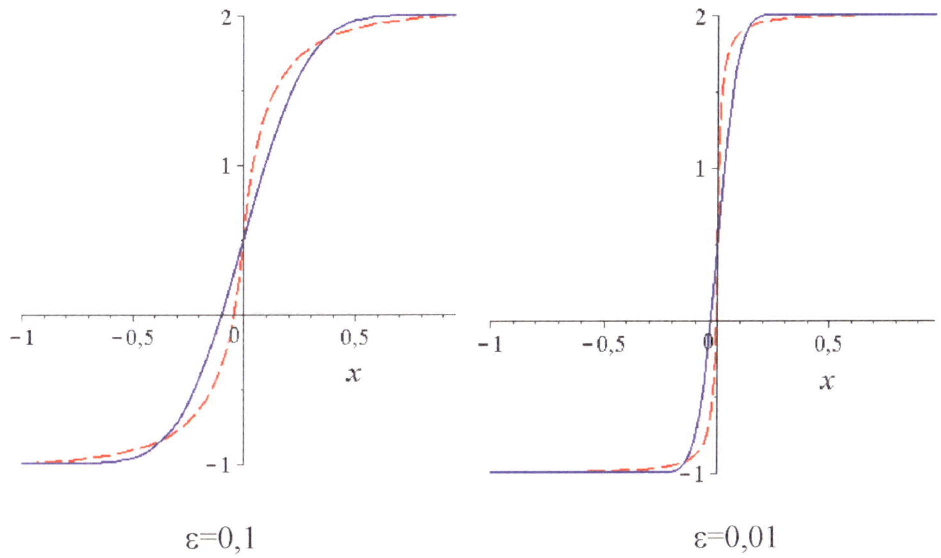

Figure 8.
Solution comparisons.

Solving this equation with respect to $U(x)$, we obtain an approximate analytical solution. **Figure 8** compares the solutions of the exact and approximate analytical solution; the solid line corresponds to the exact solution, and the dotted line corresponds to the approximate one.

Remark 1. In the given examples, the convective term is approximated by the upwind scheme. Other approximations can be used to improve.

Remark 2. In the above examples, the approximation of the term with the source is carried out constant in the considered moving segment. For improvement, other approximations can be used to obtain an improved solution.

2.3 Obtaining an analytical solution with several moving nodes

2.3.1 Moving nodes method for a one-dimensional convective-diffusion problem

Due to the importance of convective-diffusion problems, we will apply multipoint MNM to such problems [14]:

$$\frac{d\Phi}{dx} = \frac{1}{Pe}\frac{d^2\Phi}{dx^2} + S(x). \tag{20}$$

Let us take an arbitrary one node inside the segment $x \in (W, E)$.

Let us consider a difference analog of Eq. (20), in which the convective term is approximated by a one-sided difference relation.

Then the upwind scheme has the form:

$$Pe\frac{U^1 - U_W^1}{x - W} = \frac{2}{(E - W)}\left(\frac{U_E^1 - U^1}{E - x} - \frac{U^1 - U_W^1}{x - W}\right) + Pe \cdot S(x). \tag{21}$$

This schema can be rewritten like this:

$$a_P^1 U^1 = a_E^1 U_E^1 + a_W^1 U_W^1 + F^1(x), \tag{22}$$

Here

$$a_E^1 = \frac{2}{(E - W)(E - x)}, a_W^1 = \frac{Pe}{(x - W)} + \frac{2}{(E - W)(x - W)}, a_P^1 = a_E^1 + a_W^1,$$
$$F^1(x) = Pe \cdot S(x)$$

Hence, we have

$$U^1 = \frac{2(x - W)U_E^1 + (E - x)(2 + Pe(E - W))U_W^1}{(E - W)(2 + Pe(E - x))} + \frac{(x - W)(E - x)}{2 + Pe(E - x)} Pe \cdot S(x) \tag{23}$$

When $x \in (W, E)$ changes its position (let us make it moveable within the interval (W, E)), based on (23) we obtain the values of the unknown function in each position. In other words, U^1 obtained with the help of (23), will give us an approximate solution to the problem. Note that in this case, $U_W^1 = \Phi(W), U_E^1 = \Phi(E)$. The superscript corresponds to the number of nodes being moved.

Adding additional moving nodes $x_1 = \frac{x+W}{2}, x_2 = \frac{x+E}{2}$..

Now we have three moving nodes x, x_1, x_2. Note that if x changes its position, then x_1 and x_2 also changes its position.

A scheme of type (21) for a segment $[W, x]$ has the form:

$$Pe\frac{U_1^3 - U_W^3}{(x - W)/2} = \frac{2}{(x - W)}\left(\frac{U^3 - U_1^3}{x - x_1} - \frac{U_1^3 - U_W^3}{x_1 - W}\right) + Pe \cdot S(x_1). \tag{24}$$

Here $U_1^3 = U^3(x_1)$.
A scheme of type (21) for a segment $[x, E]$ has the form

$$Pe\frac{U_2^3 - U^3}{(E - x)/2} = \frac{2}{(E - x)}\left(\frac{U_E^3 - U_2^3}{E - x_2} - \frac{U_2^3 - U^3}{x_2 - x}\right) + Pe \cdot S(x_2). \tag{25}$$

Scheme upstream for a segment $[x_1, x_2]$:

$$Pe\frac{U^3 - U_1^3}{x - x_1} = \frac{2}{(x_2 - x_1)}\left(\frac{U_2^3 - U^3}{x_2 - x} - \frac{U^3 - U_1^3}{x - x_1}\right) + Pe \cdot S(x). \tag{26}$$

Here $U_2^3 = U^3(x_2)$.
In (26) we exclude U_1^3, U_2^3 using (24) and (25). Then we get the following diagram:

$$Pe\frac{U^3 - U_W^3}{\frac{(x-W)}{2} \cdot (1 + \tau_1)} = \frac{4}{(E - W)}\left(\frac{U_E^3 - U^3}{\frac{E-x}{2} \cdot (1 + \gamma_1)} - \frac{U^3 - U_W^3}{\frac{x-W}{2} \cdot (1 + \tau_1)}\right) + F^3(x) \tag{27}$$

Here we have introduced the notation.

$$\tau_1 = 2/(2 + \sigma), \gamma_1 = (2 + \theta)/2,\ \sigma = Pe(x - W), \theta = Pe(E - x),$$
$$F^3(x) = Pe \cdot S(x) + \frac{4 + Pe \cdot (E - W)}{E - W} \cdot \frac{1 - \tau_1}{1 + \tau_1} \cdot S(x_1) + \frac{4}{E - W} \cdot \frac{\gamma_1 - 1}{\gamma_1 + 1} \cdot S(x_2).$$

And $U_W^3 = \Phi(W), U_E^3 = \Phi(E)$.
(25) can be rewritten as follows:

$$a_P^3 U^3 = a_E^3 U_E^3 + a_W^3 U_W^3 + F^3(x), \tag{28}$$

where
$a_E^3 = \frac{8}{(E-W)(E-x)(1+\gamma_1)}, a_W^3 = \frac{2Pe}{(x-W)(1+\tau_1)} + \frac{8}{(E-W)(x-W)(1+\tau_1)}, a_P^3 = a_W^3 + a_E^3.$
Increase the number of moved nodes:

$x_1^- = \frac{x_1+W}{2} = \frac{x+3W}{4}, x_1^+ = \frac{x_1+x}{2} = \frac{3x+W}{4},$

$x_2^- = \frac{x_2+x}{2} = \frac{3x+E}{4}, x_2^+ = \frac{x_2+E}{2} = \frac{x+3E}{4}.$

In the difference scheme (28), the unknown function appears at three nodes: W, x, E. The function S is calculated at points x_1, x, x_2. Let us write a scheme of type (28) for each of the segments $[W, x]$ and $[x_1, x_2]$.

The scheme of type (28) for a segment has the form:

$$a_{x_1}^3 U_{x_1}^3 = a_x^3 U_x^3 + a_{W^-}^3 U_W^3 + F_-^3(x_1), \tag{29}$$

where
$a_x^3 = \frac{8}{(x-W)(x-x_1)\left(1+\gamma_1^-\right)}, a_{W^-}^3 = \frac{2Pe}{(x_1-W)\left(1+\tau_1^-\right)} + \frac{8}{(x-W)(x_1-W)\left(1+\tau_1^-\right)},$
$a_{x_1}^3 = a_x^3 + a_{W^-}^3,$
$F_-^3(x_1) = Pe \cdot S(x_1) + \frac{4+Pe\cdot(x-W)}{x-W} \cdot \frac{1-\tau_1^-}{1+\tau_1^-} \cdot S\left(x_1^-\right) + \frac{4}{x-W} \cdot \frac{\gamma_1^- -1}{\gamma_1^- +1} \cdot S\left(x_1^+\right),$

$\tau_1^- = 2/(2+\sigma^-), \gamma_1^- = (2+\theta^-)/2, \ \sigma^- = Pe\,(x_1 - W), \theta^- = Pe\,(x - x_1).$

Similarly, we write a scheme of type (29) for the segments $[x, W]$ and $[x_1, x_2]$. Excluding the obtained three systems of equations $U^3_{x_1}$ and $U^3_{x_2}$ obtain a scheme with seven movable nodes:

where

$a^7_E = \frac{2^5(1-\gamma_2)}{(E-W)(E-x)\left(1-\gamma_2^4\right)}, a^7_W = \frac{4Pe(1-\tau_2)}{(x-W)\left(1-\tau_2^4\right)} + \frac{2^5(1-\tau_2)}{(E-W)(x-W)\left(1-\tau_2^4\right)}, a^7_P = a^7_W + a^7_E.$

$\tau_2 = 4/(4+\sigma), \gamma_2 = (4+\theta)/4$

$$F^7(x) = Pe\cdot S(x) + \frac{8 + Pe\cdot(x-W)}{x-W}\cdot\frac{(1-\tau_2)^2}{1-\tau_2^4}\cdot\sum_{j=1}^{3}\sum_{i=1}^{j}\tau_2^{i-1}S\left(W + j\frac{x-W}{4}\right) -$$

$$\frac{8}{E-W}\cdot\frac{(1-\gamma_2)^2}{1-\gamma_2^4}\cdot\sum_{j=1}^{3}\sum_{i=1}^{j}\gamma_2^{i-1}S\left(x + (4-j)\frac{E-x}{4}\right).$$

Continuing in this way, we can get a scheme with $2^k - 1$ moving nodes

$$a_P^{(2^k-1)}U^{(2^k-1)} = a_E^{(2^k-1)}U_E^{(2^k-1)} + a_W^{(2^k-1)}U_W^{(2^k-1)} + F^{(2^k-1)}(x), \qquad (30)$$

where $a_E^{(2^k-1)} = \frac{2^{2k+1}(1-\gamma_k)}{(E-W)(E-x)\left(1-\gamma_k^{2^k}\right)}, a_W^{(2^k-1)} = \frac{2^{2k+1}Pe(1-\tau_k)}{(x-W)\left(1-\tau_k^{2^k}\right)} + \frac{2^{2k+1}(1-\tau_k)}{(E-W)(x-W)\left(1-\tau_k^{2^k}\right)},$

$a_P^{(2^t-1)} = a_W^{(2^t-1)} + a_E^{(2^t-1)}.\tau_k = 2^k/(2^k+\sigma), \gamma_k = (2^k+\theta)/2^k,$

$$F^{(2^k-1)}(x) = Pe\cdot S(x) + \frac{2^{k+1} + Pe\cdot(E-W)}{E-W}\frac{(1-\tau_k)^2}{1-\tau_k^{2^k}}\sum_{j=1}^{2^k-1}\sum_{i=1}^{j}\tau_k^{i-1}\cdot S\left(x + j\frac{x-W}{2^k}\right) -$$

$$\frac{2^{k+1}}{E-W}\frac{(1-\gamma_k)^2}{1-\gamma_k^{2^k}}\sum_{j=1}^{2^k-1}\sum_{i=1}^{j}\gamma_k^{i-1}\cdot S\left(x + (2^k-j)\frac{E-x}{2^k}\right).$$

Figures 9 and **10** show graphs of approximate solutions to the problem (18), obtained by (30) for $W = 0, E = 1$, with different moving nodes.

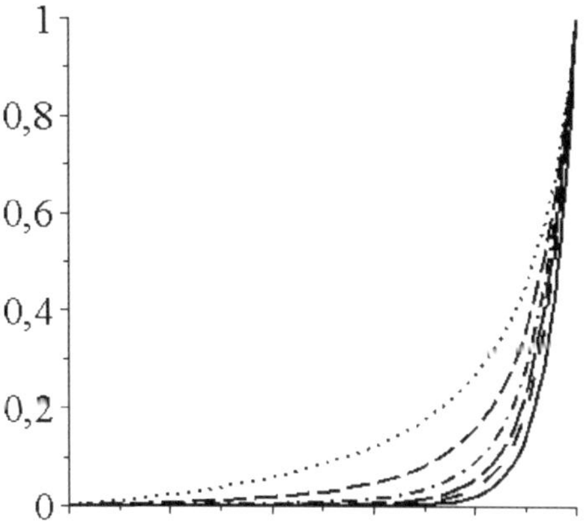

Figure 9.
$Pe = 20$, $\Phi_W = 0$, $\Phi_E = 1$, $S(x) = 0$. Approximate solutions of the problem. Dotted—at k = 1, dotted—k = 2, dotted-dotted—k = 3, long dotted—k = 4, rarely dotted—k = 5. The solid line is the exact solution.

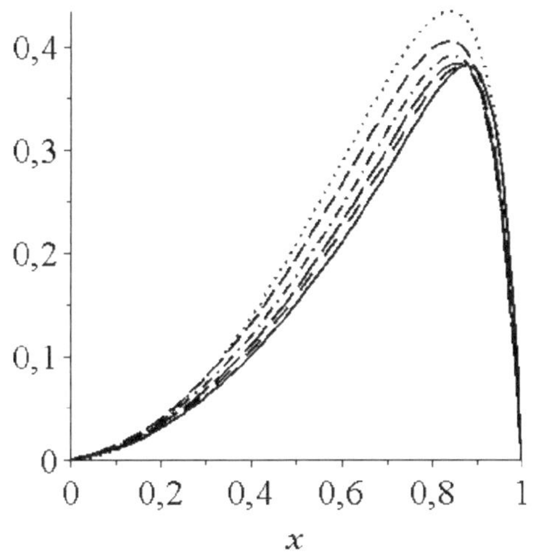

Figure 10.
$Pe = 20$, $\Phi_W = 0$, $\Phi_E = 0$, $S(x) = x$,. Approximate solutions of the problem. Dotted—at k = 1, dotted—k = 2, dotted-dotted—k = 3, long dotted—k = 4, rarely dotted—k = 5. The solid line is the exact solution.

It can be seen from the graphs that the approximate solutions give good results.

Remark. When obtaining many point-moving nodes, we proceeded from the upwind scheme. It was possible to proceed from the other three-point schemes.

2.3.2 Analytical control volume method for a one-dimensional convective-diffusion problem

It is known that differential equations are obtained on the basis of the integral conservation law. Therefore, discretization of the equations can be carried out using the approximation of integral conservation laws. This method is called the Finite Volume Method. Another name for the method is integro-interpolation.

Consider a one-dimensional convective-diffusion equation on a finite interval with boundary conditions in the form:

$$\frac{d}{dx}(\rho u \Phi) = \frac{d}{dx}\left(\Gamma \frac{d\Phi}{dx}\right) + S(x) \tag{31}$$

$$\Phi(W) = \Phi_W, \quad \Phi(E) = \Phi_E \tag{32}$$

where u is the flow velocity in the x direction, ρ is the flow density, Γ is the diffusion coefficient, $S(x)$ is a given function (source), Φ an unknown function. It follows from the continuity equation that $F = \rho u = const$.

Consider Eq. (31) into segments $[W, E]$. To obtain an approximate analytical solution to the problem using the control volume method, we take an arbitrary point $x \in [W, E]$ and control volume $[w, e]$ (**Figure 11**). Let us assume that the face w is located in the middle between the points W and x, and the face e is in the middle between the points x and E. Integrating Eq. (31) over the control volume and replacing the derivatives with the upwind scheme, we obtain the zero approximation.

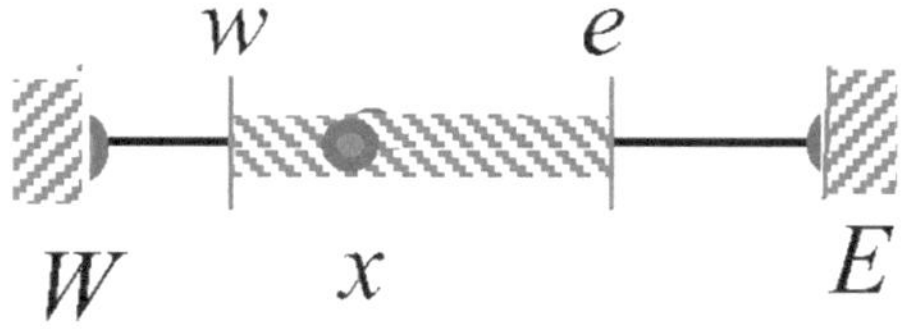

Figure 11.
Control volume [w, e].

$$(a_E + a_W)\Phi^0 = a_E\Phi_E^0 + a_W\Phi_W^0 + \frac{E - W}{2} \cdot S(x) \tag{33}$$

Here $a_E = \frac{\Gamma_e}{E-x} + \max(-F_e, 0)$; $a_W = \frac{\Gamma_w}{x-W} + \max(F_w, 0)$. Since $x \in [W, E]$ an arbitrary point from (33), we can determine Φ^0 and obtain an approximate analytical solution to problem (31).

Note that, from (33) it follows that, in the absence of a source ($S(x) \equiv 0$) on the segments $[W, E]$, the function is monotone.

To improve the approximate solution, we take additional nodes: $x_1 = \frac{x+W}{2}, x_2 = \frac{x+E}{2}$. Let us write an upwind scheme of type (33) for the segment $[W, x]$, $[x_1, x_2]$, and $[x, E]$. We get a system of three equations. We exclude the resulting system $\Phi^1(x_1)$, $\Phi^1(x_2)$ and as a result, we get an improved scheme:

$$\begin{gathered}\left[\frac{\beta_1^+}{1+\tau_1} + \frac{\alpha_1^-}{1+\gamma_1}\right]\Phi^1 = \frac{\beta_1^+}{1+\tau_1}\Phi_W^1 + \frac{\alpha_1^-}{1+\gamma_1}\Phi_E^1 + \frac{E-W}{4} \cdot S(x) + \\ \frac{1}{1+\tau_1} \cdot \frac{x-W}{2} \cdot S\left(W + \frac{x-W}{2}\right) + \frac{1}{1+\gamma_1} \cdot \frac{E-x}{2} \cdot S\left(x + \frac{E-x}{2}\right).\end{gathered} \tag{34}$$

where $\tau_1 = \frac{\beta_1^-}{\beta_1^+}, \gamma_1 = \frac{\alpha_1^+}{\alpha_1^-}$, $\beta_1^- = 2D_W + F^-, \beta_1^+ = 2D_W + F^+$, $\alpha_1^- = 2D_E + F^-$, $\alpha_1^+ = 2D_E + F^+$, $D_E = \Gamma/(E-x), D_W = \Gamma/(x-W)$, $F^- = \max(-F, 0)$, $F^+ = \max(F, 0)$.

In (34), Φ^1 is the improved value of the unknown function at the nodal point $x\left(\Phi_W^1 \equiv \Phi_W, \Phi_E^1 \equiv \Phi_E\right)$.

where in (34), the improved value of the unknown function at the nodal point is x.

Solving (34) with respect to, we obtain an improved analytical solution. Again, to improve the solution, we proceed in a similar way: we write the scheme (34) for the segment $[W, x]$, $[x_1, x_2]$ and $[x, E]$, and eliminate the unknowns at the points x_1 and x_2, and so on. Continuing this process, we get.

$$\begin{gathered}\left[\frac{(1-\tau_k)\beta_k^+}{1-\tau_k^{2^k}} + \frac{(1-\gamma_k)\alpha_k^-}{1-\gamma_k^{2^k}}\right]\Phi^k = \frac{(1-\tau_k)\beta_k^+}{1-\tau_k^{2^k}}\Phi_W^k + \frac{(1-\gamma_k)\alpha_k^-}{1-\gamma_k^{2^k}}\Phi_E^k + \frac{E-W}{2^{k+1}} \cdot S(x) \\ + \frac{1-\tau_k}{1-\tau_k^{2^k}} \cdot \frac{x-W}{2^k} \cdot \sum_{j=1}^{2^k-1}\sum_{i=1}^{j}\tau_k^{i-1}S\left(W + j\frac{x-W}{2^k}\right) \\ + \frac{1-\gamma_k}{1-\gamma_k^{2^k}} \cdot \frac{E-x}{2^k} \cdot \sum_{j=1}^{2^k-1}\sum_{i=1}^{j}\gamma_k^{i-1}S\left(x + (2^k - j)\frac{E-x}{2}\right).\end{gathered} \tag{35}$$

Here $\tau_k = \frac{\beta_k^-}{\beta_k^+}, \gamma_k = \frac{\alpha_k^+}{\alpha_k^-}$, $\beta_k^- = 2^k D_W + F^-, \beta_k^+ = 2^k D_W + F^+$, $\alpha_k^- = 2^k D_E + F^-$, $\alpha_k^+ = 2^k D_E + F^+$.

In (35), Φ^k is the improved value of the unknown function at the nodal point $x\left(\Phi_W^k \equiv \Phi_W, \Phi_E^k \equiv \Phi_E\right)$. Solving Eq. (35) with respect to Φ^k, we obtain an approximate analytical solution of the original problem.

Examples.

Figure 12 shows solutions to the problem (31) $\Gamma = const, R = \rho u/\Gamma = 20, S(x) = 0$ for segments $[0; 1]$ with boundary conditions $\Phi_W = 0$, $\Phi_E = 1$. **Figure 13** shows

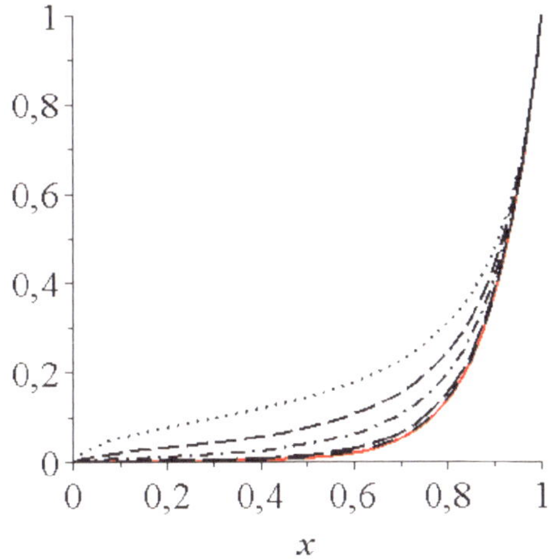

Figure 12.
Comparison of the approximate solutions for S(x) = 0. *Continuous line is exact, point—k = 0, dotted line—k = 1, dot-dotted line—k = 2, long dotted line—k = 4, rare dotted line—k = 6.*

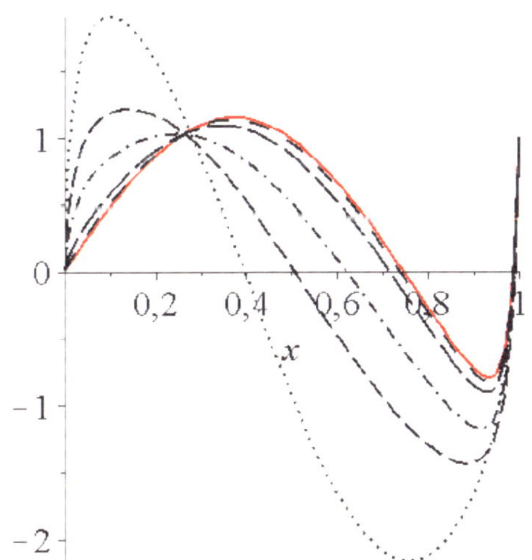

Figure 13.
Comparison of the approximate solutions for S(x) = 5cos4x. Continuous line is exact, point—k = 0, dotted line—k = 1, dot-dotted line—k = 2, long dotted line—k = 4, rare dotted line—k = 6.

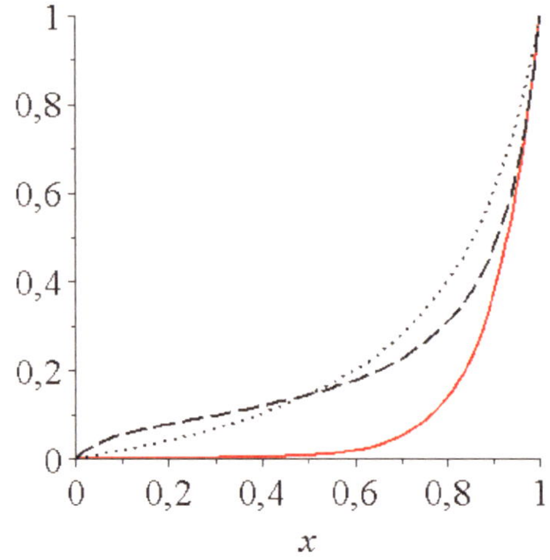

Figure 14.
Approximate solution for k = 0. Solid curves are exact solutions, point curves are the finite difference method; dotted lines—control volume method.

solutions to the problem (31) $R = 50$, $S(x) = 5\cos 4x$ for segments $[0; 1]$ with boundary conditions $\Phi_W = 0$, $\Phi_E = 1$. The graph shows that, as k increases, the approximate solutions approach the exact one.

It can be seen from the graphs that, starting from k = 6, the exact and approximate solutions visually coincide.

It is interesting to compare the analytical solution obtained by the finitely different method (30) and the control volume method (R = 10, S(x) = 0).

From **Figures 14** and **15**, it can be seen that the solution obtained by the control volume method is preferable.

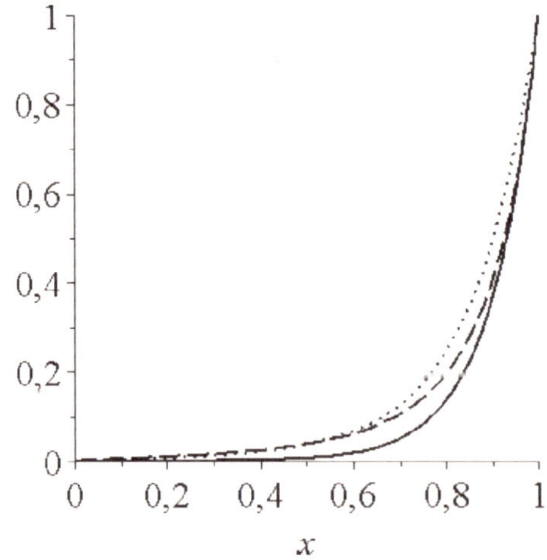

Figure 15.
Approximate solution for k = 1. Solid curves are exact solutions, point curves are the finite difference method; dotted lines—control volume method.

2.3.3 Improving accuracy with Richardson extrapolation

Using the method described, we can improve the accuracy of approximate solutions to the problem [39]. Linear combination $Q^3(x) = -\frac{1}{3}U^1(x) + \frac{4}{3}U^3(x)$ is more accurately approximates the solution. With a linear combination of $U^1(x), U^3(x)$ and $U^7(x)$ in the form $Q^7(x) = \frac{1}{45}U^1(x) - \frac{4}{9}U^3(x) + \frac{64}{45}U^7(x)$, we obtain a more refined solution to the problem [39].

Figure 16 shows graphs of approximate solutions to the problem (31) obtained by Richardson's extrapolation for $W = 0, E = 1$. The solid line in **Figure 16–19** is the exact solution.

Figures 16–19 allow us to state that Richardson's extrapolation makes it possible to obtain a more refined solution to the problem.

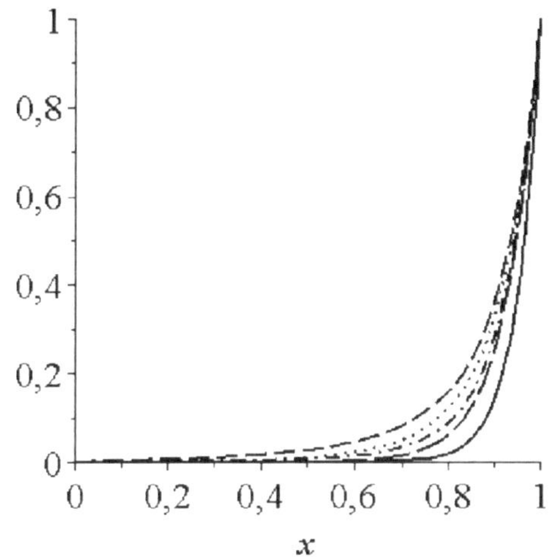

Figure 16.
$\Phi_W = 0$, $\Phi_E = 1, S(x) = 0$, $Pe = 20$. Comparisons of solutions. Dotted line is $U^3(x)$, point line—$Q^3(x)$, dot-dotted line—$U^7(x)$, long dotted line—$Q^7(x)$.

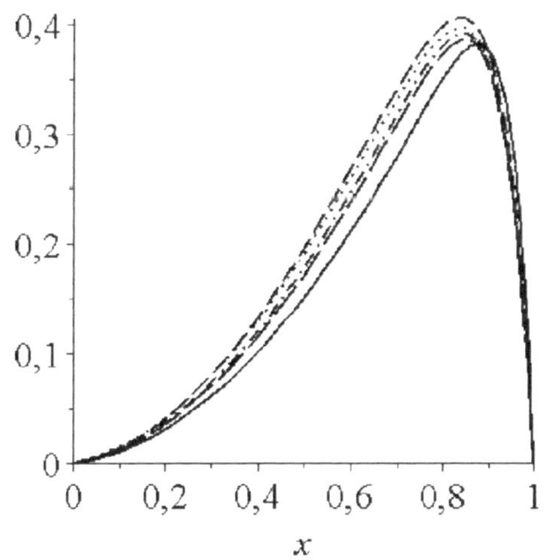

Figure 17.
$\Phi_W = 0$, $\Phi_E = 0, S(x) = x$, $Pe = 20$. Comparisons of solutions. Dotted line is $U^3(x)$, point line—$Q^3(x)$, dot-dotted line—$U^7(x)$, long dotted line—$Q^7(x)$.

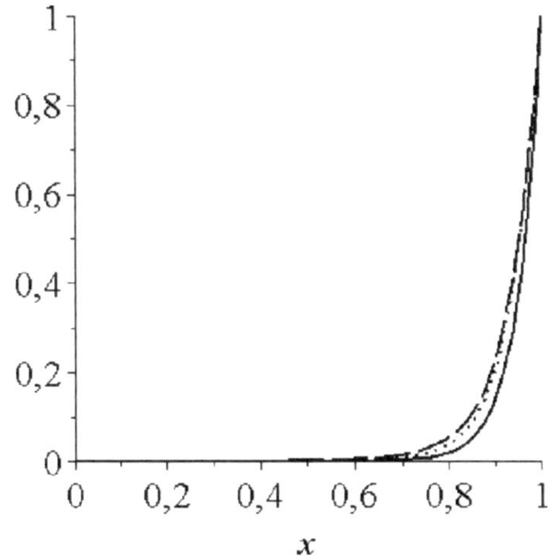

Figure 18.
$\Phi_W = 0,\ \Phi_E = 1, S(x) = 0,\ Pe = 20$. *Comparisons of solutions. Dotted line—*$U^{15}(x)$*, dotted line—*$Q^{15}(x)$.

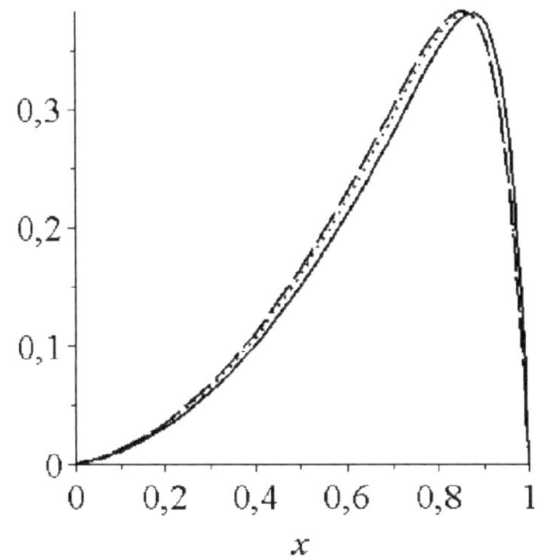

Figure 19.
$\Phi_W = 0,\ \Phi_E = 0, S(x) = x,\ Pe = 20$. *Comparisons of solutions. Dotted line—*$U^{15}(x)$*, dotted line—*$Q^{15}(x)$.

2.4 Moved node method for non-stationary problems

In the previous paragraphs, the application of the MNM for ordinary differential equations has been considered. Here we consider the application of the MNM for parabolic equations.

An example of a problem that leads to a parabolic partial differential equation is the problem of heat transfer along a long rod, described by the heat transfer (or diffusion) equation.

The problem is to find a function $U(x,t)$ in the region $\Omega = \{(x,t) \mid W \le x \le E, 0 \le t \le T\}$ satisfying the equation.

$$\frac{\partial U}{\partial t} = A\frac{\partial^2 U}{\partial x^2} + f(x,t), \quad A > 0 \tag{36}$$

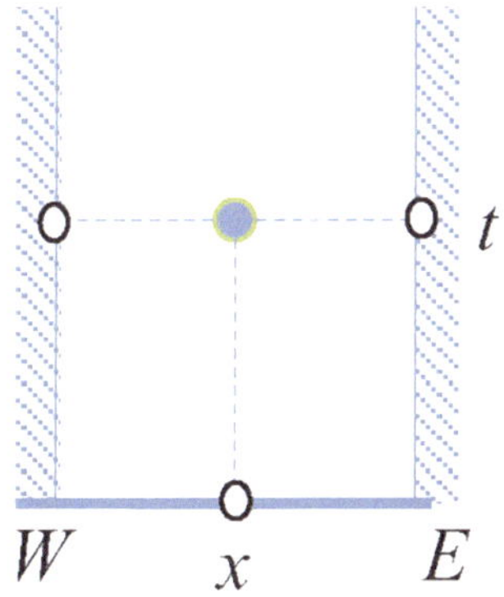

Figure 20.
The region of solution

initial condition

$$U(x,0) = U^0(x)$$

and boundary conditions of the first kind

$$U(W,t) = U_W(t); \quad U(E,t) = U_E(t).$$

Let us take an arbitrary point Ω in the area $(x,t) \in \Omega$ (**Figure 20**). We will accept this point as moving. We approximate (36) by the implicit scheme

$$\frac{Y(x,t) - U^0(t)}{t} = A\frac{2}{E-W}\left(\frac{U_E(t) - Y(x,t)}{E-x} - \frac{Y(x,t) - U_W(t)}{x-W}\right) + f(x,t), \tag{37}$$

In (37), $Y(x,t)$ is an approximate analytical solution. When the point runs through Ω, we get a solution in the area under consideration. From (37), we get

$$Y(x,t) = \frac{(E-x)(x-W)}{2At + (E-x)(x-W)}U^0(t) + \frac{2At[U_E(t)(x-W) + U_W(t)(E-x)]}{2At + (E-x)(x-W)} + \frac{(E-x)(x-W)t}{2At + (E-x)(x-W)}f(x,t). \tag{38}$$

Consider examples.

2.4.1 Test problems

Let us consider Eq. (36) $0 < x < 1$ with conditions $U^0(x) = x$, $U_W(t) = 0, U_E(t) = e^{-t}, f(x,t) = -xe^{-t}$. Exact solution of problem is $U(x,t) = xe^{-t}$. **Figure 21** presents a comparison of the exact and approximate solutions for the cross-section $x = 0,5$ and $x = 0,2$. The solid lines are the exact solution. **Figure 21** shows the closeness of the exact and approximate solutions.

Let us consider Eq. (36)$0 < x < 1$ with conditions $U^0(x) = \sin \pi x + x^2$, $U_W(t) = 0, U_E(t) = 1$, $f(x,t) = -\sin \pi x e^{-t} + \pi^2 \sin \pi x e^{-t} - 2$. Exact solution of problem

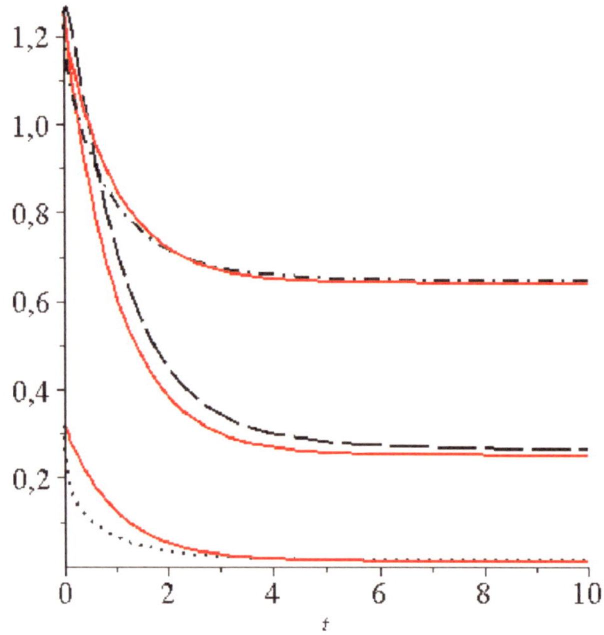

Figure 21.
Solution comparison of the exact and approximate solutions for the sections x = 0, 5 *and* x = 0.2.

presents a comparison of the exact and approximate solutions for the sections $x = 0,1$, $x = 0,5$, and $x = 0,8$. **Figure 22** shows the closeness of the exact and approximate solutions.

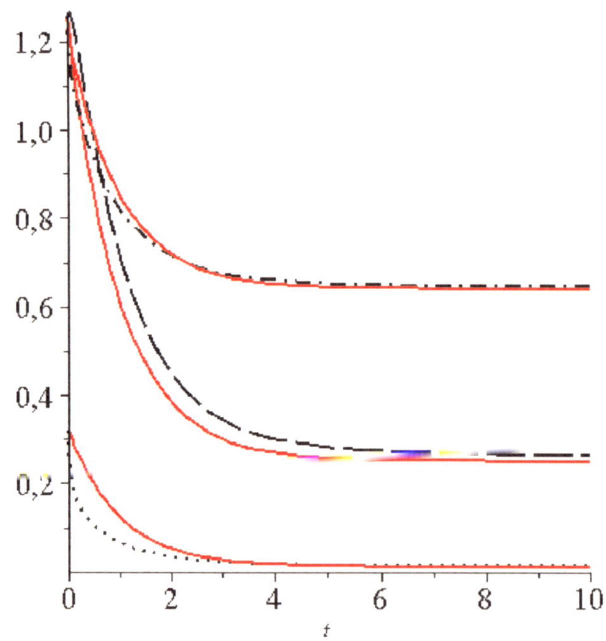

Figure 22.
Solution comparison of the exact and approximate solutions for the sections x = 0, 1, x = 0.5 *and* x = 0.8.

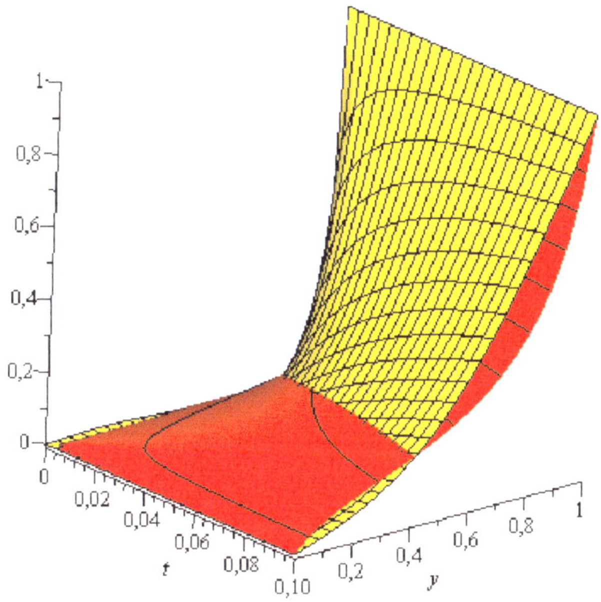

Figure 23.
Comparison of (39) and (41) (red light approx. solution).

2.4.2 Unsteady flow of a viscous fluid between parallel walls

As a practical example, consider an unsteady flow of a viscous fluid between parallel walls. Let a viscous fluid fill the entire space between horizontal planes located at a certain distance from each other. Let the lower plane be stationary all the time, and the upper one starts to move to the right at a constant speed. We neglect the action of gravity and assume that the pressure is constant everywhere. The flow is assumed to be directed parallel to the x-axis. Then the equation of motion of a viscous fluid in dimensionless variables has the form.

$$\frac{\partial u}{\partial t} = \frac{\partial^2 u}{\partial y^2} \tag{39}$$

The exact solution of the equation under the conditions:

$$u(0,y) = 0,\ u(t,0) = 0,\ u(t,1) = 1$$

looks like:

$$u = y + \frac{2}{\pi}\sum_{k=1}^{\infty}\frac{(-1)^k}{k}\sin(k\pi y)\exp\left(-k^2\pi^2 t\right) \tag{40}$$

Let us replace (39) with the difference relation:

$$\frac{u_1(t,y) - u_1(0,y)}{t-0} = \frac{2}{1-0}\left[\frac{u_1(t,1) - u_1(t,y)}{1-y} - \frac{u_1(t,y) - u_1(t,0)}{y-0}\right]$$

From here, considering the boundary conditions, we obtain an approximate solution in the form [15]:

$$u_1(t,y) = \frac{2ty}{y(1-y)+2t} \tag{41}$$

Note that, $\lim\limits_{t\to\infty} u(t,y) = \lim\limits_{t\to\infty} u_1(t,y) = y$.

There is another approach to obtaining an approximate solution to Eq. (39). We replace (39) with the following equation:

$$\frac{du_2}{dt} = \frac{2}{1-0}\left[\frac{u_2(t,1)-u_2(t,y)}{1-y} - \frac{u_2(t,y)-u_2(t,0)}{y-0}\right] \tag{42}$$

Considering Eq. (42) y as a parameter, and solving it, we get

$$u_2(t,y) = y\left(1 - \exp\left(\frac{2t}{y(1-y)}\right)\right) \tag{43}$$

This shows that partial approximation gives the best result (**Figures 23** and **24**).

2.4.3 Non-stationary convection-diffusion differential equation

Consider the equation

$$\frac{\partial\Phi}{\partial t} + \frac{\partial\Phi}{\partial x} = \frac{1}{Pe}\frac{\partial^2\Phi}{\partial x^2} + f(x,t), \tag{44}$$

Under appropriate boundary and initial conditions. We approximate Eq. (44) as follows

$$\begin{gathered}\frac{U(x,t)-U(x,0)}{t} + \frac{U(x,t)-U(W,t)}{x-W} = \\ \frac{1}{Pe}\frac{2}{(E-W)}\left(\frac{U(E,t)-U(x,t)}{E-x} - \frac{U(x,t)-U(W,t)}{x-W}\right) + f(x,t),\end{gathered} \tag{45}$$

(45) is an implicit difference scheme with a moving node. In this case, the convective term was approximated by the scheme against the flow, and the diffusion term, as usual, with the second order of accuracy.

The comparison obtained with the help of (45) of the approximate solution with the exact solution (44) under the conditions $U^0(x) = x^2 + x,\ U_W(t) = 0,\ U_E(t) = 1 + e^{-Pet},\ f(x,t) = (1+Pex)e^{-Pet} + 2x - 2/Pe$ is shown in **Figure 25**. Exact solution $(W = 0, E = 1)\ \Phi(x,t) = x^2 + e^{-Pet}x$. In **Figure 25**, the solid curves are the exact solution, the dotted curves are the approximate solution, and the graphs correspond to the sections $x = 0,1;\ \ x = 0,5;\ \ x = 0,8$. **Figure 26** same results corresponding to $t = 1;\ \ t = 5;\ \ t = 10$.

Figures 25 and **26** show the acceptability of the approximate solution for the MNM.

It should be noted that with increasing Pe the discrepancy between the exact and approximate solutions increases. On **Figures 27** and **28** compare the same problem with $Pe = 2$.

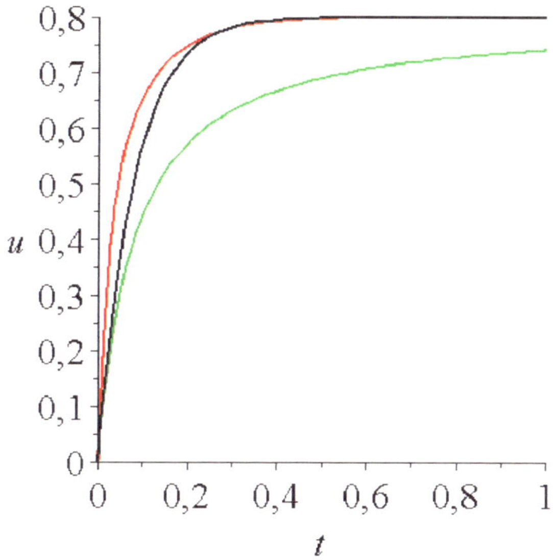

Figure 24.
Comparison of (39), (41), and (43) on the section y = 0.8. Blue line on (41) black on (43), red fine.

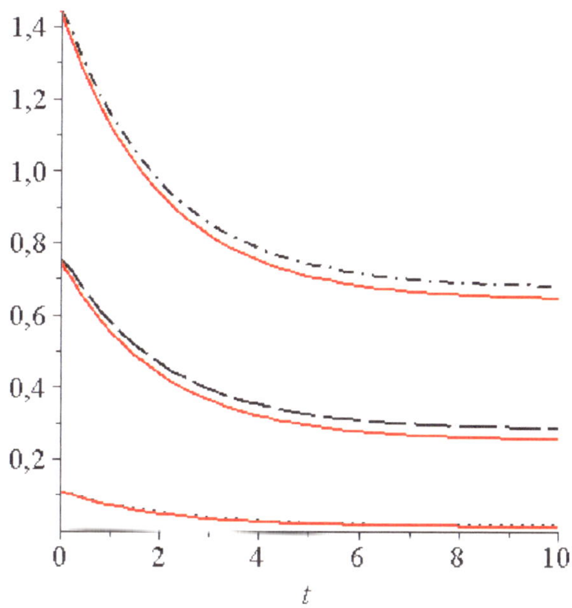

Figure 25.
Comparison solution Pe $= 0, 5$.

Thus, the MMN makes it possible to obtain an approximate analytical solution.

2.5 MNM for two-dimensional boundary value problems

Now let us consider the application of MMN to two-dimensional boundary value problems to obtain rough approximate solutions of DE.

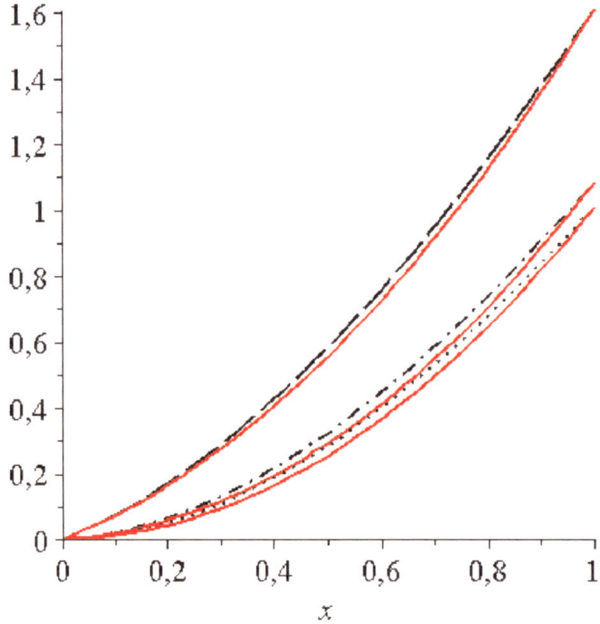

Figure 26.
Comparison solution Pe = 0, 5.

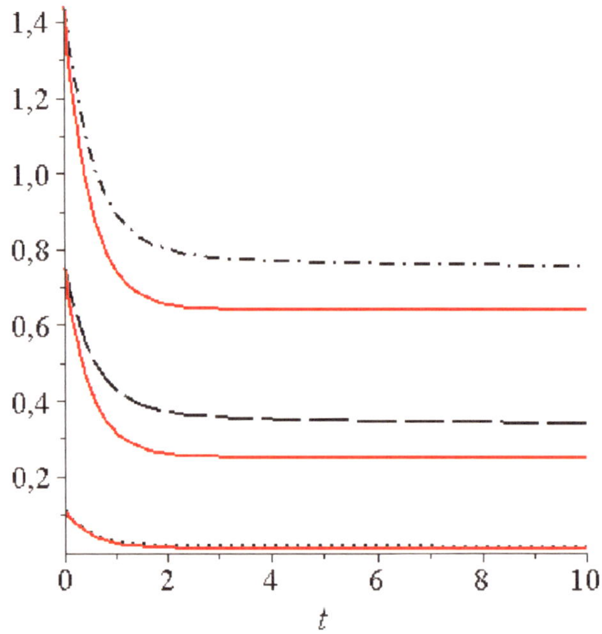

Figure 27.
Comparison solution. Pe = 2.

Consider a convex closed two-dimensional region (**Figure 29**). P point inside area. If P changes position inside the region, the boundary points E, N, W, S change their positions while being on the border of the region.

When studying stationary processes of various physical natures (oscillations, heat conduction, diffusion, hydrodynamics, etc.), one usually leads to equations of the elliptic type. The most common equation of this type is the Poisson equation.

There are various approximate-analytical and numerical methods for the equation of mathematical physics.

Consider the two-dimensional Poisson equation in a rectangle $(x,y) \in [W,E] \times [S,N]$

$$\Delta U(x,y) = f(x,y), \tag{46}$$

with boundary conditions.

$$U(W,y) = U_W(y), U(E,y) = U_E(y), U(x,S) = U_S(x), U(x,N) = U_N(x). \tag{47}$$

Take an arbitrary point in the rectangle approximation of second-order partial derivatives:

$$\frac{\partial^2 U}{\partial x^2} \approx \frac{2}{E-W}\left[\frac{U_E(y) - u(x,y)}{E-x} + \frac{u(x,y) - U_W(y)}{x-W}\right], \tag{48}$$

$$\frac{\partial^2 U}{\partial y^2} \approx \frac{2}{N-S}\left[\frac{U_N(x) - u(x,y)}{N-y} + \frac{u(x,y) - U_S(x)}{y-S}\right]. \tag{49}$$

Substituting (46) and (49) into (46), and solving, the resulting equation with respect to $u(x,y)$, we have

$$u(x,y) = \frac{1}{(E-x)(x-W) + (N-y)(y-S)} \cdot \left[\frac{(N-y)(y-S)}{E-W}((x-W)U_E + (E-x)U_W) + \right. \\ \left. + \frac{(E-x)(x-W)}{N-S}((y-S)U_N + (N-y)U_S)\right] + \frac{(E-x)(x-W)(N-y)(y-S)}{2((E-x)(x-W) + (N-y)(y-S))}f(x,y) \tag{50}$$

This is the approximate analytical solution of the Poisson equation in a rectangle. (49) satisfies the boundary conditions. Due to the fact that (48) and (49) is an approximate relation for the approximation of the second derivatives (50) is an approximate solution. Nevertheless, (50) gives an acceptable solution to many practical problems.

Consider examples.

2.5.1 Test problems

1. Consider the Laplace equation in a rectangle $[0,1] \times [0,1]$ with boundary conditions $U(0,y) = 0, U(1,y) = y, U(x,0) = 0, U(x,1) = x$. The exact solution to this problem is $U(x,y) = xy$. If we use the approximate solution (50), we obtain. In this case, the approximate solution coincides with the exact solution.

2. The function $U(x,y) = x^2 + y^2$, with boundary conditions $U(0,y) = -y^2$, $U(1,y) = 1 - y^2, U(x,0) = x^2, U(x,1) = x^2 - 1$ satisfies the Laplace equation. Relation (50) gives us an identical result.

3. The function $U(x,y) = \ln(x^2+y^2)$, with boundary conditions $U(1,y) = \ln(1-y^2)$, $U(2,y) = \ln(4+y^2), U(x,0) = \ln(x^2)$, $U(x,1) = \ln(x^2+1)$ in the region $[1,2]\times[0,1]$, satisfies the Laplace equation. An approximate solution based on (50) gives $u(x,y) = \frac{1}{(2-x)(x-1)+y(1-y)}$ $[y(1-y)((x-1)\ln(4+y^2)+(2-x)\ln(1+y^2))+(2-x)(x-1)(y\ln(1+x^2)+$ $+(1-y)\ln(x^2))]$If we compare the exact and approximate solutions in the area under consideration at points $x_i = 1+ih, y_j = jh,\ \ i,j = 1,2,\ \ldots n$ with a step $h = 0,1$ for the maximum difference, we obtain $0,0011$.

4. The function $U(x,y) = x^3 - y^3$, with boundary conditions, $U(0,y) = -y^3,\ U(1,y) = 1-y^3, U(x,0) = x^3, U(x,1) = x^3-1$ satisfies Eq. (46) for $f(x,y) = 6x - 6y$. Based on (50), the approximate solution has the form:

$$u(x,y) = \frac{xy(y-x)+y^4(1-y)+x^4(x-1)}{y(y-1)+x(x-1)}$$

The maximum absolute difference between the exact and approximate solutions calculated by points $x_i = 1+ih, y_j = jh,\ \ i,j = 1,2,\ \ldots n, h = 0,1$ is 0.048.

If we approximate the right side based on the control volume [35], the approximate solution has the form:

$$u(x,y) = \frac{xy(x-y)(1-3(x+y)+3xy)+2y^4(1-y)+2x^4(x-1)}{2[y(y-1)+x(x-1)]}$$

and the maximum absolute difference between the exact and approximate solutions calculated by points $x_i = 1+ih, y_j = jh,\ \ i,j = 1,2,\ \ldots n, h = 0,1$ is 0.024.

2.5.2 Flow in an ellipsoidal pipe

The equation describing the one-dimensional flow in an ellipsoidal tube of a viscous fluid has the form:

$$\frac{\partial^2 U}{\partial y^2} + \frac{\partial^2 U}{\partial z^2} = -\frac{\Delta p}{\mu l} \qquad (51)$$

Here u is the flow rate, μ is the flow viscosity, Δp/l (Δp/l = const) is the pressure drop. Eq. (51) is considered in the area $\frac{y^2}{a^2}+\frac{z^2}{b^2} \le 1$ (section of an ellipsoidal pipe, **Figure 30**), and the boundary condition is the no-slip condition (U = 0)

Eq. (51) is replaced by the difference

$$\frac{2}{y_E - y_W}\left(\frac{U_E-u}{y_E-y} - \frac{u-U_W}{y-y_W}\right) + \frac{2}{z_N-z_S}\left(\frac{U_N-u}{z_N-z} - \frac{u-U_S}{z-z_S}\right) = -\frac{\Delta p}{\mu l}.$$

Hence, given that

$$z_N = b\sqrt{1-y^2/a^2},\ z_S = -b\sqrt{1-y^2/a^2},\ y_E = a\sqrt{1-z^2/b^2},\ y_W = -a\sqrt{1-z^2/b^2}.$$

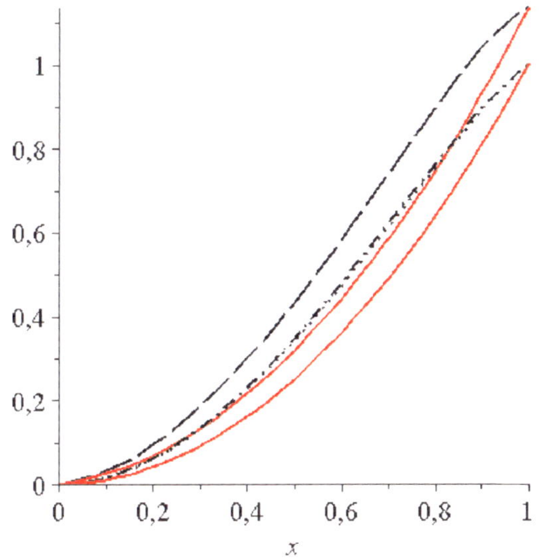

Figure 28.
Comparison solution. Pe = 2.

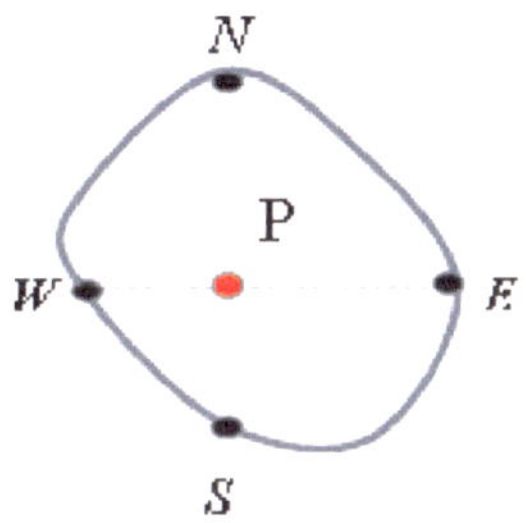

Figure 29.
The convex closed two-dimensional region.

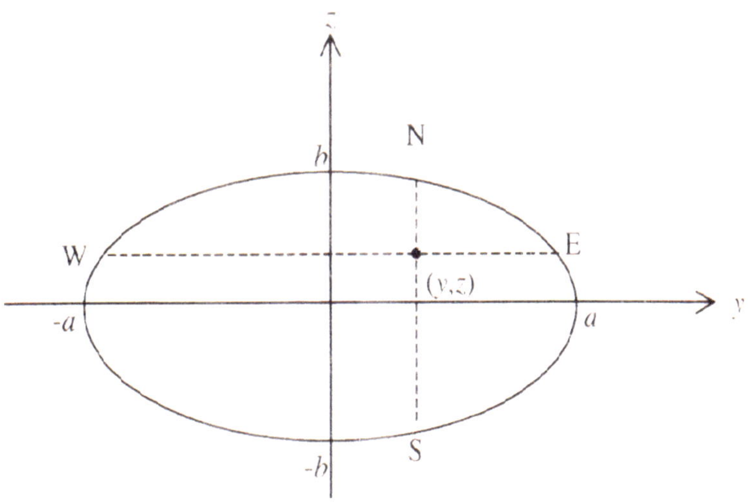

Figure 30.
Ellipsoidal pipe section.

we get

$$u = \frac{a^2 b^2}{2(a^2 + b^2)} \left(1 - \frac{y^2}{a^2} - \frac{z^2}{b^2}\right) \frac{\Delta p}{\mu l}$$

coinciding with the exact solution.

2.5.3 Two-dimensional temperature field in a solid

This problem is reduced to solving an equation $\Delta T = 0$, with boundary conditions

$$T(0,y) = 0, T(1,y) = 0, T(x,0) = T_0, T(x,1) = 0.$$

Exact solution to the problem

$$T(x,y) = \sum_{n=1}^{\infty} A_n \sin(n\pi x) sh[n\pi(y-1)]$$

where $A_n = \frac{2T_0}{n\pi} \frac{[(-1)^n - 1]}{sh(n\pi)}$.

Approximate solution

$$T(x,y) = \frac{(1-x)x(1-y)T_0}{x(1-x) + y(1-y)}.$$

The maximum absolute difference between the exact and approximate solutions calculated by points $x_i = ih, y_j = jh, \;\; i,j = 1, 2, \dots n, h = 0,01$ is 0.015.

Thus, the method presented here allows for obtaining solutions to Dirichlet problems. To improve the solution, the mesh refinement technique can be used.

To increase the accuracy, increase the number of moved nodes. When the number of nodes to be moved is four, we get.

$$\begin{aligned}
u_4(x,y) = {} & \left\{1 - \frac{4}{A+B}\left[\frac{B^2}{8}\left(\frac{x}{\left(\frac{1-x}{2}\right)^2 + B} + \frac{1-x}{\left(\frac{x}{2}\right)^2 + B}\right) + \frac{A^2}{8}\left(\frac{y}{\left(\frac{1-y}{2}\right)^2 + A} + \frac{1-y}{\left(\frac{y}{2}\right)^2 + A}\right)\right]\right\}^{-1} \\
& \times \frac{4}{A+B}\left\{\frac{B}{2}\left[\frac{1}{2}\frac{\frac{x}{2}\left(Bu_b(y) + \left(\frac{1-x}{2}\right)^2\left(yu_d\left(\frac{1+x}{2}\right) + (1-y)u_c\left(\frac{1+x}{2}\right)\right)\right)}{\left(\frac{1-x}{2}\right)^2 + B} + \right.\right. \\
& \left. + \frac{1}{4}\frac{(1-x)\left(Bu_a(y) + \left(\frac{x}{2}\right)^2\left(yu_d\left(\frac{x}{2}\right) + (1-y)u_c\left(\frac{x}{2}\right)\right)\right)}{\left(\frac{x}{2}\right)^2 + B}\right] + \\
& + \frac{A}{2}\left[\frac{1}{2}\frac{y\left(\frac{A}{2}u_d(x) + \left(\frac{1-y}{2}\right)^2\left(xu_b\left(\frac{1+y}{2}\right) + (1-x)u_a\left(\frac{1+y}{2}\right)\right)\right)}{\left(\frac{1-y}{2}\right)^2 + A} + \right. \\
& \left.\left. + \frac{1}{2}\frac{(1-y)\left(Au_c(x) + \left(\frac{y}{2}\right)^2\left(xu_b\left(\frac{y}{2}\right) + (1-x)u_a\left(\frac{y}{2}\right)\right)\right)}{\left(\frac{y}{2}\right)^2 + A}\right]\right\}
\end{aligned}$$

The maximum absolute difference between the exact and approximate solutions, calculated by points $x_i = ih, y_j = jh,\ \ i,j = 1, 2, \dots n, h = 0, 1$, is 0.14 according to the formula with one moving node, and when calculating with five moving nodes, it is 0.07.

2.5.4 *Flow in a rectangular pipe*

Eq. (51) also describes the flow of an incompressible viscous fluid in a rectangular pipe. Let us denote the height of the rectangle parallel to the axis Oz as $2h$, and the base parallel to the axis Oy as – $2\sigma h$, where σ is any positive constant. We draw the axis through the center of the rectangle and direct it downstream.

Let us transform Eq. (51) into a dimensionless form. For the scale of lengths, we take the height, h, and for the scale of speeds—the value $h^2/\mu \cdot \Delta p/l$. We introduce the following dimensionless quantities:

$$Y = y/h, \quad Z = z/h, \quad V = U\mu l/\left(h^2 \Delta p\right)$$

Substituting into (51), we obtain

$$\frac{\partial^2 V}{\partial Y^2} + \frac{\partial^2 V}{\partial Z^2} = -1 \tag{52}$$

Boundary conditions for (52)

$$V(Y, -1) = 0, \ \ V(Y, 1) = 0, \ \ V(-\sigma, Z) = 0, \ \ V(\sigma, Z) = 0 \tag{53}$$

Eq. (52) is replaced by a difference equation and taking into account the boundary condition (53) we have

$$\frac{2}{2\sigma}\left(\frac{-V}{\sigma - Y} - \frac{V}{Y + \sigma}\right) + \frac{2}{1+1}\left(\frac{-V}{1 - Z} - \frac{V}{Z + 1}\right) = -1.$$

From here we determine the approximate analytical solution:

$$V = \frac{1}{2}\frac{\left(\sigma^2 - Y^2\right)\left(1 - Z^2\right)}{1 - Z^2 + \sigma^2 - y^2} \tag{54}$$

The exact solution of the problem has the form:

$$u = \frac{16\sigma^2}{\pi^3}\sum_{n=0}^{\infty}\frac{(-1)^n}{(2n+1)^3}\left[1 - \frac{ch\left(\frac{2n+1}{2}\frac{\pi}{\sigma}Y\right)}{ch\left(\frac{2n+1}{2}\frac{\pi}{\sigma}\right)}\right]\cos\left(\frac{2n+1}{2}\frac{\pi}{\sigma}Z\right)$$

Figure 31 shows a comparison of the exact and approximate solutions on the cross-section $x = 0$ for $\sigma = 1$. The maximum absolute difference between the exact and approximate solutions is 0.045.

To increase the accuracy of the approximate solution in Eq. (52), we approximate only one of the terms. For example, we approximate Eq. (52) as follows:

$$\frac{2}{2\sigma}\left(\frac{-V}{\sigma - Y} - \frac{V}{Y + \sigma}\right) + \frac{\partial^2 V}{\partial Z^2} = -1 \tag{55}$$

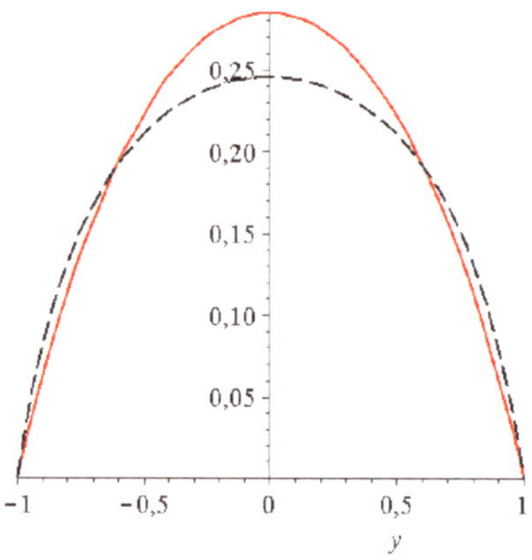

Figure 31.
Comparison of the solution on the section according to (54).

We got an ordinary differential equation, we consider the variable Y in Eq. (55) as a parameter. We solve Eq. (55) with constant coefficients, considering the boundary conditions, we find an approximate solution

$$V = C_1 \exp\left(\sqrt{k}Z\right) + C_2 \exp\left(-\sqrt{k}Z\right) + \frac{1}{k}. \tag{56}$$

Here $k = 2/((\sigma - Y)(Y + \sigma))$, $C_2 = -\frac{1}{k}\frac{\exp\left(\sqrt{k}\right) - \exp\left(-\sqrt{k}\right)}{\exp\left(2\sqrt{k}\right) - \exp\left(-2\sqrt{k}\right)}$, $C_1 = -C_2 - \exp\left(2\sqrt{k}\right) - \frac{1}{k}\exp\left(-\sqrt{k}\right)$.

Figure 31 shows a comparison of the exact approximate solution obtained based on (56) on the cross-section $x = 0$ at $\sigma = 1$. A comparison of **Figures 31** and **32** shows that the calculation by formula (56) gives a more accurate result. The maximum absolute difference between the exact and approximate solutions is equal to that obtained by (56) and equals 0.024. In **Figures 31** and **32** solid curves are the exact solution.

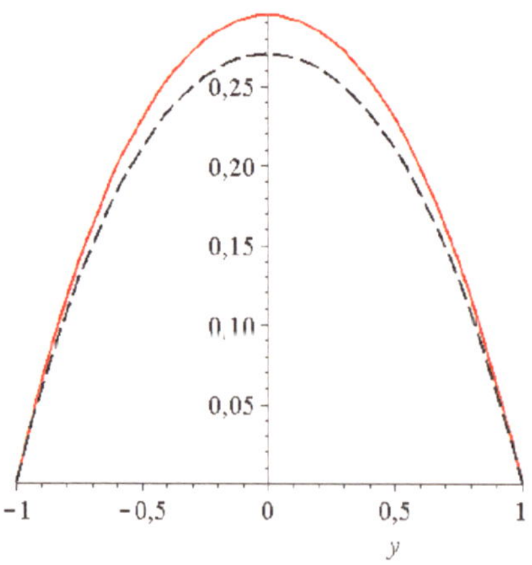

Figure 32.
Comparison of the solution on the section according to (56).

Figure 33.
Coordinate systems and the region of solution.

2.5.5 Flow at the inlet section of the pipe

With appropriate simplifications, the flow of a viscous incompressible fluid in a dimensionless form is described by the following differential equation:

$$u\frac{\partial u}{\partial x} = -\frac{1}{\mathrm{Re}}N + \frac{1}{\mathrm{Re}}\left(\frac{\partial^2 u}{\partial y^2} + \frac{\partial^2 u}{\partial x^2}\right), \tag{57}$$

Here, N = −12 is the pressure drop, Re is the Reynolds number. The equations are considered in the area $D: -0,5 < x < 0.5, \quad 0 < y < L$. (**Figure 33**). Boundary conditions for (57):

$$u(0, y) = 1; \; u(L, y) = 1,5(1 - 4y^2);$$
$$u(x, -0,5) = 0; \; u(x, 0.5) = 0.$$

The convective term is linearizable

$$u\frac{\partial u}{\partial x} \approx \frac{\partial u}{\partial x}.$$

Approximating in (57) by the liquid volume $(y + 0,5)/2 < y < (y - 0,5)/2$, we obtain an ordinary equation, solving which we obtain an approximate solution:

$$u = C_1 \exp(k_1 x) + C_2 \exp(k_2 x) - \frac{1 - 4y^2}{8}N, \tag{58}$$

where $k_{1,2} = \frac{\mathrm{Re}}{2}\left(1 \pm \sqrt{1 + \frac{32}{\mathrm{Re}^2(1-4y^2)}}\right)$.

For comparison, solutions (57) were also made with the numerical method.

Figure 34 shows the velocity profiles obtained on the basis of an approximate solution. The solid curve to the section $x = 0,1$, and the pointed curve to $x = 0,5$, the dotted one corresponds to the section $x = 3$. **Figure 35** shows a comparison of the approximate and numerical solution of Eq. (57). The solid lines correspond to the solution (58), and the dotted lines correspond to the numerical solution (velocity profiles are given for the cross-section $x = 0,1$ and $x = 0,5$).

2.6 Solution of the flow problem in the combined region

Exact solution. Let a liquid flow in a flat pipe partially filled with a porous medium. The lower part of the horizontal pipe is filled with a porous medium of height h

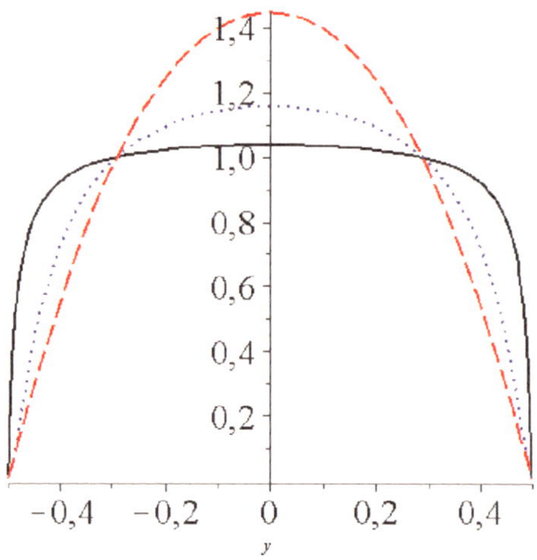

Figure 34.
Approximate solution based on (58). Velocity profiles corresponding to sections $x = 0, 1; 0, 5; 3$. Re = 1, L = 5.

(pipe height H). Considering the flow to be one-dimensional and stationary, we obtain from the Rakhmatulin equation [16, 17], we obtain

$$\mu \frac{du}{dy}\left(f\frac{du}{dy}\right) - Ku = f\frac{dp}{dx}. \tag{59}$$

In (59) for the parameter K, we use the Kozeny-Karman relation as adopted in porous media:

$$K = \frac{\mu \cdot f^2}{k}. \tag{60}$$

where the $k = \frac{d^2 f^3}{150(1-f)^2}$, permeability, d is the characteristic size of the porous medium.

Let us pass to dimensionless variables assuming $u = \overline{u}U,\ y = \overline{y}H, x = \overline{x}H,\ p = \frac{\rho U^2}{\mathrm{Re}}\overline{p}$. Then Eq. (59) in dimensionless form for $f = const$, has the form:

$$\frac{d^2\overline{u}}{d\overline{y}^2} - A\overline{u} = \frac{d\overline{p}}{d\overline{x}}. \tag{61}$$

Here $A = 180(H/d)^2(1-f)^2/f^2$.

In the free zone, the one-dimensional flow satisfies the equation

$$\frac{d^2\overline{u}}{d\overline{y}^2} = \frac{d\overline{p}}{d\overline{x}}. \tag{62}$$

In the future, in Eqs. (61) and (62), we release the dash above the variables.

Eq. (61) is considered when $0 < y < h_0$, and Eq. (62) $h_0 < y < 1$. Equations are solved under the following boundary conditions.

No-slip conditions for Eq. (61) to the lower walls, and for Eq. (62) to the upper walls:

$$u(0) = 0, \; u(1) = 0. \tag{63}$$

In the inner boundary region, we set the conditions for the continuity of the flow and the equality of the shear stress:

$$u(h_0 - 0) = u(h_0 + 0), \; \frac{du(h_0 - 0)}{dy} = \frac{du(h_0 + 0)}{dy}. \tag{64}$$

It is easy to obtain an analytical solution of (61) and (62) under the given boundary conditions. **Figure 36** shows an analytical solution. The dimensionless pressure difference is adopted $\frac{d\overline{p}}{dx} = -12$, so that it corresponds to the flow without a porous layer. The dotted line corresponds to the solution obtained with a porosity of 0.3, and the dotted-dotted line is 0.5.

Numerical solution. Consider eq. (61) for the entire region and set

$$f = \begin{cases} \varepsilon \; npu \; 0 < y < h_0 \\ 1 \; npu \; h_0 \leq y < 1 \end{cases}. \tag{65}$$

In this case, Eq. (61) in the pure region takes the form (62). Thus, Eq. (61) can be used in the entire area, with porosity (65), while the interboundary conditions are satisfied automatically (in the porous layer, the porosity is taken equal to ε). For this purpose, a finite-difference approximation of Eq. (61) was compiled and calculated using the sweep method in the combined region. **Figure 37** presents the results of numerical calculations (solid curves are the analytical solution, and point data are the numerical results). This shows that it is possible to perform a thorough calculation without highlighting the interboundary condition.

Approximate analytical solution using a moving node. Using the moving node method, one can find an approximate analytical solution to the problem.

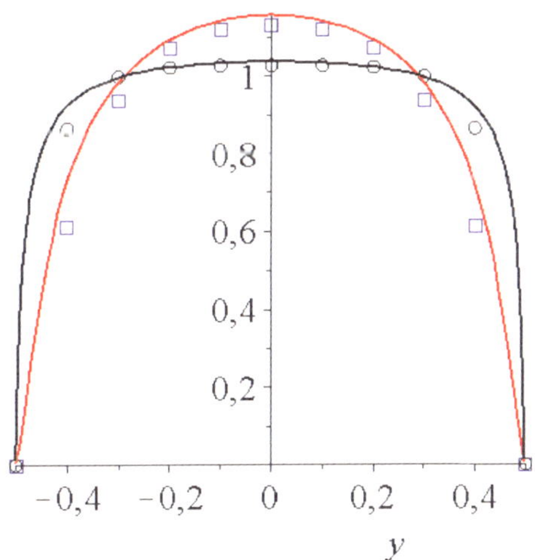

Figure 35.
Comparison of approximate and numerical solution. Re = 1, L = 5.

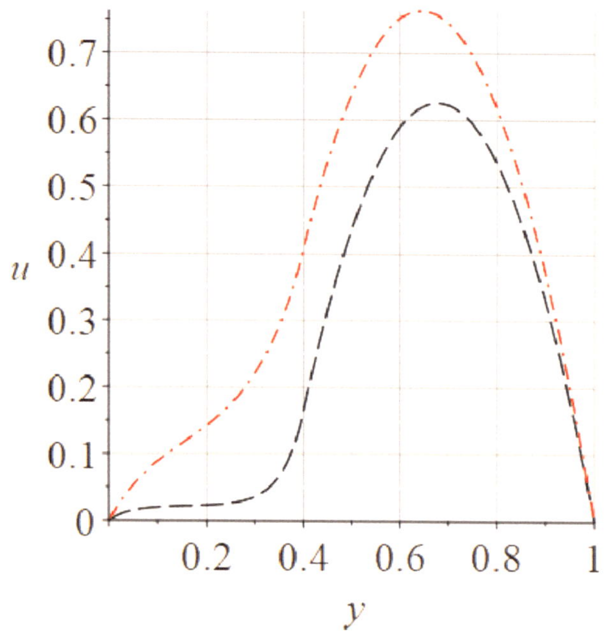

Figure 36.
Exact solution: velocity distributions for different porosity values.

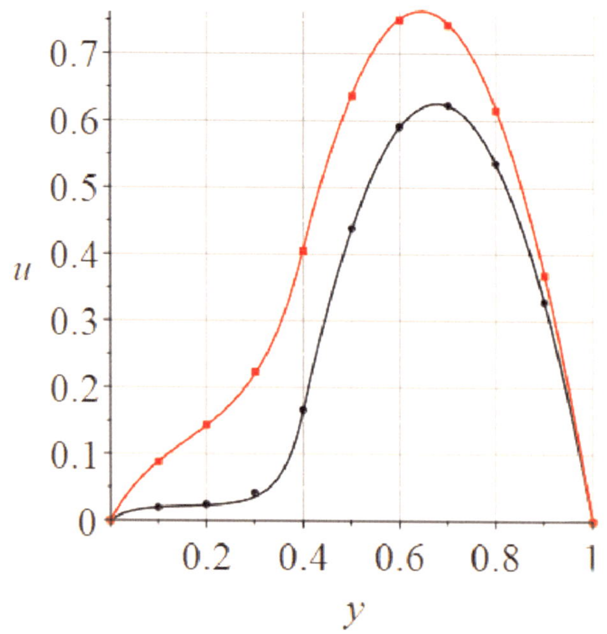

Figure 37.
Comparison of exact and numerical results for different porosity values.

Eqs. (61) and (62) is approximated by difference relations:

$$\frac{2}{h_0}\left(\frac{u_G - u}{h_0 - y} - \frac{u}{y}\right) - Au = \frac{dp}{dx}. \tag{66}$$

$$\frac{2}{1-h_0}\left(\frac{-u}{1-y}-\frac{u-u_G}{y-h_0}\right)=\frac{dp}{dx}. \tag{67}$$

In the difference Eqs. (66) and (67) no-slip boundary conditions are used. In Eqs. (66) and (67) u_G— the value of the unknown function on the inner boundary. To find u_G, we use the second interboundary condition (64). We put in (66) $y \to h_0 - 0,$ then we have

$$\frac{2}{h_0}\left(\frac{du}{dy}|_{h_0-0}-\frac{u}{h_0}\right)-Au_G=\frac{dp}{dx}. \tag{68}$$

If in (67) $y \to h_0 + 0$ then we have:

$$\frac{2}{1-h_0}\left(\frac{-u_G}{1-h_0}-\frac{du}{dy}|_{h_0+0}\right)=\frac{dp}{dx}. \tag{69}$$

Using (64), we obtain

$$u_G=-\frac{h_0(1-h_0)}{2+h_0^2(1-h_0)}\frac{dp}{dx}. \tag{70}$$

Using (70) from (66) and (67) we determine the distribution of velocities in the porous

$$u=-\frac{y}{2+Ay(h_0-y)}\left(h_0-y+\frac{2(1-h_0)}{2+h_0^2(1-h_0)}\right)\frac{dp}{dx}. \tag{71}$$

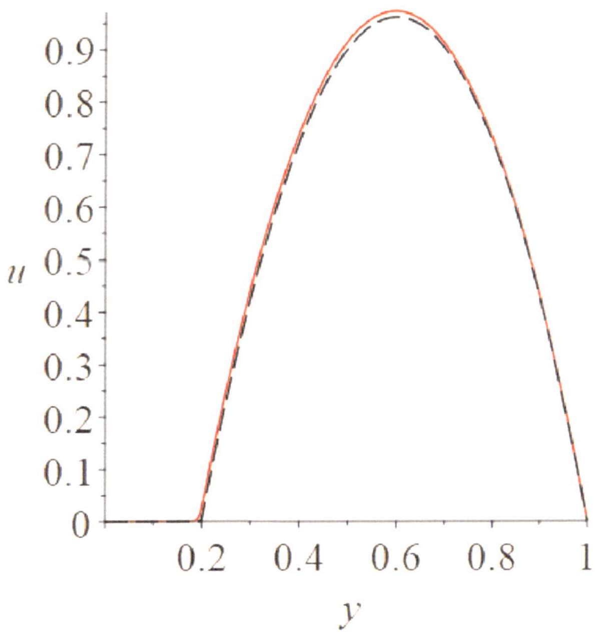

Figure 38.
Solution comparison.

and free zone

$$u = -\frac{1-y}{1-h_0}\left(\frac{(1-h_0)(y-h_0)}{2} + \frac{h_0(1-h_0)}{2+Ah_0^2(1-h_0)}\right)\frac{dp}{dx}. \tag{72}$$

Figure 38 compares the exact and approximate solutions (the solid line is the exact solution, and the dotted line is the approximate one obtained using (71) and (72) at $A = 40000, h_0 = 0,2$).

3. Application of the moving node method

In the first chapter, we considered MNM for some boundary value problems in order to obtain an approximate analytical solution. This chapter focuses on some uses of moved nodes. With the help of multipoint moving nodes, improved schemes are built for the convective-diffusion problem. It is proposed to improve the accuracy of schemes using the Richardson extrapolation method. Some properties of schemes are also presented for research with the help of MNM

3.1 Obtaining discrete compact schemes for the convective-diffusion problem of MNN

Of great interest is the construction and analysis of the discretization of a singularly perturbed ordinary differential equation of the second order. The equation of convection-diffusion is basic in modeling fluid flow at high Reynolds numbers and in convective mass exchange at high Peclet numbers. Many works are devoted to this subject [25, 40–45].

Numerical solutions of the convection-diffusion equation often show numerical fluctuations. In practical calculations, many authors have observed parasitic oscillations at high Peclet numbers when the central approximation for the convective term is used. On the other hand, the upwind scheme usually leads to unpleasant artificial numerical diffusion.

This dilemma is central to the numerical solutions of convection-diffusion problems.

In order to compute approximate solutions to a partial differential equation, some form of local approximation must be used. This means that the decision values at each node are used to generate an approximate decision value. With finite differences, one usually tries to make the local area as compact as possible, for example, using only neighboring nodes when updating on a node.

If we consider the approximation of the convection-diffusion problem on a uniform grid, we can observe that most of the literature deals with the choice between schemes in a three-point pattern: (W,P,E). To obtain an approximation of a high order of accuracy, it is necessary to increase the number of points of the computational pattern.

Here we use the structure described in the first chapter to derive a new finite difference scheme [14]. Although such a procedure cannot be easily generalized to partial differential equations with variable coefficients.

Consider the DE of convection-diffusion

$$\frac{d\Phi}{dx} = \frac{1}{Pe}\frac{d^2\Phi}{dx^2} + S(x), \tag{73}$$

with boundary conditions.

$$\Phi(0) = \Phi_0, \Phi(1) = \Phi_1 \tag{74}$$

where Pe is the Peclet number ($Pe = \rho v L/\Gamma$), (v is the velocity, ρ – density, L is the length scale, Γ is the diffusion coefficient, x is the dimensionless coordinate, S(x) is the source.

On [0,1] we introduce a non-uniform grid

$$\Omega = \{x_i, i = 0, 1, 2, \dots, N, 0 = x_0 < x_1 < \dots < x_{i-1} < x_i < x_{i+1} < \dots < x_N = 1\}.$$

In the first chapter, with the help of moving nodes, an analytical solution to problem (73), (74) was constructed. When constructing compact circuits, we rely on a circuit against the flow, which is monotonic for any Peclet numbers.

Let us rewrite the scheme against the flow (21) for the segment (W,E)

$$Pe\frac{U^1 - U^1_W}{x - W} = \frac{2}{(E - W)}\left(\frac{U^1_E - U^1}{E - x} - \frac{U^1 - U^1_W}{x - W}\right) + Pe \cdot S(x). \tag{75}$$

In (75), the equation relates the unknown function at three points: W, x, E, i.e. Eq. (75) is written in a three-point pattern. Now let us write Eq. (75) for an arbitrary internal node x_i, which is connected with neighboring nodes x_{i-1}, x_{i+1}. Then

$$Pe\frac{U^1_i - U^1_{i-1}}{x_i - x_{i-1}} = \frac{2}{(x_{i+1} - x_{i-1})}\left(\frac{U^1_{i+1} - U^1_i}{x_{i+1} - x_i} - \frac{U^1_i - U^1_{i-1}}{x_i - x_{i-1}}\right) + Pe \cdot S(x_i). \tag{76}$$

Here, $i = 1, 2, \dots, i, \dots, N - 1$ and U^1_i means the approximate value of the unknown function at the node x_i.

This schema can be rewritten like this:

$$a^1_P U^1_P = a^1_E U^1_E + a^1_W U^1_W + F^1_i, \tag{77}$$

But now

$$a^1_E = \frac{2}{(x_{i+1} - x_{i-1})(x_{i+1} - x_i)}, a^1_W = \frac{Pe}{(x_i - x_{i-1})} + \frac{2}{(x_{i+1} - x_{i-1})(x_{i+1} - x_i)}, \tag{78}$$
$$a^1_P = a^1_E + a^1_W, F^1_i = Pe \cdot S(x_i)$$

To increase accuracy, based on three moving nodes (28),

$$a^3_P U^3_P = a^3_E U^3_E + a^3_W U^3_W + F^3_i, \tag{79}$$

here $a_E^3 = \frac{8}{(x_{i+1}-x_{i-1})(x_{i+1}-x_i)(1+\gamma_1)}, a_W^3 = \frac{2Pe}{(x_i-x_{i-1})(1+\tau_1)} + \frac{8}{(x_{i+1}-x_{i-1})(x_i-x_{i-1})(1+\tau_1)},$
$a_P^3 = a_W^3 + a_E^3,\ \theta = Pe\,(x_{i+1}-x_i), \sigma = Pe\,(x_i - x_{i-1}),\ \tau_1 = 2/(2+\sigma), \gamma_1 = (2+\theta)/2$
$F_i^3 = Pe \cdot S(x_i) + \frac{4+Pe\cdot(x_{i+1}-x_{i-1})}{x_{i+1}-x_{i-1}} \cdot \frac{1-\tau_1}{1+\tau_1} \cdot S(x_{i-1/2}) + \frac{4}{x_{i+1}-x_{i-1}} \cdot \frac{\gamma_1-1}{\gamma_1+1} \cdot S(x_{i+1/2})$

$$x_{i-1/2} = 0,5(x_{i-1}+x_i), x_{i+1/2} = 0,5(x_i + x_{i+1}).$$

Based on with $2^k - 1$ moving nodes (31), we have

$$a_P^{(2^k-1)} U_P^{(2^k-1)} = a_E^{(2^k-1)} U_E^{(2^k-1)} + a_W^{(2^k-1)} U_W^{(2^k-1)} + F_i^{(2^k-1)}, \tag{80}$$

where

$$a_E^{(2^k-1)} = \frac{2^{2k+1}(1-\gamma_k)}{(x_{i+1}-x_{i-1})(x_{i+1}-x_i)\left(1-\gamma_k^{2^k}\right)}, a_W^{(2^k-1)} = \frac{2^{2k+1}Pe(1-\tau_k)}{(x_i - x_{i-1})\left(1-\tau_k^{2^k}\right)} +$$
$$\frac{2^{2k+1}(1-\tau_k)}{(x_{i+1}-x_{i-1})(x_i-x_{i-1})\left(1-\tau_k^{2^k}\right)},$$

$$a_P^{(2^k-1)} = a_W^{(2^k-1)} + a_E^{(2^k-1)}.\tau_k = 2^k/\left(2^k+\sigma\right), \gamma_k = \left(2^k+\theta\right)/2^k,$$

$$F_i^{(2^k-1)} = Pe \cdot S(x_i) + \frac{2^{k+1} + Pe\cdot(x_{i+1}-x_{i-1})}{x_{i+1}-x_{i-1}} \frac{(1-\tau_k)^2}{1-\tau_k^{2^k}} \sum_{j=1}^{2^k-1} \sum_{i=1}^{j} \tau_k^{i-1} \cdot S\left(x_i + j\frac{x_i - x_{i-1}}{2^k}\right) -$$

$$\frac{2^{k+1}}{x_{i+1}-x_{i-1}} \frac{(1-\gamma_k)^2}{1-\gamma_k^{2^k}} \sum_{j=1}^{2^k-1} \sum_{i=1}^{j} \gamma_k^{i-1} \cdot S\left(x_i + \left(2^k - j\right)\frac{x_{i+1}-x_i}{2^k}\right).$$

Let us consider numerical experiments.

Figures 39 and **40** show graphs for solving problem (73), (74) for $Pe = 50$ on segments $[0; 1]$ with boundary conditions $\Phi_0 = 0,\ \Phi_1 = 1$. **Figure 39** corresponds to $S(x) = 5\cos 4x$, and the graphs in **Figure 40** are obtained with

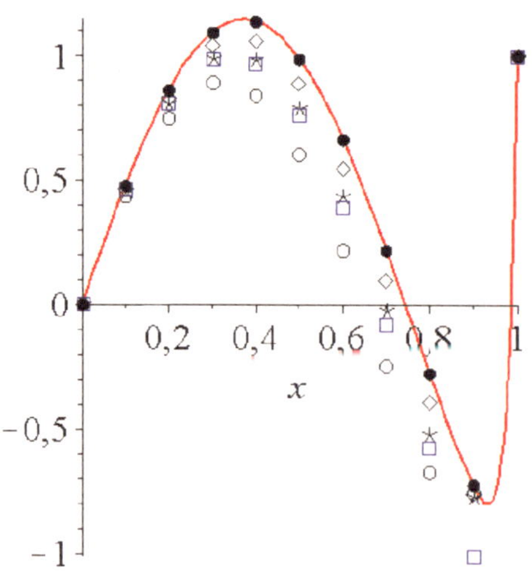

Figure 39.
Comparison of various schemes with source term $S(x) = 5\cos 4x$.

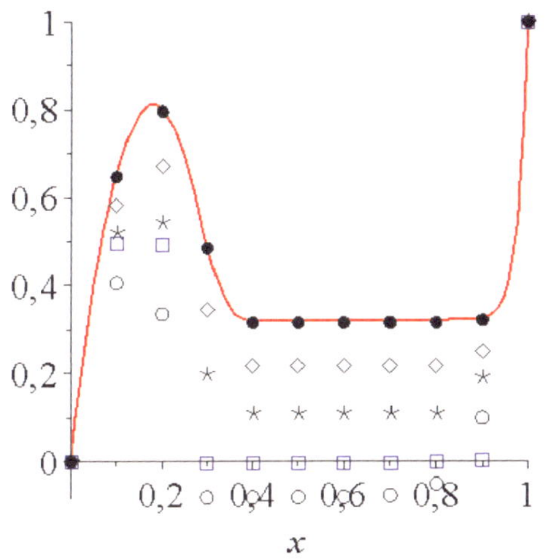

Figure 40.
Comparison of various schemes with source term (81).

$$S(x) = \begin{cases} 10 - 50x \quad if \quad x \le 0,3 \\ 50x - 20 \quad if \quad 0,3 < x < 0,4 \\ 0 \quad if \quad 0,4 < x \end{cases} \tag{81}$$

The numerical results are obtained for $h = 0,1$ and the grid Peclet number is equal to 5. The solid lines are plots of the exact solutions of the problem. Circle symbols are obtained for the upwind scheme, rectangles according to the Patankar scheme, asterisks according to (79), diamonds according to (80) at $k = 2$, and circles according to (80) at $k = 7$.

Table 1 shows the root-mean-square errors $\sigma = \sqrt{\sum_1^N (\Phi(x_i) - U_i)^2 / N}$ for the considered schemes. Φi is exact solution at nodal points, Ui is numerical solution obtained by the considered schemes, N number of nodes.

From **Figures 39** and **40**, and from the **Table 1**, it is clear that the proposed schemes give good results.

3.2 Construction of compact schemes of the convective-diffusion problem based on the finite volume method

The finite volume method is one of the methods that can give a good approximate solution to the problem. Here we explore the application of the finite volume method to solve the convection-diffusion equation for constructing compact schemes.

Scheme	Upwind	Patankar	(79)	(80), k = 2	(80), k = 7
$S(x) = 5\cos 4x$	0.361	0.300	0.192	0.096	0.004
$S(x)$ with (81)	0.282	0.199	0.154	0.077	0.003

Table 1.
The root-mean-square errors.

The basic strategy of all finite volume methods is to write the differential equation in a conservative form, integrate it over small domains (called "cells" or "finite volumes"), and transform each such integral over the cell boundary.

Our goal is to construct a qualitative scheme for the problem (82)

$$\frac{d}{dx}(\rho u \Phi) = \frac{d}{dx}\left(\Gamma \frac{d\Phi}{dx}\right) + S(x) \tag{82}$$

$$\Phi(0) = \Phi_0, \quad \Phi(1) = \Phi_1 \tag{83}$$

based on the control volume method. The procedure for obtaining a scheme is similar to that described in paragraph 3.1.

On [0,1] we introduce a non-uniform grid

$$\Omega = \{x_i, i = 0, 1, 2, \ldots, N, 0 = x_0 < x_1 < \ldots < x_{i-1} < x_i < x_{i+1} < \ldots < x_N = 1\}.$$

In the first chapter, with the help of moving nodes, an analytical solution to problem (82), (83) was constructed using the control volume method in the form.

$$\begin{aligned}
&\left[\frac{(1-\tau_k)\beta_k^+}{1-\tau_k^{2^k}} + \frac{(1-\gamma_k)\alpha_k^-}{1-\gamma_k^{2^k}}\right]U^k = \frac{(1-\tau_k)\beta_k^+}{1-\tau_k^{2^k}}U_W^k + \frac{(1-\gamma_k)\alpha_k^-}{1-\gamma_k^{2^k}}U_E^k + \frac{E-W}{2^{k+1}}\cdot S(x) \\
&+\frac{1-\tau_k}{1-\tau_k^{2^k}}\cdot\frac{x-W}{2^k}\cdot\sum_{j=1}^{2^k-1}\sum_{i=1}^{j}\tau_k^{i-1}S\left(W + j\frac{x-W}{2^k}\right) \\
&+\frac{1-\gamma_k}{1-\gamma_k^{2^k}}\cdot\frac{E-x}{2^k}\cdot\sum_{j=1}^{2^k-1}\sum_{i=1}^{j}\gamma_k^{i-1}S\left(x + (2^k - j)\frac{E-x}{2}\right).
\end{aligned} \tag{84}$$

Here $\tau_k = \frac{\beta_k^-}{\beta_k^+}, \gamma_k = \frac{\alpha_k^+}{\alpha_k^-}$, $\beta_k^- = 2^k D_W + F^-, \beta_k^+ = 2^k D_W + F^+$, $\alpha_k^- = 2^k D_E + F^-$, $\alpha_k^+ = 2^k D_E + F^+$, $F^- = \max(-F, 0)$, $F^+ = \max(F, 0)$, $D_E = \Gamma/(E-x)$, $D_W = \Gamma/(x-W)$.

Now let us write Eq. (84) for an arbitrary internal node x_i, which is connected with neighboring nodes x_{i-1}, x_{i+1}.

Then

$$\begin{aligned}
&\left[\frac{(1-\tau_k)\beta_k^+}{1-\tau_k^{2^k}} + \frac{(1-\gamma_k)\alpha_k^-}{1-\gamma_k^{2^k}}\right]U_P{}^k = \frac{(1-\tau_k)\beta_k^+}{1-\tau_k^{2^k}}U_W^k + \frac{(1-\gamma_k)\alpha_k^-}{1-\gamma_k^{2^k}}U_E^k + \frac{x_{i+1}-x_{i-1}}{2^{k+1}}\cdot S(x_i) + \\
&\frac{1-\tau_k}{1-\tau_k^{2^k}}\cdot\frac{x_i-x_{i-1}}{2^k}\cdot\sum_{j=1}^{2^k-1}\sum_{m=1}^{j}\tau_k^{m-1}S\left(x_{i-1} + j\frac{x_i-x_{i-1}}{2^k}\right) + \\
&\frac{1-\gamma_k}{1-\gamma_k^{2^k}}\cdot\frac{x_{i+1}-x_i}{2^k}\cdot\sum_{j=1}^{2^k-1}\sum_{m=1}^{j}\gamma_k^{m-1}S\left(x_i + (2^k - j)\frac{x_{i+1}-x_i}{2}\right).
\end{aligned} \tag{85}$$

What does it have to do with $D_E = \Gamma/(x_{i+1} - x_i), D_W = \Gamma/(x_i - x_{i-1})$.

3.3 Improving the accuracy of circuits using the Richardson extrapolation method

The Richardson extrapolation method is used to solve grid problems on a sequence of grids. The method consists in carrying out calculations for the same circuit, with different steps. Then we have several grid solutions. On the basis of the grid solutions, some linear combination is compiled. The resulting linear combination has a higher order of accuracy.

Creation of new schemes using Richardson extrapolation based on the schemes given in paragraph 3.1.

The accuracy of scheme (76), with a uniform arrangement of grid nodes, is $O(h)$. Scheme (79) has order $O(h/2)$. For a linear combination $Q^3(x_i) = -\frac{1}{3}U^1(x_i) + \frac{4}{3}U^3(x_i)$, we get an approximation error for a uniform grid $O(h^2)$. A linear combination of $U^1(x_i), U^3(x_i)$ and $U^7(x_i)$ in the form $Q^7(x_i) = \frac{1}{45}U^1(x_i) - \frac{4}{9}U^3(x_i) + \frac{64}{45}U^7(x_i)$ has an approximation order of $O(h^4)$. Consider $N = 10$, $S(x) = x^2$, $Pe = 30$. **Table 2** shows the absolute difference between the exact and approximate solutions according to the schemes.

Table 3 shows the root-mean-square error $\sigma = \sqrt{\sum_1^N (\Phi(x_i) - U_i)^2/N}$ for the considered schemes. $\Phi(x_i)$ the exact solution at the nodal points, U_i is the numerical solution obtained by the considered schemes.

Figures 41 and **42** show numerical solutions for $\Phi_W = 0$, $\Phi_E = 0$.

From the graphs in **Figures 41** and **42**, and from **Tables 2** and **3**, it is clear that the Richardson linear combination allows you to get a more improved circuit.

x	0.1	0.2	0.3	0.4	0.5	0.6	0.7	0.8	0.9
$U^1(x_i)$	0.001	0.004	0.007	0.011	0.017	0.025	0.039	0.073	0.160
$U^3(x_i)$	0.001	0.002	0.004	0.006	0.008	0.012	0.018	0.034	0.089
$U^7(x_i)$	0.000	0.001	0.002	0.003	0.005	0.007	0.010	0.019	0.046
$Q^3(x_i)$	0.000	0.001	0.002	0.004	0.006	0.007	0.010	0.021	0.065
$Q^7(x_i)$	0.000	0.001	0.001	0.002	0.003	0.005	0.007	0.014	0.030

Table 2.
The absolute difference between the exact and approximate solutions.

Schemes	$U^1(x)$	$U^3(x)$	$U^7(x)$	$Q^3(x)$	$Q^7(x)$
$S = x^2$, Pe = 50, $\Phi_W = 0$, $\Phi_E = 1$	0.047	0.023	0.011	0.015	0.006
S = 10, Pe = 50, $\Phi_W = 0$, $\Phi_E = 1$	0.033	0.017	0.008	0.011	0.005
$S = x^2$, Pe = 100, $\Phi_W = 0$, $\Phi_E = 1$	0.034	0.014	0.006	0.008	0.003
$S = 5\cos(4\pi x)$, Pe = 50, $\Phi_W = 0$, $\Phi_E = 1$	0.213	0.120	0.061	0.090	0.038

Table 3.
Comparison by the root-mean-square errors.

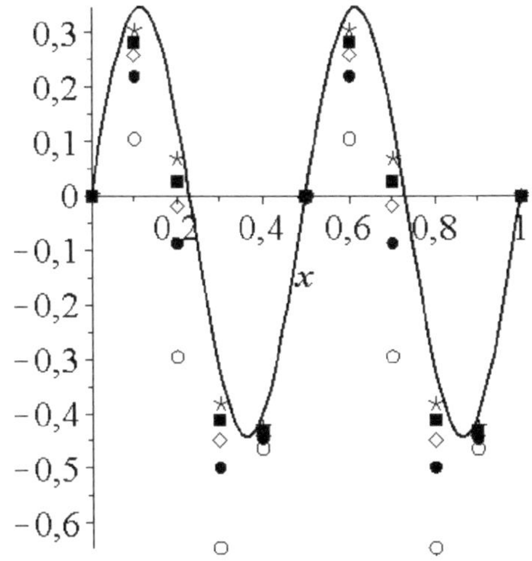

Figure 41.
Pe = 100, S = 5cos4πx Solid curve exact solution, circle obtained by scheme U^1*, circle by* U^3*, solid rectangle by* U^7*, diamond by* Q^3*, star by* Q^7*.*

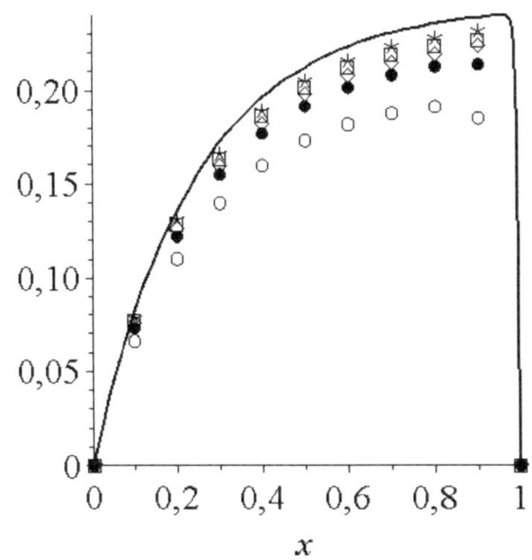

Figure 42.
Pe = 100, S = exp.(−4x). Solid curve exact solution, circle obtained by scheme U^1*, circle by* U^3*, solid rectangle by* U^7*, diamond by* Q^3*, star by* Q^7*.*

3.4 Influence of the choice of profile on the face of the control volume on the quality of difference schemes

When obtaining discrete analogs of the convective-diffusion problems given above, on the basis of multipoint PUs, it was possible to construct better compact circuits in a three-point template. However, there is another approach to improve the quality of the scheme based on the choice of the decision profile.

Since the work of Leonard, in order to improve the results of the numerical solution, attempts have been made to improve the algorithm, which is built in a five-point pattern.

In all the above schemes (except for the scheme against the flow), the conditions of boundedness and non-negativity of the coefficients are violated.

Here it is proposed to improve the scheme based on the choice of the solution profile on the edge of the control volume in the three-point template of the convective-diffusion problem. The upwind scheme with one-sided differences is taken as the initial scheme. The QUICK scheme uses quadratic upwind interpolation to determine the convective flow. Here we use the solution obtained by the upwind scheme based on the method of moving nodes.

MNM for simple cases allows one to obtain an analytical representation of the solution between the nodal points of the boundary value problem. Based on this representation, it is possible to construct a better discrete scheme.

We integrate (73) over the control volume $[w, e]$

$$\Phi_e - \Phi_w = \frac{1}{Pe}\left(\frac{d\Phi}{dx}\right)_e - \frac{1}{Pe}\left(\frac{d\Phi}{dx}\right)_w + \int_w^e S(x)dx.$$

Replacing the derivatives with difference relations, we have

$$\Phi_e - \Phi_w = \frac{1}{Pe}\frac{\Phi_E - \Phi_P}{x_E - x_P} - \frac{1}{Pe}\frac{\Phi_P - \Phi_W}{x_P - x_W} + (x_e - x_w)f_P. \tag{86}$$

Here $f_P = \frac{1}{x_e - x_w}\int_w^e S(x)dx$. Depending on the type of function profile Φ on the control volume, different schemes are obtained.

Let the profile Φ be piecewise constant in each control volume. Then, assuming $\Phi_e = \Phi_P, \Phi_w = \Phi_W$, we have an upwind scheme:

$$\Phi_P - \Phi_W = \frac{1}{Pe}\frac{\Phi_E - \Phi_P}{x_E - x_P} - \frac{1}{Pe}\frac{\Phi_P - \Phi_W}{x_P - x_W} + (x_e - x_w)f_P. \tag{87}$$

If the profile Φ is linear between the nodes and the edges of the control volume are located in the middle between the node points, we have a scheme with central differences:

$$\frac{\Phi_E + \Phi_P}{2} - \frac{\Phi_P + \Phi_W}{2} = \frac{1}{Pe}\frac{\Phi_E - \Phi_P}{x_E - x_P} - \frac{1}{Pe}\frac{\Phi_P - \Phi_W}{x_P - x_W} + (x_e - x_w)f_P. \tag{88}$$

To improve the accuracy of circuits, many authors recommended various circuits. All these schemes are multipoint (more than three). Here is a way to improve three-point circuits.

From (87) we get

$$\begin{aligned}\Phi_P &= \frac{x_P - x_W}{\mathrm{Pe}(x_E - x_P)(x_P - x_W) + x_E - x_W}\Phi_E + \frac{(x_E - x_P)(1 + \mathrm{Pe}(x_P - x_W))}{\mathrm{Pe}(x_E - x_P)(x_P - x_W) + x_E - x_W}\Phi_W + \\ &= \frac{(x_P - x_W)\Phi_E + (x_E - x_P)(1 + Pe(x_P - x_W))\Phi_W}{\mathrm{Pe}(x_E - x_P)(x_P - x_W) + x_E - x_W}\end{aligned} \tag{89}$$

If the nodes x_E and x_W are fixed, and the node x_P is movable, we get a profile Φ_P between the nodes x_E and x_W. This profile is used in (86) to determine Φ_e and Φ_w.

To improve scheme (87), we proceed as follows. Eq. (89) connects at three nodes (x_W, x_P, x_E), if we apply Eq. (89) for nodes (x_W, x_w, x_P), we have

$$\Phi_w = \frac{2+R_h}{4+R_h}\Phi_P + \frac{2}{4+R_h}\Phi_E + \frac{R_h}{2(R_h+4)}\frac{h}{4}f_w. \tag{90}$$

Similarly, for nodes (x_P, x_e, x_E), we have

$$\Phi_e = \frac{2+R_h}{4+R_h}\Phi_P + \frac{2}{4+R_h}\Phi_E + \frac{R_h h}{2(R_h+4)}f_e. \tag{91}$$

Substituting (90) and (91) into (86) we have

$$\left[\frac{R_h^2}{4+R_h}+2\right]\Phi_P = \left[1+\frac{2+R_h}{4+R_h}\right]\Phi_W + \left[1-\frac{2R_h}{4+R_h}\right]\Phi_E + hR_h f_P - \frac{h\cdot R_h^2}{2(4+R_h)}(f_e - f_w). \tag{92}$$

The condition $R_h < 4$ is ensured by the positivity of the coefficients and the stability of the scheme (92).

Proceeding similarly as in the derivation of (92), but using (92) for the profile, for a uniform step we obtain

$$\left[\frac{(4+R_h)^2-16}{(4+R_h)^2+16}\right]\Phi_P = \left[\frac{1}{R_h} - \frac{16}{(4+R_h)^2+16}\right]\Phi_E + \left[\frac{(4+R_h)^2}{(4+R_h)^2+16}+\frac{1}{R_h}\right]\Phi_W + hS(P) + \frac{h\left(8R_h+8R_h^2\right)}{2\left[(4+R_h)^2+16\right]}\cdot(S(x_w)-S(x_e)), \tag{93}$$

Test problems

1. Consider the equation

$$\frac{du}{dx} = \frac{1}{Pe}\frac{d^2u}{dx^2} + \sin \pi x.$$

with boundary conditions $u(0) = u(1) = 0$. **Table 4** shows the maximum absolute differences of the schemes calculated at the nodal points (u is the exact solution of the problem, u_1 is the solution obtained according to the upwind scheme, u_2 is according to the power law, u_3 according to the Leonard scheme, u_4 according to (92) and u_5- according to the scheme (93).

Pe	R_h	**max**$\lvert u-u_1\rvert$	**max**$\lvert u-u_2\rvert$	**max**$\lvert u-u_3\rvert$	**max**$\lvert u-u_4\rvert$	**max**$\lvert u-u_5\rvert$
100	10	0.0526	0.03770	0.1801	0.03701	0.00077
1000	100	0.0470	0.0464	0.2732	0.01607	0.00927

Table 4.
The maximum absolute differences.

2. Consider the equation

$$\frac{du}{dx} = \frac{1}{\text{Pe}}\frac{d^2u}{dx^2} + s(x),$$

with boundary conditions $u(0) = 0,\ u(1) = 1$, with source

$$s(x) = \begin{cases} 10 - 50x, & 0 \le x \le 0.3, \\ 50x - 20, & 0.3 < x \le 0.4, \\ 0, & 0.4 < x \le 1 \end{cases}$$

Figure 43 shows that scheme (93) gives the best results. Leonard's scheme gives an incorrect solution near the right boundary. Scheme (92) also exhibits a slight non-monotonicity. This is due to the fact that scheme (92) is stable for $R_k < 4$.

Figure 44 Shows that for large grid Peclet numbers, the upstream and Patankar schemes give close results. Scheme (93) gives the best results. This can also be seen in **Table 5**, which compares the considered schemes (SDS—central scheme).

3. Two-dimensional case. Consider the equation

$$\frac{\partial g}{\partial x} = \frac{1}{\text{Re}}\left(\frac{\partial^2 g}{\partial x^2} + \frac{\partial^2 g}{\partial y^2}\right) + s(x, y).$$

Exact solution $g = 6y^{10}(1 - y^{10})(1 - x^3) + 6x^3y(1 - y)$. The equations are solved in the area $[0, 1] \times [0, 1]$. The source term is defined so that the given function is a

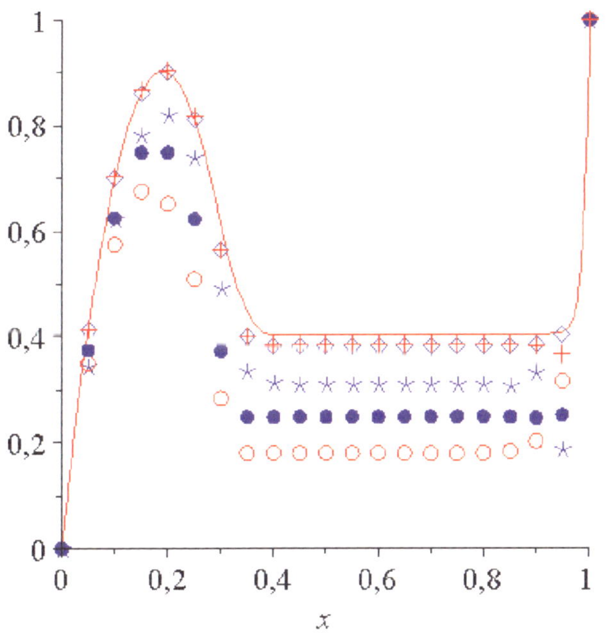

Figure 43.
Comparison of various schemes. Pe $= 100$, $R_h = 5$. *The solid line is the exact solution, the circle is the upwind scheme, the circle is the Patankar scheme, the asterisk is the Leonard scheme, + is the scheme (92), the diamond is according to (93).*

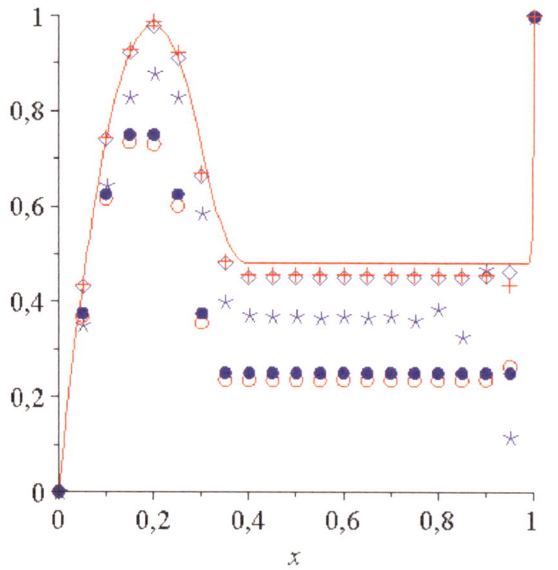

Figure 44.
Comparison of various schemes. Re = 500, $R_h = 25$. *The solid line is the exact solution, the circle is the upwind scheme, the circle is the Patankar scheme, the asterisk is the Leonard scheme, + is the scheme (92), the diamond is according to (93).*

Scheme	Re	h	R_h	$\max\lvert u - u_p \rvert$	$\frac{\sum \lvert u_i - (u_p)_i \rvert}{\sum u_i}$
Upwind	100	1/40	2.5	0.1627	0.2116
	100	1/20	5	0.3258	0.4224
	500	1/20	25	0.3650	0.4101
Power	100	1/40	2.5	0.0833	0.1057
	100	1/20	5	0.2385	0.3025
	500	1/20	25	0.3454	0.3868
(8)	100	1/40	2.5	0.0164	0.0169
	100	1/20	5	0.0460	0.0401
	500	1/20	25	0.0531	0.0398
(12)	100	1/40	2.5	0.0129	0.00840
	100	1/20	5	0.0452	0.0358
	500	1/20	25	0.0571	0.0404
QUICK	100	1/40	2.5	0.0700	0.0701
	100	1/20	5	0.2231	0.1931
	500	1/20	25	0.3653	0.2055
CDS	100	1/40	2.5	0.1237	0.0062
	100	1/20	5	0.3033	0.0467
	500	1/20	25	0.5136	0.1355

Table 5.
Comparison of circuits with respect to grid Peclet number.

Scheme	Re = 100, n = 5, h = 0,1		Re = 500, n = 5, h = 0,1		Re = 1000, n = 10, h = 0,1	
	$\max\lvert g-g_p\rvert$	$\frac{\sum\lvert g-g_p\rvert}{\sum\lvert g\rvert}$	$\max\lvert g-g_p\rvert$	$\frac{\sum\lvert g-g_p\rvert}{\sum\lvert g\rvert}$	$\max\lvert g-g_p\rvert$	$\frac{\sum\lvert g-g_p\rvert}{\sum\lvert g\rvert}$
Upwind	0.150	0.074	0.169	0.074	0.186	0.129
CDS	0.074	0.023	0.035	0.018	0.470	0.382
Power	0.130	0.061	0.165	0.071	0.184	0.127
QUICK	0.063	0.017	0.020	0.0051	0.097	0.016
(8)	0.035	0.019	0.013	0.008	0.057	0.023
(12)	0.033	0.016	0.008	0.005	0.060	0.015
VONOS	0.055	0.016	0.019	0.005	0.073	0.015

Table 6.
Results of calculations of errors according to the schemes.

solution to the equation. The boundary conditions were determined based on the exact solution. **Table 6** shows the results of calculations according to the schemes.

From **Table 6**, it is clear that the proposed schemes show the best results.

3.5 Schema improvement with flow equality

MNM can improve the quality of the scheme. We demonstrate this method based on the upwind scheme (87) written in the form:

$$\frac{\Phi_P-\Phi_W}{x_P-x_W}=\frac{2}{Pe(x_E-x_W)}\left(\frac{\Phi_E-\Phi_P}{x_E-x_P}-\frac{\Phi-\Phi_W}{x_P-x_W}\right)+S(x_P). \tag{94}$$

In (94) we pass to the limit at $x_E\to x_P$ and, assuming the existence of the limit, we have

$$\frac{\Phi_P-\Phi_W}{x_P-x_W}=\frac{2}{Pe(x_P-x_W)}\left(\frac{d\Phi_P^-}{dx_P}-\frac{\Phi_P-\Phi_W}{x_P-x_W}\right)+S(x_P).$$

Here, $d\Phi_P^-/dx_P$ is the left-hand derivative of the unknown function at the point x_P. From here

$$\frac{d\Phi_P^-}{dx_P}=\frac{2+Pe(x_P-x_W)}{2}\cdot\frac{\Phi_P-\Phi_W}{x_P-x_W}-\frac{Pe(x_P-x_W)}{2}\cdot S(x_P), \tag{95}$$

Similarly, taking an arbitrary point $x\in(x_P,x_E)$ and passing to the limit $x\to x_P$, we find

$$\frac{d\Phi_P^+}{dx_P}=\frac{2}{2+Pe(x_E-x_P)}\cdot\frac{\Phi_E-\Phi_P}{x_E-x_P}+\frac{Pe(x_E-x_P)}{2+Pe(x_E-x_P)}\cdot S(x_P),$$

By equating $d\Phi^+/dx=d\Phi^-/dx$ the flows, we get an improved scheme:

$$c_P\Phi_P=a_P\Phi_W+b_P\Phi_E+d_PS(x_P) \tag{96}$$

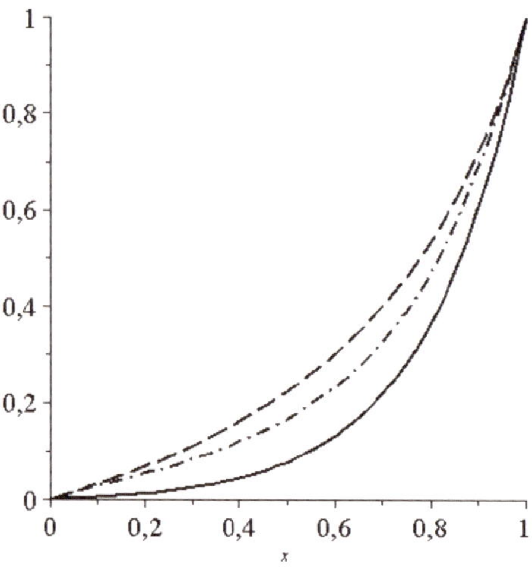

Figure 45.
Comparison of schemes. The solid curve is the exact solution, the dotted line according to (87), the dotted line according to (96).

where
$a_P = \frac{2+Pe(x_P - x_W)}{(x_P - x_W)}$, $b_P = \frac{2}{[2+Pe(x_E - x_P)](x_E - x_P)}$, $c_P = a_P + b_P$, .

Figure 45 shows a comparison of the exact solution and the schemes according to (87) and (96) for $Pe = 5$, with one moving node ($S(x) = 0$). It can be seen from the graph that the solution is improving. Numerical diffusion decreases.

3.6 Investigation of the scheme by the MNM

At this point, we are dealing with monotonicity and MMN approximation of the circuit. On the basis of the analytical form of the approximate solution of the problem between the nodes, which is obtained on the basis of the MMN, it is possible to investigate monotonicity and the type of approximation of the scheme.

3.6.1 Investagation of monotonicity

Scheme with central-difference approximation of the convective term. Consider Eq. (73). Take a segment $[x_{i-1}, x_{i+1}] \subset [0, 1]$ and any point $x \equiv x_i \in (x_{i-1}, x_{i+1})$. Consider the grid analog (73)

$$\frac{u_{i+1} - u_{i-1}}{x_{i+1} - x_{i-1}} = \frac{2}{Pe(x_{i+1} - x_{i-1})}\left(\frac{u_{i+1} - u}{x_{i+1} - x} - \frac{u - u_{i-1}}{x - x_{i-1}}\right) + S(x) \tag{97}$$

If we set $x = (x_{i+1} + x_{i-1})/2$, we have a central-difference approximation. Here, u_{i+1} is the approximate value of the solution at the point x_{i+1}, u is the approximate value of the solution at the point x. To obtain a physically plausible solution in simple cases, we set $S(x) = 0$.

From (97) we find

$$u = \frac{(x - x_{i-1})(2 - Pe(x_{i+1} - x))u_{i+1} + (x_{i+1} - x)(2 + Pe(x - x_{i-1}))u_{i-1}}{2(x_{i+1} - x_{i-1})}. \tag{98}$$

By changing x the values on the interval (x_{i-1}, x_{i+1}), we can determine the behavior of the solution. For given values $x_{i+1}, x_{i-1}, u_{i-1}, u_{i+1}$(98) is a parabola.

From (98) one can get

$$\frac{u - u_{i-1}}{u_{i+1} - u_{i-1}} = \frac{(x - x_{i-1})(2 - Pe(x_{i+1} - x))}{2(x_{i+1} - x_{i-1})}. \tag{99}$$

A physically plausible solution is obtained if $0 \le \frac{u - u_{i-1}}{u_{i+1} - u_{i-1}} \le 1$. This condition imposes a restriction $2 - Pe(x_{i+1} - x) \ge 0$. This condition is the condition of monotonicity of the central-difference scheme for a non-uniform grid. In the case of a uniform grid, we have $2 \ge Pe \cdot h$. This condition is the well-known monotonicity condition [46]. For a coarse grid (N = 2, one movable node) at Pe = 5, the solution of exact and approximate solutions are shown in **Figure 46**.

In **Figure 46**, the solid curve represents the exact solution, while the dotted one represents the approximate solution obtained on the basis of (99). It can be seen from the graph that scheme (99) does not give a physically plausible analytical solution. That is why scheme (99) for large Peclet numbers gives an oscillatory numerical solution. A plausible solution should have the same qualitative character as the exact solution. When solving numerically, scheme (97) is implemented using a sweep, and for the stability of the sweep, the nodes are selected so that $2 - Pe(x_{i+1} - x_i) \ge 0$. For example, for a coarse grid (one nodal point), the credibility condition gives $x_i \ge 0, 6$. Indeed, for Pe = 5, it $2 - Pe(1 - x) \ge 0$ follows that $x_i \ge 0, 6$ (see **Figure 46**).

For Pe = 2, comparisons of the solutions are shown in **Figure 47**, which gives a physically plausible solution.

Upwind scheme. Let us consider a difference analog of Eq. (73), in which the convective term is approximated by a one-sided difference relation (without a source)

$$\frac{u - u_{i-1}}{x - x_{i-1}} = \frac{2}{Pe(x_{i+1} - x_{i-1})} \left(\frac{u_{i+1} - u}{x_{i+1} - x} - \frac{u - u_{i-1}}{x - x_{i-1}} \right).$$

From here we get

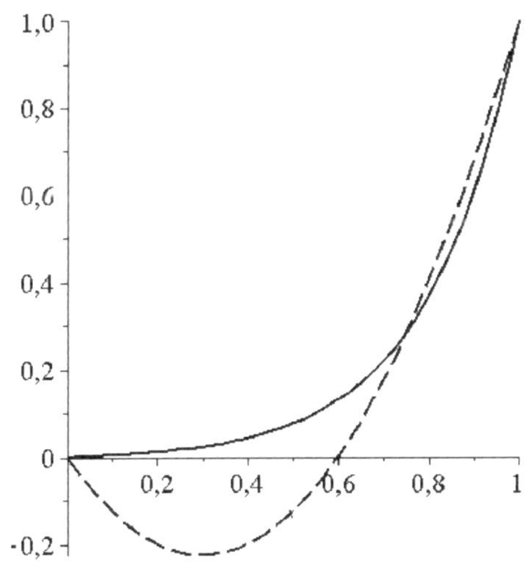

Figure 46.
Comparison of solutions in a coarse grid. The dotted curve is approximate, the solid curve is exact, Pe = 5 (Φ_0 = 0, Φ_1 = 1).

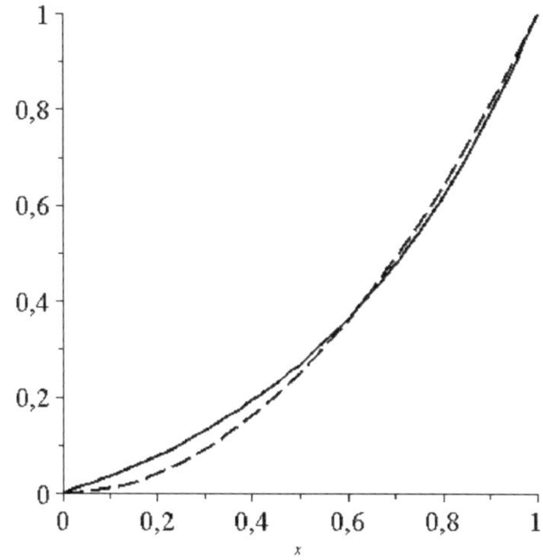

Figure 47.
Comparison of the solution in a coarse grid. The dotted curve is approximate, the solid curve is exact, Pe = 2 (Φ_0 = 0, Φ_1 = 1).

$$u = \frac{2(x - x_{i-1})u_{i+1} + (x_{i+1} - x)(2 + Pe(x_{i+1} - x_{i-1}))u_{i-1}}{(x_{i+1} - x_{i-1})(2 + Pe(x_{i+1} - x))}$$

or

$$\frac{u - u_{i-1}}{u_{i+1} - u_{i-1}} = \frac{2(x - x_{i-1})}{(x_{i+1} - x_{i-1})(2 + Pe(x_{i+1} - x))}. \qquad (100)$$

Since, the right side of relation (100) into segments is a hyperbola and therefore we have $0 \leq \frac{u-u_{i-1}}{u_{i+1}-u_{i-1}} \leq 1$. Those the upstream circuit is always monotonic. **Figure 48** shows a comparison of the exact and approximate analytical solutions (Pe = 5). However, numerical diffusion occurs.

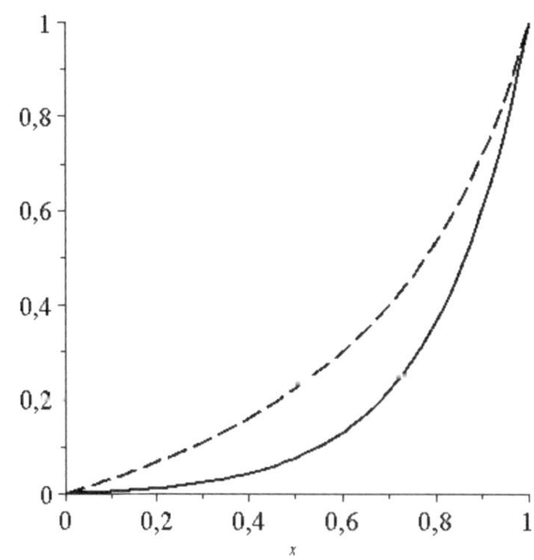

Figure 48.
Comparison of the solution in a coarse grid. The dotted curve is approximate, and the solid curve is exact (Φ_0 = 0, Φ_1 = 1).

3.7 An explicit expression of the approximation error of ordinary differential equations based on the moved node method

Here discusses the issue of the possibility of calculating the approximation error. When replacing differential equations with discrete ones, one of the key issues is the closeness of the discrete solution to the exact solution. For the difference solution to the problem, a grid area is formed. The discrete solution is determined at the nodal points. Traditionally, in questions of replacing a differential equation with a descriptive one, one usually indicates the degree of approximation of the $O(h^p)$ type. Here h is the grid step.

However, it is possible to calculate the approximation error at nodal points based on the method of moving nodes. The method of moving nodes allows for obtaining an approximate analytical expression. On the basis of the approximate form, it is possible to calculate the approximation error. On the other hand, at each node one can construct a differential analog of the difference equation. Using simple examples, the calculation of approximation errors is demonstrated and schemes of the collocation type are constructed.

3.7.1 Introduction

Here describes the application of the moving nodes method to the calculation of the approximation error. When a two-point boundary value problem is solved by different methods, the question of the degree of approximation usually appears. The closeness of the exact and approximation of the solution, and the quality of the difference scheme are evaluated based on the degree of this parameter. With such an analysis, other parameters (the coefficients of the differential equation) are not explicitly involved in the approximation error expression. Obtaining an explicit expression for the approximation error makes it possible to analyze it.

Consider the simplest ordinary differential equation with boundary conditions

$$\frac{d^2u}{dx^2} = C, \quad u(0) = 0, \quad u(1) = 1 \tag{101}$$

where C—const.

Create a uniform grid on segments [0,1] with step h. A uniform grid on a segment $x \in [0, 1]$ with step h has the form:

$$\overline{\omega}_h = \{x_k = hk, \ k = 0, 1, \dots, N, h \cdot N = 1\}$$

Let us replace the second-order derivative with the difference relation:

$$\frac{U_{i+1} - 2U_i + U_{i-1}}{h^2} = C, \quad 1 \le i \le N-1, U_0 = 0, \ U_N = 1 \tag{102}$$

Difference scheme (102) traditionally has order $O(h^2)$. However, if we solve system (102) by the Tomas algorithm, we obtain a numerical solution that coincides with the exact analytical solution for any grid steps h at the grid nodes. Those. scheme (102) approximates (101) exactly.

3.7.2 Methodology

Let us have a differential equation

$$Lu = f, \tag{103}$$

where L is a differential operator, f is a known function, and u is an unknown function. (103) the equation is considered in some domain D with appropriate boundary conditions. The differential Eq. (103) is replaced by the difference equation:

$$L_h u_h = f_h, \tag{104}$$

where L_h is the difference operator, u_h is the unknown grid function, and f_h is the approximation of the function f at the grid nodes.

Usually, the approximation error is given as [2, 3]:

$$Q_h = L_h[u]_h - f_h, \tag{105}$$

where $[u]_h$ is the exact solution of (103) at the grid nodes. Using the Taylor series, from (105) one obtains that, $Q_h = O(h^m)$, where h is the grid step and m is the degree of approximation.

You can determine an explicit approximation error if you use the method of a moving node, which allows you to extend the definition to the entire area D. This allows you to introduce an approximation error like this:

$$R_h = L_h\{u\}_h - f_h. \tag{106}$$

Here $\{u\}_h$ is a predefined continuous function by means of a moveable node. The approximate calculation of the approximation error of type (106) is demonstrated using simple examples.

3.7.3 Results and discussion

As an application of the above approach, consider examples.

1. Consider a simple boundary value problem:

$$\frac{d^2u}{dx^2} = f(x), \quad u(0) = u_a, \quad u(1) = u_b \tag{107}$$

Let us build a non-uniform grid on segments [0; 1]:

$$\overline{\omega}_h = \{0 = x_0, < x_1 < \dots < x_{N-1} < x_N = 1, k = 0, 1, \dots, N\}$$

In the non-uniform grid, we replace (107) with the difference problem:

$$\frac{2}{x_{i+1} - x_{i-1}} \left(\frac{U_{i+1} - U_i}{x_{i+1} - x_i} - \frac{U_i - U_{i-1}}{x_i - x_{i-1}} \right) = f(x_i), \quad i = 1, 2, \dots, N-1. \tag{108}$$

Here U_i is the grid solution of the problem. From here

$$U_i = \frac{U_{i+1}(x_i - x_{i-1}) + U_{i-1}(x_{i+1} - x_i)}{x_{i+1} - x_{i-1}} - \frac{1}{2}f(x_i)(x_i - x_{i-1})(x_{i+1} - x_i), \quad i = 1, 2, \dots, N - 1. \tag{109}$$

We redefine the value of the function at non-nodal points as follows. To do this, we consider in (109) $x_{i+1}, x_{i-1}, U_{i-1}, U_{i+1}$, to be fixed, and x_i to be moved, and the function f(x) to be smooth. Thus, we will complete the grid function on each segment (x_{i-1}, x_{i+1}). From (109) we get

$$U_i''(x_i) = -\frac{1}{2}f''(x_i)(x_{i+1} - x_i)(x_i - x_{i-1}) - f'(x_i)(x_{i+1} + x_{i-1} - 2x_i) + f(x_i) \tag{110}$$

Then the approximation error for the nodal points looks like this:

$$R_h(x_i) = -\frac{1}{2}f''(x_i)(x_{i+1} - x_i)(x_i - x_{i-1}) - f'(x_i)(x_{i+1} + x_{i-1} - 2x_i) \tag{111}$$

If the grid is uniform for the approximation error, we obtain the expression

$$R_h(x_i) = -\frac{1}{2}f''(x_i)h^2, \; i = 1, 2, \dots, N - 1. \tag{112}$$

If on the segments (x_{i-1}, x_{i+1}) the function constant approximation error is identically equal to zero and we get the exact solution.

Based on expression (110), the following conclusion can be drawn.

Given a two-point boundary value problem

$$\frac{d^2u}{dx^2} = f^*(x), \quad u(0) = u_a, \; u(1) = u_b$$

and $f^*(x)$ can be represented as

$$f^*(x_i) = -\frac{1}{2}f''(x_i)(x_{i+1} - x_i)(x_i - x_{i-1}) - f'(x_i)(x_{i+1} + x_{i-1} - 2x_i) + f(x_i)$$

then the difference scheme

$$\frac{2}{x_{i+1} - x_{i-1}}\left(\frac{U_{i+1} \quad U_i}{x_{i+1} - x_i} - \frac{U_i - U_{i-1}}{x_i - x_{i-1}}\right) = f(x_i), \quad i = 1, 2, \dots, N - 1,$$

gives a grid solution coinciding with the exact solution at the nodal points.

If there is only one internal node point (the node being moved is one), then an approximate analytical solution can be obtained. Indeed, if we rewrite scheme (108) for one moving node, we have

$$2\left(\frac{U_i - U_{i-1}}{1 - x} - \frac{U(x) - U_a}{x}\right) = f(x_i). \tag{113}$$

From here we obtain an approximate analytical solution:

$$U(x) = \frac{U_b x + U_a(1-x)}{x_{i+1} - x_{i-1}} - \frac{1}{2} f(x_i)(1-x)x. \tag{114}$$

In this case, (114) represents the exact solution to the problem (107).
if we put

$$f^*(x) = -\frac{1}{2} f''(x)(1-x)x - f'(x)(1-2x) + f(x).$$

The form of the approximation error (111) allows the construction of new schemes of the collocation type. Indeed, if in problem (108) we replace the right side with the expression

$$f(x_i) + A(x_i - x_{i-1})(x_{i+1} - x_i),$$

Here A is still an unknown constant. Parameter A is determined so that the approximation error (111) for a uniform step at node x_i is equal to zero, i.e. collocation type scheme. Then we have

$$A = \frac{1}{4} f''(x_i)$$

2. Consider a stationary equation in which only convection and diffusion are present without a source.

$$\varepsilon v'' + v' = 0, \tag{115}$$

with boundary conditions $v(0) = 0, v(1) = 1$.
There are various schemes for the difference solution (115). Based on the moving node technique, it is possible to explicitly express local errors in the approximation of differential equations. Using the moving node method, we will show the efficient calculation of local approximation errors for the model problem (115).

Scheme with central-difference approximation of the convective term. Take a segment (x_{i-1}, x_{i+1}) and any point $x \in (x_{i-1}, x_{i+1})$. Consider the different analog (115).

$$\frac{2\varepsilon}{x_{i+1} - x_{i-1}} \left(\frac{u_{i+1} - u}{x_{i+1} - x} - \frac{u - u_{i-1}}{x - x_{i-1}} \right) + \frac{u_{i+1} - u_{i-1}}{x_{i+1} - x_{i-1}} = 0 \tag{116}$$

At $x = (x_{i-1} + x_{i+1})/2$, we have a central difference approximation. Here, u is the approximate value of the solution at point x.
From (116) we find.

$$u = \frac{(x - x_{i-1})(2\varepsilon + x_{i+1} - x)u_{i+1} + (x_{i+1} - x)(2\varepsilon - x + x_{i-1})u_{i-1}}{2\varepsilon(x_{i+1} - x_{i-1})}. \tag{117}$$

From here we get,

$$u' = \frac{2\varepsilon + x_{i+1} + x_{i-1} - 2x}{2\varepsilon} \frac{u_{i+1} - u_{i-1}}{x_{i+1} - x_{i-1}}, \tag{118}$$

$$u'' = -\frac{1}{\varepsilon}\frac{u_{i+1} - u_{i-1}}{x_{i+1} - x_{i-1}}. \tag{119}$$

If the difference solution at nodal points is known, then formula (117) makes it possible to determine the unknown at points that are not nodal.

Using formulas (118) and (119), the derivatives are restored at any point of the segment. Multiplying (119) by and adding with (118), we obtain.

$$\varepsilon u'' + u' = \Psi_1, \tag{120}$$

where

$$\Psi_1 = \frac{x_{i+1} + x_{i-1} - 2x}{2\varepsilon}\frac{u_{i+1} - u_{i-1}}{x_{i+1} - x_{i-1}}.$$

Eq. (120) can be called a differential analog of the difference Eq. (16); difference Eq. (116) is a collocation-type scheme.

Using (119), the approximation error can be written as.

$$\Psi_1 = -\frac{x_{i+1} + x_{i-1} - 2x}{2}u''.$$

Then Eq. (120) takes the form

$$\left(\varepsilon + \frac{x_{i+1} + x_{i-1} - 2x}{2}\right)u'' + u' = 0. \tag{121}$$

Thus, difference Eq. (116) exactly approximates differential Eq. (121) on the segment $[x_{i-1}, x_{i+1}]$.

Comparison of Eqs. (115) and (121) shows that when Eq. (115) is approximated by scheme (116), scheme diffusion appears with a variable coefficient $(x_{i+1} + x_{i-1} - 2x)/2$.

Upwind Scheme. Let us consider the difference analog of Eq. (115), in which the convective term is approximated by the one-sided difference relation.

$$\frac{2\varepsilon}{x_{i+1} - x_{i-1}}\left(\frac{u_{i+1} - u}{x_{i+1} - x} - \frac{u - u_{i-1}}{x - x_{i-1}}\right) + \frac{u_{i+1} - u}{x_{i+1} - x} = 0. \tag{122}$$

From here we get

$$u = \frac{(x - x_{i-1})(2\varepsilon + x_{i+1} - x_{i-1})}{(x_{i+1} - x_{i-1})(2\varepsilon + x - x_{i-1})}\frac{u_{i+1} + 2\varepsilon(x_{i+1} - x)u_{i-1}}{} \tag{123}$$

Determine the first and second derivatives:

$$u' = \frac{2\varepsilon(2\varepsilon + x_{i+1} - x_{i-1})}{(2\varepsilon + x - x_{i-1})^2}\frac{u_{i+1} - u_{i-1}}{x_{i+1} - x_{i-1}}, \tag{124}$$

$$u'' = \frac{-4\varepsilon(2\varepsilon + x_{i+1} - x_{i-1})}{(2\varepsilon + x - x_{i-1})^3} \frac{u_{i+1} - u_{i-1}}{x_{i+1} - x_{i-1}} \tag{125}$$

Let us calculate the approximation error.

$$\Psi_2 = \frac{2\varepsilon(x - x_{i-1})(2\varepsilon + x_{i+1} - x_{i-1})}{(2\varepsilon + x - x_{i-1})^3} \frac{u_{i+1} - u_{i-1}}{x_{i+1} - x_{i-1}}$$

The differential analog of scheme (122) has the form.

$$\left(\varepsilon + \frac{x - x_{i-1}}{2}\right)u'' + u' = 0, \tag{126}$$

those with a scheme against the flow, we have a scheme diffusion with a coefficient $(x_{i+1} - x)/2$. Based on (123)—is a hyperbola, which is monotone on the segment, i.e. scheme (122) is monotonic.

Based on the form of the differential analog (126), we can conclude that the differential equation

$$\left(\varepsilon + \frac{x}{2}\right)u'' + u' = 0 \tag{127}$$

is exactly approximated by the scheme

$$2\varepsilon\left(\frac{u_b - u}{1 - x} + \frac{u - u_a}{x}\right) + \frac{u_b - u}{1 - x} = 0 \tag{128}$$

Thus solving (128) with respect to u, we obtain the exact solution of differential Eq. (127).

3.8 On convergence of MNM

Let us show the convergence of MNM on model problems.

1. Consider the Cauchy problem

$$\frac{du}{dx} = -u, \quad u(0) = 1. \tag{129}$$

Let us replace the derivative with the forward difference,

$$\frac{du}{dx} \approx \frac{U_1(x) - U_1(0)}{x - 0} = \frac{U_1(x) - 1}{x}, \tag{130}$$

In (130) $U_1(x)$ the approximate value of the unknown function at the moving point is if there is only one moving node.

Using (130) we write the difference Eq. (129)

$$\frac{U_1(x) - 1}{x} = -U_1(x), \tag{131}$$

Take, now, two moving x nodes and $x/2$. For these points, we write difference equations of the type (131)

$$\frac{U_2(x/2)-1}{x/2}=-U_2(x/2),\qquad \frac{U_2(x)-U_2(x/2)}{x-x/2}=-U_2(x), \tag{132}$$

Eliminating these equations $U_2(x/2)$, we get

$$U_2(x)=\frac{1}{(1+x/2)^2}.$$

For three moved nodes $x/3$, $2x/3$ and x we get

$$U_3(x)=\frac{1}{(1+x/3)^3}.$$

If the number of nodes n, we get

$$U_n(x)=\frac{1}{(1+x/n)^n}. \tag{133}$$

If we strive for the number of nodes to infinity, we get

$$\lim_{n\to\infty}U_n(x)=\lim_{n\to\infty}\frac{1}{(1+x/n)^n}=e^{-x}.$$

Thus, we obtain the exact solution to problem (129).
2. Consider the problem

$$\frac{d\Phi}{dx}=\frac{1}{Pe}\frac{d^2\Phi}{dx^2},\qquad \Phi(0)=0,\ \ \Phi(1)=1. \tag{134}$$

For this problem, the difference scheme with 2^k-1 moving nodes has the form (29):

$$a_P^{(2^k-1)}U^{(2^k-1)}=a_E^{(2^k-1)}U_E^{(2^k-1)}+a_W^{(2^k-1)}U_W^{(2^k-1)} \tag{135}$$

where
$a_E^{(2^k-1)}=\frac{2^{2k+1}(1-\gamma_k)}{(1-x)\left(1-\gamma_k^{2k}\right)}, a_W^{(2^k-1)}=\frac{2^{2k+1}Pe(1-\tau_k)}{x\left(1-\tau_k^{2k}\right)}+\frac{2^{2k+1}(1-\tau_k)}{x\left(1-\tau_k^{2k}\right)}, a_P^{(2^k-1)}=a_W^{(2^k-1)}+a_E^{(2^k-1)}.$
$\tau_k=2^k/\left(2^k+\sigma\right), \gamma_k=\left(2^k+\theta\right)/2^k,\ \theta=Pe\,(1-x).$
If we find from (135) $U^{(2^k-1)}$ and pass to the limit at $k\to\infty$, we have

$$\lim_{k\to\infty}U^{(2^k-1)}(x)=\frac{e^{Pe\,x}-1}{e^{Pe}-1}.$$

The obtained limit coincides with the exact solution.

4. Conclusions

Summary: Derivation of approximate analytical solutions of differential equations by the moving nodes method

- The method of moving nodes allows one to obtain approximate analytical solutions for boundary value problems of mathematical physics.
- This is especially true in engineering applications, to obtain a rough analytical representation of the solution. The analytical method has its advantages over the numerical ones for its subsequent use and analysis of the structure of the solution.
- To refine the solution of differential equations, you can achieve this by adding the number of nodes to be moved.
- The examples given show the possibilities of applying and using the method of moving nodes for applied problems.
- Using the method of moving nodes based on the upwind scheme, compact schemes with high resolution for convective-diffusion problems are constructed.

Summary: Application of the moving node method

- Using the method of moving nodes based on the control volume method, compact schemes with high resolution for convective-diffusion problems are constructed.
- Using a combination of moving node methods and Richardson's extrapolation, compact, high-resolution schemes for convective-diffusion problems are constructed.
- Choices of the influence of the profile on the faces of the control volume are studied.
- The possibilities of using movable nodes for the analysis of schemes are shown.
- Based on the method of moving nodes, the possibilities of finding errors in the approximation of differential equations are shown.
- For simple problems, the convergence of the moving nodes method is given.

Author details

Umurdin Dalabaev[1*] and Malika Ikramova[2]

1 The University of World Economy and Diplomacy, Tashkent, Uzbekistan

2 Scientific Research Institute of Irrigation and Water Problems of the Ministry of Water Resources of UZ, Uzbekistan

*Address all correspondence to: udalabaev@mail.ru

References

[1] Anderson D, Tannekhil D, Pletcher R. Vychislitel'naya gidromekhanika i teploobmen. Moskov.: Mir. 1990

[2] Marchuk GI. Metody vychislitel'noy matematiki/G.I. Marchuk. M.: Nauka; 1977

[3] Marchuk GI, Shaydurov VV. Povysheniye tochnosti resheniy raznostnykh skhem. M.: Nauka, glavnaya redaktsiya fiziko-matematicheskoy literatury; 1979

[4] Na TS. Vychislitel'nyye metody resheniya prikladnykh granichnykh zadach. M.: Mir; 1982. 294 s

[5] Paskonov VM, Polezhayev VI, Chudov LA. Chislennoye modelirovaniye protsessov teplo-i massoobmena. M.: Nauka; 1984. 286 s

[6] Samarskiy AA. Vvedeniye v teoriyu raznostnykh skhem/A.A. Moskov, Nauka; 1971

[7] Fletcher K. Vychislitel'nyye metody v dinamike zhidkostey. Moskov, Mir; 1991

[8] Tikhonov AN, Samarskiy AA. Uravneniya matematicheskoy fiziki, Uchebnoye posobiye dlya vuzov. — 5-ye izd., stereotip. M.: Nauka; 1977. p. 735 s.: il

[9] Doolan ER, Miller JJH, Schilders WHA. Uniform Numerical Methods for Problems with Initial and Boundary Layers. Dublin: Boole Press; 1980

[10] Il'in AM. Raznostnaya skhema dlya differentsial'nogo uravneniya s malym parametrom pri starshey proizvodnoy. Matem. zametki. 1969;**6** (2):237-248

[11] Patankar S. Chislennyye metody resheniya zadach teploobmena i dinamiki zhidkosti. M.: Energoatomizdat; 1984. 152 s

[12] Samarskiy AA, Andreyev VB. Raznostnyye metody dlya ellipticheskikh uravneniy. M.: Nauka; 1976

[13] Samarskiy AA, Vabishchevich PN. Chislennyye metody reshcheniya zadach konvektsii-diffuzii. Izd. stereotip. M.: Knizhnyy dom «LIBROKOM»; 2015. 248 s

[14] Umurdin D. Increasing the accuracy of the difference scheme using the Richardson extrapolation based on the movable node method. Academic Journal of Applied Mathematical Sciences. DOI: 10.32861/ajams.68.204.212. Available from: https://arpgweb.com/journal/journal/17

[15] Dalabaev U. Application of the method of moving nodes to the solution of applied boundary problems. Bulletin of the Institute of Mathematics—Tashkent. 2018;**6**. C. 5-9

[16] Dalabaev U. The stability of the difference scheme for the Equation of Rahmatulin. Malaysian Journal of Mathematical Sciences—Malaysia. 2009; **3**(1):1-11

[17] Dalabaev U. On the lattice viscous flow. Turkish Journal of Physics—Turkiya. 1997;**21**(5):649-654

[18] Darvish MS. A new high-resolution scheme based on the normalized variable formulation. Numerical Heat Transfer, Part B. 1993;**24**:353-373

[19] Darwish M, Asmar D, Moukalled F. A comparative assessment within a multigrid environment of segregated pressure-based algorithms for fluid flow

at all speeds. Numerical Heat Transfer, Part B. 2003;**45**:49-74

[20] Ferreira VG, Kurokawa FA, Queiroz RAB, Kaibara MK, Oishi CM, Cuminato JA, et al. Assessment of a high-order finite difference upwind scheme for the simulation of convection-diffusion problems. International Journal for Numerical Methods in Fluids. 2009;**60** (1):1-26

[21] Ferreira VG, de Queiroz RAB, GAB L, Cuenca RG, Oishi CM, Azevedo JLF, et al. A bounded upwinding scheme for computing convection-dominated transport problems. Computers & Fluids. 2012;**57**:208-224

[22] Gaskell PH, Lau AKC. Curvature–compensated convective transport: SMART, a new boundedness preserving transport algorithm. International Journal for Numerical Methods in Fluids. 1988;**8**:617641

[23] Hayase TA, Humphrey JAC, Greif R. Consistently formulated QUICK scheme for fast and stable convergence using finite–volume iterative calculation procedure. Journal of Computational Physics. 1992;**98**

[24] Leer BV. Towards the ultimate conservation difference scheme V. A second order sequel to Godunov's method. Journal of Computational Physics. 1977;**23**:101-136

[25] Leer BV. Towards the ultimate conservative difference scheme. II. Monotonicity and conservation combined in a second order scheme. Journal of Computational Physics. 1974; **14**:361-370

[26] Leonard BP. A stable and accurate convective modelling procedure based on quadratic upstream interpolation. Computer Methods in Applied Mechanics and Engineering. 1979;**19**: 59-98

[27] Leonard BP. The ULTIMATE conservative difference scheme applied to unsteady one-dimensional advection. Computer Methods in Applied Mechanics and Engineering. 1991;**88**: 17-74

[28] Leonard BP. A Survey of finite difference with upwind for numerical modeling of the incompressible convective diffusion equation. In: Taylor C, Morgan K, editors. Computational Techniques in Transient and Turbulent Flows. Swansea, UK: Prineridge Press; 1981. pp. 1-35

[29] Li B, Chen Z, Huan G. Control volume function approximation methods and their applications to modeling porous media flow. Advances in Water Resources. 2003;**26**:435-444

[30] Lin CH, Lin CA. Simple high–order bounded convection scheme to model discontinuities. AIAA Journal. 1997;**35** (3):563-565

[31] Patankar SV, Spolding DB. Heat and Mass Transfer in Boundary Layers. Cambridge University Press; 1970. 255 pp

[32] Raithby GD. A critical evaluation of upstream differencing applied to problems involving fluid flow. Computer Methods in Applied Mechanics and Engineering. 1976;**9**:75-103

[33] Shyy W. A study of finite difference approximations to steady–state, convection–dominated flow problems. Journal of Computational Physics. 1995; **57**:415-438

[34] Shyy W, Thakur S, Wright J. Second-order upwind and central difference scheme for recirculating flow

computation. AIAA Journal. 1999;**30**: 923-932

[35] Dalabaev U. Raznostno-analiticheskiy metod priblizhennogo resheniya zadachi Dirikhle. Sinergiya nauk. 2018;**21**:344-349. Available from: http://synergy-journal.ru/archive/article 1949

[36] Dalabayev U. Primeneniye metoda peremeshchayemykh uzlov k issledovaniyu monotonnosti raznostnoy skhemy i yego uluchsheniye dlya odnomernoy konvektivno-diffuzionnoy zadachi. Problemy vychislitel'noy i prikladnoy matematiki. 2019;**6**(24): 44-52

[37] Targ SM. Osnovnyye zadachi teorii laminarnykh techeniy. Moskov.; 1951. 420 s

[38] Dalabaev U. Difference -analytical method of the one-dimensional convection-diffusion equation. IJISET—International Journal of Innovative Science, Engineering & Technology. 2016;**3**(1):234-239

[39] Dalabaev U. Computing technology of a method of control volume for obtaining of the approximate analytical solution one-dimensional convection-diffusion problems. Open Access Library Journal. 2018;**5**:e4962

[40] Gao W, Li H, Jian Y. An oscillation–free high order TVD/CBC-based upwind scheme for convection discretization. Numerical Algorithms. 2012;**59**:29-50

[41] Moukalled F, Mangani L. Darwish M, The Finite Volume Method in Computational Fluid Dynamics. Springer International Publishing Switzerland; 2016

[42] Zho JA. Low-diffusive and oscillation free convection scheme. Communications in Applied Numerical Methods. 1991;7:225-232

[43] Zhu J, Rodi W. A low-dispersion and bounded convection scheme. Computer Methods in Applied Mechanics and Engineering. 1991;**92**:87-96

[44] Yeoh GH, Tu J. Computational techniques for multi–phase flows. Basics and Applications. 2019:619

[45] Yu B, TaoWQ, Zhang DS,Wang QW. Discussion on numerical stability and boundedness of convective discretized scheme. Numerical Heat Transfer, Part B. 2001;**40**(4):343-365

[46] Loytsyanskiy LG. Mekhanika zhidkosti i gaza. M.: Nauka; 1973. 736 s

Chapter 16

On the Analytical Properties of Prime Numbers

Shazali Abdalla Fadul

Abstract

In this work we have studied the prime numbers in the model $P = am + 1, m, a > 1 \in \mathbb{N}$. and the number in the form $q = ma^m + bm + 1$ in particular, we provided tests for hem. This is considered a generalization of the work José María Grau and Antonio M. Oller-marcén prove that if $C_m(a) = ma^m + 1$ is a generalized Cullen number then $m^{a^m} \equiv (-1)^a (mod\ C_m(a))$. In a second paper published in 2014, they also presented a test for Broth's numbers in Form $kp^n + 1$ where $k < p^n$. These results are basically a generalization of the work of W. Bosma and H.C Williams who studied the cases, especially when $p = 2, 3$, as well as a generalization of the primitive MillerRabin test. In this study in particular, we presented a test for numbers in the form $ma^m + bm + 1$ in the form of a polynomial that highlights the properties of these numbers as well as a test for the Fermat and Mersinner numbers and $p = ab + 1\ a, b > 1 \in \mathbb{N}$ and $p = qa + 1$ where *q is prime odd* are special cases of the number $ma^m + bm + 1$ when b takes a specific value. For example, we proved if $p = qa + 1$ where q is odd prime and $a > 1 \in \mathbb{N}$ where $\pi_j = \frac{1}{q}\binom{q}{j}$ then $\sum_{j=1}^{q-2} \pi_j(-C_m(a))^{q-j-1}(q - a^m) \equiv \chi_{(m,q-a^m)}(mod\ p\ )$ Components of proof Binomial theorem Fermat's Litter Theorem Elementary algebra.

Keywords: broth numbers, Cullen number, polynomial, Fermat number, Mersinne numbers

1. Introduction

No algorithm that produces prime numbers in explicit forms, or rather, this goal was not reached, mathematicians resorted to an alternative method to discover prime numbers, which are primitive tests since Fermat's era or before, and this method proved its effectiveness to the extent that many prime numbers were discovered The Great Until. Euler studied Fermat's prime numbers and discovered some of them. Cullen, Broth and Mersinne also studied those numbers, as well as Pedro Berrizbeitia, Wieb Bosma and A. Schönhage. The results that we reached in this study are in the same way as those who followed the work of [1–3]. In a paper published on March 11, 2011 MO [3] prove the following result. $C_m(a)$ is a prime where $C_m(a) = ma^m + 1$ then

$m^{a^m} \equiv (-1)^a (mod\ C_m(a))$ And in a paper published on July 10, 4102 using the same ideas found in MO [2], they proved [3] the following result. Let $N = kp^n + 1$.

where is p is odd prime and $k < p^n$ Assume that $a \in \mathbb{Z}$ is a p-th power non-residue modulo $\boldsymbol{N}$, then $\boldsymbol{N}$ is a prime if only if $\phi_p\left(a^{\frac{N-1}{p}}\right) \equiv 0(mod\ N)$. The numbers in form $C_m(a) = ma^m + 1$ are called Cullen numbers, first studied by Cullen in 1905. And the numbers in the form of $kp^n + 1$ are called the Broth numbers and we call the number primes the form $M_P = 2^P - 1$ mersenne number discovered in 2005 by Martin nowak the largest prime number of Mersenne $M_{25964951}$ and 42 in the list. We know about Mersenne's number if M_p it is not prime then there is a prime number $q = 2pr + 1$ where M_p/q example M_{11} of a non-prime. Also there is a relationship between Mersenne prime and the perfect numbers. And number in form $F_n = 2^{2^n} + 1$ are called Fermat numbers were first studied by Pierre de Fermat, The importance of these numbers lies in providing the large prime numbers of the known. All the large prime numbers are in the form $ma^n + b$ or $a^n + b$, for example, in 2021, $2525532.73^{2525532} + 1$ was discovered the largest prime number defined by Tom Greer. There is a program in the Internet called [Prime Grid] The goal of discovering this is a kind of numbers See [*https* : *primegrid.com*] Researchers use several techniques in the study such as preliminary tests and high-precision computers. Prove Broth if $N = k2^n + 1$ where K is odd and $k < 2^n$ if $a^{\frac{N-1}{2}} \equiv -1(mod\ N)$ same $a \in \mathbb{Z}$ then N is a prime. The next important step was made in 1914 by Pocklington his result is the first generalization of Proth's theorem suitable for numbers of the form prove Pocklington if $N = Kp^n + 1$ where K is odd and $k < p^n$ if for same $a \in \mathbb{Z}$ $a^{N-1} \equiv -1(mod\ N)$ *and* $g.c.d\left(a^{\frac{N-1}{p}} - 1, p\right)$ then N is prime. There are many works that discuss Broth's theorem and numbers. Case $p = 3$ studied by W. Bosma [4] and A. Guthmann [5] Also, for a discussion on the Broth numbers, see H.C Williams [6, 7] P. Berrizbeitia [8, 9].

The purpose of this work is to study the numbers in model $ma^m + bm + 1$ and $p = ba + 1$ where $a, b > 1 \in \mathbb{Z}$ and $p = aq + 1, a, q \in \mathbb{N}$ where q is an odd prime number, In addition, tests for Fermat and Mersenne numbers are presented and the study of the relationship between two prime numbers and a polynomial with finite properties. From the results we obtained we proved, for example, *if* p and q are prime numbers, $p = qa + 1$ where $q, a > 1 \in \mathbb{N}$ and where $C_m(a) = ma^m + 1$ and $\pi_j = \frac{1}{q}\binom{q}{j}$ then

$$\sum_{j=1}^{q-2} \pi_j(-C_m(a))^{q-j-1}(q - a^m) \equiv \chi_{(m,q-a^m)}(mod\ p) \tag{1}$$

Our approach to the proof differs from the one in [2, 3]. We explicitly relied on the binomial theorem, elementary algebra, and Fermat's litter theorem. A deductive method of analysis using basic operations in elementary algebra.

2. Proof of the theorem 1

In this section, we prove theorem 1. Components of the proof Elementary algebra basic operations such as subtraction from both sides and extraction of the common factor with the binomial theorem form the foundations of the proof. Theorem 1 is an expression of a polynomial that shows the properties of numbers in the form $p = ma^m + bm + 1$,

$a, m, b > 1 \in \mathbb{N}$, so it can be used as a test to reveal the prime number in the form $ma^m + bm + 1$. In addition to that, it is used to prove the results in the next section where it plays an essential role in the proofs.

THEOREM 1. if M is a prime where $\mathcal{M} = ma^m + bm + 1\, and\, a, m > 1 \in \mathbb{N}, b \neq 0 \in \mathbb{Z}$ then

$$\begin{cases} \eta_{(\lambda)} \equiv \chi_{(x,y)} (mod\ \mathcal{M}) & if\ \mathcal{M}\ and\ \lambda\ is\ a\ prime\ where \\ \eta_{(\lambda)} \equiv \chi_{(\lambda,x,y)} (mod\ \mathcal{M}) & if\ \mathcal{M}\ is\ a\ prime \end{cases} \tag{2}$$

Proof. let $\mathcal{M} = ma^m + \delta\, where\, m, a, \mathrm{n} > 1 \in \mathbb{N}$ and $\delta = bm + 1$ where $b \in \mathbb{Z}$ according to the binomial theorem, we find that

$$(\mathcal{M} + (-\delta))^n - m^n = (ma^m)^n - m^n = \sum_{j=0}^{n-1} \binom{n}{j} \mathcal{M}^{n-j} (-\delta)^j + (-\delta)^n - m^n \tag{3}$$

Then

$$m^n (a^{mn} - 1) - ((-bm - 1)^n - m^n) = \sum_{j=1}^{n-1} \binom{n}{j} \mathcal{M}^{n-j} (-bm - 1)^j \tag{4}$$

Note that

$$\sum_{j=0}^{n-1} \binom{n}{j} \mathcal{M}^{n-j} (-bm - 1)^j \equiv 0 (mod\ \mathcal{M}) \tag{5}$$

Then from (4) and (5) we have that

$$a^{nm} - 1 - ((-bm - 1)^n - m^n) \equiv 0 (mod\ \mathcal{M}) \tag{6}$$

Now we conclude that from Eq. (6)

$$a^{nm} \equiv 1 (mod\ \mathcal{M})\ if\ and\ only\ if\ (-bm - 1)^n \equiv m^n (mod\ \mathcal{M}) \tag{7}$$

Suppose that $n = \frac{\mathcal{M}-1}{m}$ and $\mathcal{M}$ is a prime so

$$a^{\mathcal{M}-1} \equiv 1 (mod\ \mathcal{M})\ if\ and\ only\ if\ (-bm - 1)^{\frac{\mathcal{M}-1}{m}} \equiv m^{\frac{\mathcal{M}-1}{m}} (mod\ \mathcal{M}) \tag{8}$$

From the assumption $\mathcal{M}$ is a prime from Fermat's litter theorem see [Kenneth H. Rosen 2 pp. 161] we have that if $\mathcal{M}\ is\ a\ prime$ then $a^{\mathcal{M}-1} \equiv 1\ (mod \mathcal{M})$ where $a > 1$. This means if $\mathcal{M}$ is a prime number, then from (8) we find that

$$(-bm-1)^{\frac{\mathcal{M}-1}{m}} \equiv m^{\frac{\mathcal{M}-1}{m}} (mod\ \mathcal{M}) \tag{9}$$

Then

$$(bm+1)^{\frac{\mathcal{M}-1}{m}} \equiv (-m)^{\frac{\mathcal{M}-1}{m}} (mod\ \mathcal{M}) \tag{10}$$

Let be $\lambda = \frac{\mathcal{M}-1}{m}$ $\lambda = a^m + b$ then from binomial theorem we have that

$$\begin{aligned}(bm+1)^{\lambda} - (-m)^{\lambda} &= \sum_{j=0}^{\lambda} \binom{\lambda}{j} (bm)^{\lambda-1} - (-m)^{\lambda} \\ &= (bm)^{\lambda} - (-m)^{\lambda} + \sum_{j=1}^{\lambda-1} \binom{\lambda}{j} (bm)^{\lambda-1} + 1\end{aligned} \tag{11}$$

$\mathcal{M} = ma^m + bm + 1$ then $\lambda = \frac{\mathcal{M}-1}{m} = a^m + b$ According to the binomial theorem, if λ is a prime number, then $\binom{\lambda}{j} \equiv 0(mod\ \mathcal{M})$ means $\binom{\lambda}{j}$ is divisible by λ for every $2 \leq \lambda \leq \lambda - 1$. So suppose λ is a prime number and $\pi_j = \frac{1}{\lambda}\binom{\lambda}{j}$ follows from that $\binom{\lambda}{j}(bm) = \pi_j b(\mathcal{M}-1)$. so from (11) we have that

$$\begin{aligned}(bm+1)^{\lambda} - (-m)^{\lambda} &= (bm)^{\lambda} - (-m)^{\lambda} + 1 \\ &\quad + \sum_{j=1}^{\lambda-1} \pi_j b^{\lambda-j} m^{\lambda-j-1} (\mathcal{M}-1) \\ &= (bm)^{\lambda} - (-m)^{\lambda} + 1 + \left(\sum_{j=1}^{\lambda-1} \pi_j b^{\lambda-j} m^{\lambda-j-1} \right) \mathcal{M} \\ &\quad - \sum_{j}^{\lambda-1} \pi_j b^{\lambda-j} m^{\lambda-j-1}\end{aligned} \tag{12}$$

From Eq. (10) $(bm+1)^{\lambda} \equiv (-m)^{\lambda} (mod\ \mathcal{M})$. and Notice the Eq. (12) on the right-hand side consisting of two terms the first multiplied by $\mathcal{M}$ and the second empty of $\mathcal{M}$. Then we have that

$$\left(\sum_{j=1}^{\lambda-2} \pi_j b^{\lambda-j} m^{\lambda-j-1} \right) \mathcal{M} \equiv 0(mod\ \mathcal{M}) \tag{13}$$

And

$$(bm)^{\lambda} - (-m)^{\lambda} - \sum_{j=1}^{\lambda-2} \pi_j b^{\lambda-j} m^{\lambda-j-1} - b + 1 \equiv 0(mod\ \mathcal{M}) \tag{14}$$

Then

$$\sum_{j=1}^{\lambda-2} \pi_j b^{\lambda-j} m^{\lambda-j-1} \equiv (bm)^{\lambda} - (-m)^{\lambda} - b + 1(mod\ \mathcal{M}) \tag{15}$$

Let be $\eta_{(\lambda)}(x,y) = \sum_{j=1}^{\lambda-2} \pi_j x^{\lambda-j} y^{\lambda-j-1}$ and $\chi_{(x,y)} = (bm)^{\lambda} - (-m)^{\lambda} - b + 1$ So we have that

$$\eta_{(\lambda)}(x,y) \equiv \chi_{(x,y)}(mod\ \mathcal{M}) \tag{16}$$

This is the first case of proof when λ is a prime. The proof of the second case is similar to the case of the first and there is no fundamental difference, according to the binomial theorem let be $\lambda > 2 \in \mathbb{N}$ then from (12) we have

$$\begin{aligned} \lambda\left((bm+1)^{\lambda} - (-m)^{\lambda}\right) &= \lambda\left((bm)^{\lambda} - (-m)^{\lambda}\right) + \lambda + \sum_{j=1}^{\lambda-1} \binom{\lambda}{j} b^{\lambda-j} m^{\lambda-j-1}(\mathcal{M}-1) \\ &= \lambda\left((bm)^{\lambda} - (-m)^{\lambda}\right) + \lambda + \left(\sum_{j=1}^{\lambda-1} \binom{\lambda}{j} b^{\lambda-j} m^{\lambda-j-1}\right)\mathcal{M} \\ &\quad - \sum_{j=1}^{\lambda-2} \binom{\lambda}{j} b^{\lambda-j} m^{\lambda-j-1} - \lambda b \end{aligned} \tag{17}$$

Note that

$$\left(\sum_{j=1}^{\lambda-1} \binom{\lambda}{j} b^{\lambda-j} m^{\lambda-j-1}\right)\mathcal{M} \equiv (mod\ \mathcal{M}) \tag{18}$$

And from (10) we have that

$$\lambda\left((bm+1)^{\lambda} - (-m)^{\lambda}\right) \equiv 0(mod\ \mathcal{M}) \tag{19}$$

Then

$$\sum_{j=1}^{\lambda-2} \binom{\lambda}{j} b^{\lambda-j} m^{\lambda-j-1} \equiv \lambda(bm)^{\lambda} - \lambda(-m)^{\lambda} + \lambda - b\lambda(mod\ \mathcal{M}\ ) \tag{20}$$

Let be $\eta_{(\lambda)}(x,y) = \sum_{j=1}^{\lambda-2} \binom{\lambda}{j} x^{\lambda-j} y^{\lambda-j-1}$ and $\chi_{(\lambda,y,x)} = \lambda(bm)^{\lambda} - \lambda(-m)^{\lambda} + \lambda - \lambda b$ so we have

$$\eta_{(\lambda)}(x,y) \equiv \chi_{(\lambda,x,y)}(mod\ \mathcal{M}) \tag{21}$$

Then

$$\begin{cases} \eta_{(\lambda)}(x,y) \equiv \chi_{(y,x)}(mod\ \mathcal{M}) & if\ \lambda\ and\ \mathcal{M}\ is\ aprime \\ \eta_{(\lambda)}(x,y) \equiv \chi_{(\lambda,y,x)}(mod\ \mathcal{M}) & if\ \mathcal{M}\ is\ a\ prime \end{cases} \tag{22}$$

REMARK 1: We note that the proof has little complexity, as we explicitly relied on the binomial theorem and elementary algebra to obtain Eq. (12). After that, Fermat's Litter Theorem was used, which is a theorem dating back to the year 1610.In 1610 Fermat wrote in a letter to Frenicle, that whenever p is prime p divides $a^{p-1} - 1$ for all integers a not divisible p, a result now known as Fremat's little theorem, As equivalent formulation is the assertion that p divide $a^p - a$ for all integers a , whenever p is prime. The question naturally arose as to whether the prime are the only integer exceeding that satisfy this criterion, but Carmichael pointed out in 1910 that 561 = 11 × 17 × 3 divides a $a^{560} \equiv 1(mod\ 561)$ now. A composite integer which satisfies $a^{n-1} \equiv 1(mod\ n)$ for all positive integers a with g.c.d(a, n) = 1 is called a Carmichael number. For a related discussion see Kenneth H. Rose page (55). This means that Theorem 1 is not a definitive test, but it fails at the numbers Carmichael, but on the one hand we find that it is more general than those [2, 10] because of the variables m, a, b in the number $ma^m + bm + 1$. And we will explain this by proving results for Mersenne and Fermat numbers, which are special cases when the variable b takes a certain value.

THEOREM 2. *if* $\lambda = a^m + (-1)^\sigma$ and $q = ma^m + (-1)^\sigma m + 1$ where q is a prime and $a, m > 1 \in \mathbb{N}$ then

$$\begin{cases} \Psi_{(m)} \equiv \lambda - (-1)^\sigma \lambda (mod\ q\ ) \\ \quad if\ \sigma = 1\ and\ a > 1 \in \mathbb{N}\ and\ if\ \sigma = 2\ then\ a\ is\ odd \\ \Psi_{(m)} \equiv 2\lambda m^\lambda + \lambda - (-1)^\sigma \lambda (mod\ q\ )\ if\ \sigma = 2\ and\ a\ is\ even \end{cases} \tag{23}$$

Proof. Let be $b = (-1)^\sigma$ From theorem 1 we have

$$\sum_{j=1}^{\lambda-2} \binom{\lambda}{j} ((-1)^\sigma)^{\lambda-j} m^{\lambda-j-1} \equiv \lambda((-1)^\sigma m)^\lambda - \lambda(-m)^\lambda + \lambda - (-1)^\sigma \lambda (\text{mod q}) \tag{24}$$

Let be

$$\Psi_{(m)} = \sum_{j=1}^{\lambda-2} \binom{\lambda}{j} ((-1)^\sigma)^{\lambda-j} m^{\lambda-j-1} \tag{25}$$

From Eq. (24) we get the following

$$\begin{cases} \Psi_{(m)} \equiv \lambda - \lambda(-1)^\sigma (mod\ q) \\ if\ \sigma = 1\ and\ a > 1 \in \mathbb{N}\ and\ if\ \sigma = 2\ a\ is\ odd \\ \Psi_{(m)} \equiv 2\lambda m^\lambda + \lambda - \lambda(-1)^\sigma (mod\ q) \\ if\ \sigma = 2\ and\ a\ is\ even \end{cases} \tag{26}$$

LEMMA 1. let be $p = 3^m - 2$ and $\mathcal{M} = m3^m - 2m + 1$ where p and $\mathcal{M}$ is a prime number then

$$\sum_{j=1}^{p-2} \pi_j (-2)^{p-j} m^{p-j-1} \equiv \left(2_1^p\right)(-m)^p + 3(mod\ \mathcal{M}) \tag{27}$$

Proof. Let be in theorem 1 $a = 3\ and\ b = -2\ where\ p\ is\ a\ prime$ then we get $\lambda = 3^m - 2\ and\ \mathcal{M} = m3^m - 2m + 1$ and we have

$$\sum_{j=1}^{p-2} \pi_j(-2)^{p-j} m^{p-j-1} \equiv \binom{2^p}{1}(-m)^p + 3(mod\ \mathcal{M}) \tag{28}$$

Then

$$\sum_{j=1}^{M_p-2} \pi_j(-1)^{M_p-j} p^{M_p-j-1} \equiv 2(mod\ \mathcal{M}) \tag{29}$$

LEMMA 2. If F_n fermat number and $p = 2^n F_n + 1$ where p is a prime then

$$\sum_{j=1}^{F_n-2} \pi_j (2^n)^{F_n-j-1} \equiv 2(2^n)^{F_n}(mod\ p) \tag{30}$$

Proof. Let be in theorem 1 $b = 1$ *and* $a = 2$ *and* $m = 2^n$ then we get $\lambda = a^m + b = 2^{2^n} + 1$ and $\mathcal{M} = ma^m + bm + 1 = 2^{2^n+n} + 2^n + 1 = p$ where

$$\sum_{j=1}^{F_n-2} \pi_j (2^n)^{F_n-j-1} \equiv 2(2^n)^{F_n}(mod\ p) \tag{31}$$

REMARK 2: Fermat,s numbers $F_n = 2^{2^n} + 1$ are named after pierre de farmat because he was the first to stud these numbers guess that all fermat numbers are prime

$$3, 5, 17, 57, 65537, \ldots . \tag{32}$$

But this conjecture was denied by Euler's proved the Fermat number F_5 is not prime

$$F_5 = 2^{2^5} + 1 = 4294967297 = 641 \times 6700417 \tag{33}$$

These numbers was named $2^P - 1$ Mersnne numbers, so in Ref. to Marin Meresenne, who began studying them by 2020 he discovered fifty –one prime numbers. There is a program called (the big search for Mersenne prime on the internet). Many prime numbers of Meresnne numbers have been discovered, we know about $M_2, M_3, M_5, M_7, M_{17}, M_{19}, M_{31}, M_{521} \ldots, M_{1279}, M_{110305}, M_{132049}, M_{25964951}$ all prime numbers M_{11} is not prime number and they give good results from fermat numbers that only four digits of it have been discovered so far. We know about Fermat numbers, if F_n is not prime, then there is $b = k2^{n+2} + 1$ where F_n/b, and likewise abuot Mersenne numbers, if M_p is not prime, there is $q = 2pr + 1$ where M_p/q and q is prime. From a computational point of view, we find that the results that we have reached are more robust and generalizable those of the results mentioned. Firstly, this is due to the existing ideas and properties is those results. This is represented in highlighting an integral relationship between two prime numbers and more prime numbers. We notice in the LEMMA 1 that the prime numbers and the $p = 3^m - 2$ and the number in form $\mathcal{M} = m3^m - 2m + 1$ numbers in form combine in one result, and also the properties of the Mersnne numbers. Such a correlation does not exist [5, 6] as well as with the ratio of the Fermat numbers also meet with the numbers in LEMMA 2 and this shows relationship between the Fermat numbers and those numbers. In addition to that, the result are expressed a polynomial that highlights the properties of those numbers, and it can also be used as a primitive test to discover those numbers . For a discussion of such issues see [3, 8, 9, 11–14] on there are several numbers studied $A3^n \pm 1, k2^n + 1, 2^n \pm 1$ and close to these formulas.

3. Prime numbers In form $p = am + 1$

In this section we study the prime numbers in form $p = am + 1\, where\ a, m > 1 \in \mathbb{N}$ which is a special case of the prime numbers form $a^m + bm + 1$, when substituting b a certain value, we study the properties of numbers $p = am + 1, a, m \in \mathbb{N}$ and $q = bp + 1\, where\ b, p > 1 \in \mathbb{N}$, as well as relationship between polynomial. These polynomial numbers show us special properties of this numbers $am + 1$ as well as the numbers $p = qa + 1\, where\ q\ is\ a\ prime\ number$ of the properties, polynomial can be used as a primitive testing algorithm for those numbers. The proof depend mainly on THEOREM 1. In this section, we explain and realize that there is more than one variable in the theorem, for example, $ma^m + bm + 1$ variable $m, a > 1 \in \mathbb{N}\ and\ b \in \mathbb{Z}, b \neq 0$, because of this, have many distinctive properties. We prove THEOREM 3 is this section and THEOREM 4 by directly changing the value of that variable without any the complexity mentions in particular the basic operations and the binomial theorem are the other extreme in the proofs.

THEOREM 3. *if* $p = qm + 1$ where p, q *is a prime odd* and $a, m > 1 \in \mathbb{N}$ where $C_m(a) = ma^m + 1$ and $\pi_j = \frac{1}{q}\binom{q}{j}$ then

$$\sum_{j=1}^{q-2} \pi_j(-C_m(a))^{q-j-1}(q - a^m) \equiv \chi_{(m,q-a^m)}(mod\ p) \tag{34}$$

Proof. let be in theorem $b = q - a^m\ where\ q > 2\ is\ prime\ odd$ then $\mathcal{M} = ma^m + (q - a^m)m + 1 = qm + 1 = p$ therefore also $\lambda = a^m + b = a^m + q - a^m = q$ so we have that

$$\eta_{(\lambda)}(m, \mathrm{q} - a^m) = \sum_{j=1}^{q-2} \pi_j(q - a^m)^{q-j}m^{q-j-1} \tag{35}$$

And

$$\chi_{(m,q-a^m)} = (mq - ma^m)^\lambda - (-m)^\lambda - q + a^m + 1 \tag{36}$$

Note that $m(q - a^m) = mq - ma^m = p - ma^m - 1\ le\ be\ C_m(a) = ma^m + 1$ then we have

$$\eta_{(q)}(m, q - a^m) = \sum_{j=1}^{q-2} \pi_j(p + (-C_m(a)))^{q-j-1}(q - a^m) \tag{37}$$

Note that from binomial theorem

$$\begin{aligned}&(p+(-C_m(a)))^{q-j-1}\\&=\sum_{k=0}^{q-j-2}\binom{q-j-1}{k}p^{q-j-1-k}(-C_m(a))^k(q-a^m)\\&+(-C_m(a))^{q-j-1}(q-a^m)\,for\ all\ 1\le j\le q-1\end{aligned}\tag{38}$$

Let be

$$\psi_m=\sum_{k=1}^{q-j-2}\binom{q-j-1}{k}p^{q-j-1-k}(-C_m(a))^k\tag{39}$$

Then from (37) and (38) and (39) we have

$$\eta_{(q)}(m,q-a^m)=\sum_{j=1}^{q-2}\pi_j\psi_m+\sum_{j=1}^{q-2}\pi_j(-C_m(a))^{q-j-1}(q-a^m)\tag{40}$$

Note that $\psi_m\equiv 0(mod\,p)$ then also

$$\sum_{j=1}^{q-2}\pi_j\psi_m\equiv 0(mod\,p)\ all\ j=1,2,3\dots .q-1\tag{41}$$

And we know from Eq. (16)

$$\eta_{(q)}(m,q-a^m)\equiv\chi_{(m,q-a^m)}(mod\,p)\tag{42}$$

Now we conclude that from Eqs. (37), (39), and (41) we have

$$\sum_{j=1}^{q-2}\pi_j(-C_m(a))^{q-j-1}(q-a^m)\equiv\chi_{(m,q-a^m)}(mod\,p)\tag{43}$$

From (16) we knew $b=q-a^m$ *then* note that

$$\begin{aligned}\chi_{(m,q-a^m)}&-(mq-ma^m)^q-(-m)^q+1-q+a^m\\&=(p+(-ma^m-1))^q-(-m)^q-q+a^m+1\\&=\sum_{j=0}^{q-1}\binom{q}{j}p^{q-j}(-ma^m-1)^j+(-ma^m-1)^q-(-m)^q-q+a^m+1\end{aligned}\tag{44}$$

Note that

$$\sum_{j=0}^{q-1}\binom{q}{j}p^{q-j}(-ma^m-1)^j\equiv 0(mod\,p)\tag{45}$$

If the terms multiplied by variable p are excluded, we get the following

$$\chi_{(m,q-a^m)} = (-ma^m - 1)^q - (-m)^q - q + a^m + 1 \tag{46}$$

Therefore

$$\sum_{j=1}^{q-2} \pi_j(-C_m(a))^{q-j-1}(q - a^m) \equiv \chi_{(m,q-a^m)}(mod\ p) \tag{47}$$

LEMMA 3. *if* $p = 2q + 1$ and q *is a prime odd* then

$$\sum_{j=1}^{q-2} \pi_j(-2a^2 - 1)^{q-j-1}(q - a^m) \equiv \chi_{(2,q-a^2)}(mod\ p) \tag{48}$$

Proof. Let be in theorem 3 $m = 2$

LEMMA 4. *if* $p = qm + 1$ *and* q *is prime odd and* $m > 1 \in \mathbb{N}$ then

$$\sum_{j=1}^{q-2} \pi_j(-m2^m - 1)^{q-j-1}(q - 2^m) \equiv \chi_{(m,q-2^m)}(mod\ p) \tag{49}$$

Proof. Le be in theorem 3 $a = 2$

REMARK 3: if we make a comparison between the results found in [2, 3] about the generalized Cullen numbers and the results that we reached here, in fact, we find that these results are more generalized than those, and also rich in properties than those. The first general and the Cullen numbers in particular, and this is considered one of the properties of the prime numbers of the number in an adjective, as well as this relationship in the form a polynomial that combines the Cullen numbers and the prime number in general $P = qa + 1$ *where* q *is prime odd note* $P \in \mathbb{P} - \{2, 3, 5\}$. Such ideas do not exist in [5, 6, 7, … .12]. we also note that polynomials can be used as a primitive test to discover prime numbers.

THEOREM 4. *if* $q > 2, m > 1 \in \mathbb{N}$ *and* $p = qm + 1$ and p is prime then

$$\sum_{j=1}^{q-2} \binom{q}{j} (-C_m(a))^{q-j-1}(q - a^m) \equiv \chi_{(q,m,q-a^m)}(mod\ p) \tag{50}$$

Proof. we will prove this theorem with the same ideas as in the proof THEOREM 3 now according THEOREM 1 we have that

$$\eta_{(\lambda)}(b, m) = \sum_{j\ 1}^{\lambda-2} \binom{\lambda}{j} b^{\lambda-j} m^{\lambda-j-1} \tag{51}$$

And

$$\chi_{(\lambda,b,m)} = \lambda(bm)^\lambda - \lambda(-m)^\lambda + \lambda - b\lambda \tag{52}$$

Then let be $b = q - a^m$ *and* $m, a, q > 1 \in \mathbb{N}$ where $\lambda = a^m + b = a^m + q - a^m = q$ then we have that

$$\eta_{(q)}(q-a^m,m)=\sum_{j=1}^{q-2}\binom{q}{j}(q-a^m)^{q-j}m^{q-j-1}$$
$$=\sum_{j=1}^{q-2}\binom{q}{j}(p+(-C_m(a)))^{q-j-1}(q-a^m) \tag{53}$$

Then from binomial theorem we conclude that

$$\eta_{(q)}(q-a^m,m)=\sum_{j=1}^{q-2}\binom{q}{j}\sum_{k=0}^{q-j-2}\binom{q-j-1}{j}p^{q-j-1-k}(-C_m(a))^k(q-a^m)$$
$$+\sum_{j=1}^{q-2}\binom{q}{j}(-C_m(a))^{q-j-1}(q-a^m) \tag{54}$$

Note that

$$\sum_{j=1}^{q-2}\binom{q}{j}\sum_{k=0}^{q-j-2}\binom{q-j-1}{j}p^{q-j-1-k}(-C_m(a))^k(q-a^m)\equiv 0(mod\,p) \tag{55}$$

But from theorem 1 we kwon that

$$\eta_{(q)}(q-a^m,m)\equiv\chi_{(\,q,q-a^m,m)}(mod\,p) \tag{56}$$

Then from (54)–(56) we get that

$$\sum_{j=1}^{q-2}\binom{q}{j}(-C_m(a))^{q-j-1}(q-a^m)\equiv\chi_{(\,q,q-a^m,m)}(mod\,p) \tag{57}$$

Note that

$$\chi_{(q,q-a^m,m)}=q(qm-ma^m)^q-q(-m)^q+q-q^2+qa^m \tag{58}$$

Now we have from binomial theorem

$$q(qm-ma^m)^q=q\sum_{j=0}^{q-1}\binom{q}{j}p^{q-j}(-ma^m-1)^j+q(-ma^m-1)^q \tag{59}$$

Note that

$$\sum_{j=0}^{q-1}\binom{q}{j}p^{q-j}(-ma^m-1)^j\equiv 0(mod\,p) \tag{60}$$

Now we are going to eliminate all terms in congruence (60) because it is divisible be p so after that we get the value of $\chi_{(q,q-a^m,m)}$ as follows So we from (56)–(60) we get that

$$\chi_{(q,q-a^m,m)} = q(-ma^m - 1)^q - q(-m)^q + q - q^2 + qa^m \tag{61}$$

Then

$$\sum_{j=1}^{q-2} \binom{q}{j} (-C_m(a))^{q-j-1}(q - a^m) \equiv \chi_{(q,q-a^m,m)}(mod\ p) \tag{62}$$

LEMMA 5. *if* $p = 6m + 1$ and $a > 1 \in \mathbb{N}$ then

$$\sum_{j=1}^{4} \binom{6}{j} (-C_m(a))^{5-j}(6 - a^m) \equiv \chi_{(6,m,6-a^m)}(mod\ p\ ) \tag{63}$$

Proof. Let be in theorem 4 $q = 6$

4. Conclusion

We notice theorem 1 that shows us the relationship between the numbers in the form $a^m + b$ and $ma^m + bm + 1$. This is represented by a polynomial that combines these two numbers. One of the benefits of this relationship is that polynomial can be used as a primitive test, as well as clarifying the properties that those numbers have. But from an abstract arithmetic point of view, we find that theorem 3 is in fact more general than the theorem 1, and this is due to theorem 3 combining all the prime numbers. These results are in Cullen numbers and those in [5, 6] we note that these results are more generalized and differ from those in the form of a gem, and this appears and ideas used differ from those in [5, 6 7, ...12,13]. In general, we studied all numbers in from $p = ba + 1$ where $a, b > 1 \in \mathbb{N}$, as well as Mersnne numbers and Cullen numbers, Fermat numbers. We showed about these numbers that have properties and a relationship between them. We proved this relationship in the form of a polynomial that combines two types of prime number or more. We note that only prime numbers in form $p = ba + 1\ where\ a, b > 1 \in \mathbb{N}$ have been studied no more prime numbers in form $p = ba + m\ where\ a, b, m > 1 \in \mathbb{N}$ have not been studied.

Author details

Shazali Abdalla Fadul
Faculty of Mathematics Sciences and Statistics, AL-Neelain University, Khartoum, Sudan

*Address all correspondence to: shazlyabdullah3@gmail.com

References

[1] Rose HS. A Course in Number Theory. 2nd ed. Clarendon Press; 14 Dec 1995. ISBN-10: 0198523769

[2] Grau JM, Oller-Marcén AM. An ~O (log 2(N)) time primality test for generalized Cullen numbers. Mathematics of Computation. 2011;**80** (276):2315-2323. DOI: 10.1090/S0025-5718-2011-02489-0 MR 2813363 6

[3] María GJ, Oller-Marcén Antonio M, Daniel S. A primality test for Kp^n+1 numbers. Mathematics of Computation. 2015;**84**(291):505-512

[4] Bosma W. Cubic Reciprocity and Explicit Primality Tests for $h.3^k \pm 1$ High Primes and Misdemeanours: Lectures in Honour of the 60th Birthday of Hugh Cowie

[5] Pocklington HC. The determination of the prime or composite nature of large numbers by fermat's theorem. Proceedings of the Cambridge Philosophical Society. 1914;**18**:29-30

[6] Williams HC. A note on the primality of $6^{2^n}+1$ and $10^{2^n}+1$. The Fibonacci Quarterly. 1988;**26**(4):296-305 MR 967648

[7] Williams HC, Zarnke CR. Some prime numbers of the forms $2A3^n+1$ and $2A3^n-1$. Mathematics of Computation. 1972;**26**:995-998. MR 314747. DOI: 10.1090/S0025-5718-1972-0314747-X

[8] Berry PBTG, Tena-Ayuso J. A generalization of Proth's theorem. Acta Arithmetica. 2003;**110**(2):107-115. MR 2008078. DOI: 10.4064/aa110-2-1

[9] Berrizbeitia P, Iskra B. Deterministic primality test for numbers of the form $A^2 3^n+1$ $n>3$ odd. Proceedings of the American Mathematical Society. 2002; **130**(2):363-365. MR 1862113. DOI: 10.1090/S00029939-01-06100-7

[10] James J. Tattersall Elementary Number Theory in Nine Chapters. Cambridge University Press; 2005. ISBN: 0521850142, 9780521850148

[11] Berrizbeitia P, Berry TG. Generalized strong pseudoprime tests and applications. Journal of Symbolic Computation. 2000;**30**(2):151-160. MR 1777169. DOI: 10.1006/jsco.1999.0343

[12] Berrizbeitia P, Olivieri A. A generalization of Miller's primality theorem. Proceedings of the American Mathematical Society. 2008;**136**(9): 3095-3104. MR 2407072. DOI: 10.1090/S0002-9939-08-09303-9

[13] Crandall R, Pomerance C. Prime Numbers. A Computational Perspective. 2nd ed. New York: Springer; 2005 MR 2156291

[14] Guthmann A. Effective primality tests for integers of the forms $N=k2^n+1$ and $N=k2^m.3^n+1$. BIT. 1992;**32**(3): 529-534. MR 1179238. DOI: 10.1007/BF02074886